T5-ARV-579

THE PHARMACOLOGY OF NERVE AND MUSCLE IN TISSUE CULTURE

The Pharmacology of Nerve and Muscle in Tissue Culture

Alan L. Harvey

Department of Physiology and Pharmacology,
University of Strathclyde, Glasgow, UK

Alan R. Liss, Inc., New York

First published in the United States of America 1984 by
Alan R. Liss, Inc., 150 Fifth Avenue, New York, NY 10011
Printed and bound in Great Britain

Library of Congress Cataloging in Publication Data

Harvey, Alan L., 1950-
The pharmacology of nerve and muscle in tissue culture

Bibliography: p.
Includes index.
1. Pharmacology, Experimental. 2. Drugs–Physiological effect. 3. Neurons. 4. Muscle cells. 5. Tissues culture. I. Title. [DNLM: 1. Nervous system–Physiology. 2. Nervous system–Drug effects. 3. Muscles–Physiology. 4. Muscles–Drug effects. 5. Tissue culture. WL 102 H341p]
RM301.H38 1983 615'.77'0724 83-23878
ISBN 0-8451-3011-0

CONTENTS

PREFACE

The techniques of tissue culture were introduced at the beginning of this century. They have become more and more popular as it is realized that they are not as difficult or as esoteric as some early protagonists liked to maintain. Most of the work performed with culture methods has simply concerned cell growth and survival. Biologists have long used culture approaches to provide a simple system in which to study cell division and multiplication. Any pharmacology done on cultured tissue was largely toxicological or as part of a screening programme for potential anti-cancer drugs.

In the last decade there has been a great increase in the use of excitable cells in tissue culture. Nerves and muscles from a wide variety of sources can maintain their highly differentiated properties in culture. Such cultures offer an attractive preparation for use in physiological and pharmacological investigations. Consequently, a vast amount of work has been produced, and this book is an attempt to review it. It is hoped that this will introduce physiologists and pharmacologists to the potential of culture methods for their experiments and also indicate to more traditional tissue culture users further possible areas of interest. By being more comprehensive in scope and by trying to concentrate largely on drug actions, I hope that the present volume usefully extends the treatment of the subject begun earlier in the excellent works by Crain (1976) and Nelson and Lieberman (1981).

I would like to thank the many authors who have supplied me with details of their work and allowed me to use some of their results, and I would like to thank my colleagues who read earlier drafts of some of the chapters. Finally, I would like to repay two 'bills': my special thanks go to Bill Dryden for introducing me to tissue culture, and to Bill Bowman for introducing me to pharmacology.

1 AN INTRODUCTION TO TISSUE CULTURE AND ITS USE IN THE STUDY OF NERVE AND MUSCLE

What is Tissue Culture?

Tissue culture is a means of keeping animal cells alive for more than a few hours after their removal from the animal. Different types of tissue culture can be classified according to the degree of disruption of the tissue before it is placed in culture. There are four broad classes: organ culture, explant culture, cell culture and continuous cell line culture.

Organ cultures are the least disrupted type of culture. Whole organs, or large sections of them, are placed in a warm nutritive solution and maintained for periods of only a few days. As the normal blood flow to the organ is obviously disrupted, exchange of nutrients and metabolites depends on simple diffusion. Hence, surface cells may survive but cells within the mass of the organ culture soon die. Release of degradation products may subsequently affect the functioning of the surviving cells, making interpretation of pharmacological results difficult. Nevertheless, in some instances, successful use has been made of organ cultures to study the actions of drugs (e.g. Wildenthal, 1974).

Explant cultures are really mini-organ cultures because they are made by cutting the starting tissue into small pieces, usually approximately 0.5-1 mm cubes. One or several of these tissue blocks are anchored to a culture dish where they can be maintained, often for several weeks. Explants share the drawbacks of organ cultures, although to a lesser extent. Usually experiments are performed, not on the explant itself, but on cells that have migrated from it. The uncertainty is whether these cells are typical of the starting tissue or are artificially selected by the culture technique. There is also difficulty in making explants consistently. However, some tissues appear to survive better as explants than as other forms of culture, perhaps because the dense group of cells in the explant beneficially modifies the environment. For example, longer survival and better development were found with explants of human skeletal muscle compared with dissociated cell cultures from the same tissue (Harvey *et al.*, 1979, 1980).

Cell cultures are made by dissociating the starting tissue into its component cells. Enzymes such as trypsin and collagenase are most frequently used but often mechanical dissociation in solutions lacking Ca^{2+} and Mg^{2+} is equally successful. With consistent techniques applied to similar starting material it is possible to obtain cell suspensions with little batch-to-batch variation, and by always plating the same density of cells, any one culture should be reasonably similar to any other. Hence, with cell cultures it is possible to prepare large numbers of replicates.

The cells tend to grow in monolayers so that transfer of waste products and nutrients should be adequate. Cells in monolayer cultures are more visible under the microscope than cells in organ or explant cultures, and are more accessible to recording electrodes and to pipettes for iontophoretic application of drugs. Against these advantages are the problems introduced by the loss of normal tissue geometry and cell-to-cell interactions.

Continuous cell lines are formed by adaptation of some cells to permanent growth and multiplication in culture. Usually such cells are transformed by a virus or a carcinogenic chemical and are genetically different from cells of the starting population. Cell lines derived from single parent cells are known as clonal cell lines; although they have abnormal chromosome numbers, they have the advantage that each cell in the culture is genetically identical.

Cell lines offer advantages of simplicity and of mass production so that they are useful for biochemical studies. Difficulties arise when attempts are made to relate properties of cells in cell line cultures to cells in the corresponding non-cultured tissue. However, if the property being examined is of fundamental interest (for example, an ionic conductance or drug receptor mechanism), it may not matter that it is not precisely identical to that of the parent tissue. The cell line can be regarded as a model system that is much more convenient to study.

Development of Tissue Culture

Although several workers had made attempts to maintain tissues *in vitro*, the first true cultures are usually regarded as the explants of frog embryo nervous tissue that were grown by R.G. Harrison (1907, 1910). In these cultures, fragments of medullary tissue were placed in drops of clotted lymph and maintained for several days. Harrison observed the outgrowth of long nerve processes from cells within the explant (Figure 1.1).

Harrison's techniques were soon adapted for other excitable tissues

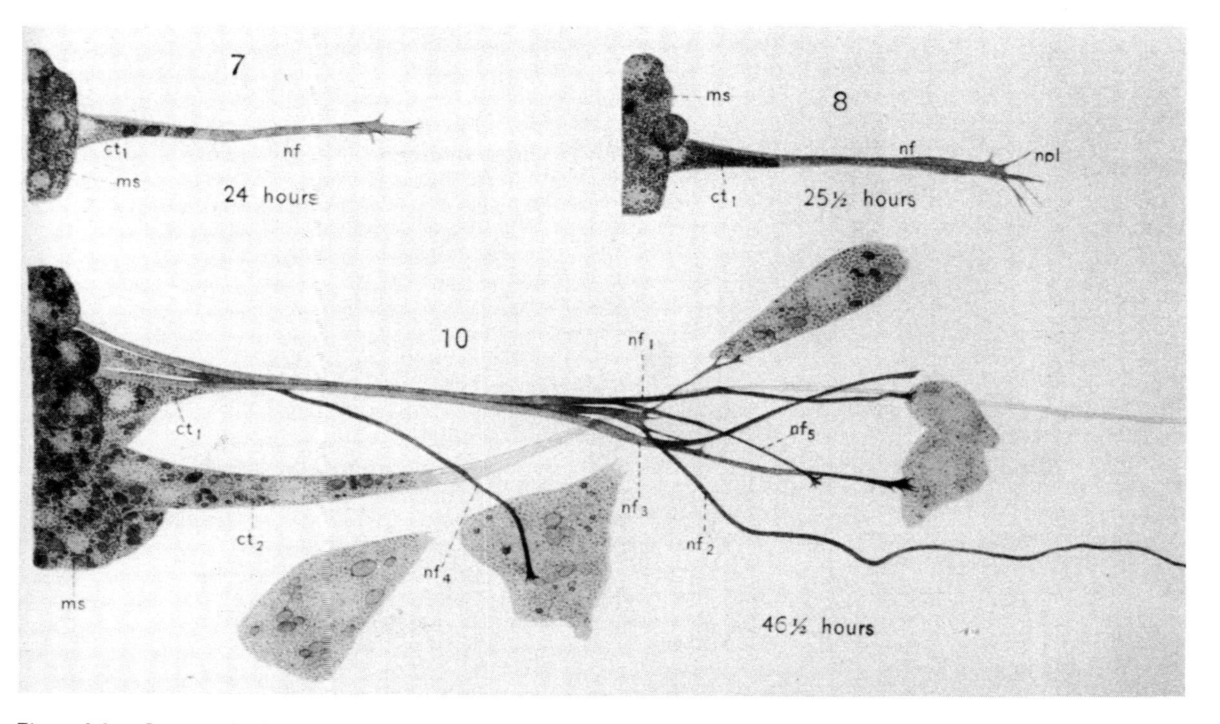

Figure 1.1 Outgrowth of Living Nerve Processes in an Explant Culture of Medullary Tube from Frog Embryo. From Harrison (1910).

from other species. Mammalian nerves were grown by Ingebrigtsen (1913); cardiac muscle by Burrows (1910); skeletal muscle from frog embryo had also been grown by Harrison (1907, 1910) and was later cultured from chick embryo by Lewis and Lewis (1917b); and smooth muscle was grown by Champy (1913/14).

Unfortunately, the early days of tissue culture were before the advent of antibiotics and technical problems tended to dominate discussions of the potential use of the method. The idea of the intrinsic difficulty of tissue culture techniques is still (mistakenly) widespread today. Some reasons for this myth have been discussed by Witkowski (1979). There are several general accounts of tissue culture methods, incuding the standard works by Rothblat and Cristofalo (1972) and Paul (1975), and the more recent reviews by Prasad (1981) and Schlapfer (1981).

General Advantages of Tissue Culture

Tissue culture offers several experimental advantages for pharmacological studies, including good visualization of individual cells, ready accessibility of cells to applied drugs, lack of connective tissue which can prevent penetration of target cells by microelectrodes, and simplification of experimental conditions by removing indirect influences such as blood flow, metabolism and hormonal effects. Additionally, tissue culture methods provide a means to study changes in drug effects during development, and they are also useful when very long exposures to contant low levels of drugs are desired. Finally, tissue culture allows more economic use to be made of human tissue, so that drugs can be assessed experimentally on man rather than on other species.

Despite these unique advantages, many studies that have involved tissue culture experiments have not made full use of the technique's potential. Rather, there has been a tendency to use culture for culture's sake: to grow a particular cell, or to prove that what is known to happen *in vivo* also takes place in culture. While there is obviously a need to establish the validity of the culture model, more imaginative uses of the technique can be made so that information can be gathered which would not otherwise be accessible.

Examples of the power of the tissue culture technique in resolving biological problems are provided by the very first experiments on cultured cells by Harrison (1907, 1910) and by later work on skeletal muscle myogenesis. Prior to Harrison's invention of tissue culture, nerves could only be examined closely in fixed and stained preparations, causing controversy about the origin of nerve axons. With his elegantly simple experiments, Harrison demonstrated that living nerve cells, maintained in culture, produced elongated processes (see Figure 1.1). As Harrison (1910) wrote, 'Such an empirical determination must have more weight than any amount of *a priori* argumentation upon the subject.'

In early studies on myogenesis *in vivo* there was controversy about the origin of multinuclear skeletal muscle fibres (see Murray, 1960, for references). Some authors maintained that such cells arose from nuclear division rather than by fusion of uninuclear cells. Tissue culture studies helped to clarify the situation (for references, see Harvey, 1980). Time lapse cinematography of cells in culture has shown fusion taking place, and morphological examination of multinuclear cells in culture has never revealed the presence of dividing nuclei. Skeletal muscle fibres can develop in culture from a single myoblast under cloning conditions. Multinuclear myotubes in culture do not synthesise DNA, although DNA synthesis occurs in myoblasts. Exposure of early cultures to concentrations of nitrogen mustard sufficient to inhibit DNA synthesis has no effect on cell fusion. Finally, hybrid myotubes are formed if myoblasts from mouse and chick embryos are mixed in monolayer culture, demonstrating that fusion must have taken place.

Recording Techniques

Early studies on cells in tissue culture were limited to morphological examination, but with the development of techniques for recording electrical activity of membranes of excitable cells it was found that cultured cells were suitable for electrophysiological experiments. Indeed, cultured cells can offer distinct advantages for electrophysiologists in addition to their excellent visibility and accessibility. For example, voltage clamp recording techniques require that the area of the cell membrane under study is maintained at a precise value of membrane potential. This can be difficult to achieve in cells whose complic-cated geometry leads to uneven dissipation of injected current. With manipulations in culture it is often possible to create cells that are more suitable for voltage clamping. Examples are the very large nerve-derived tumour cells (neuroblastoma), spherical 'myoballs' and 'myosacs' prepared from skeletal muscle, and aggregates of cardiac cells in which there is electrical continuity between individual cells. Use of voltage clamp methods allows the precise characterization of the properties of membrane ion channels and, coupled with fluctuation analysis, the determination of the fundamental properties of conductance and the lifetime of receptor-operated ion channels. These techniques and their application to cultured cells have been reviewed in depth by Smith *et al.* (1981) and Lecar and Sachs (1981).

An electrophysiological technique for which cultured cells are especially suited is the so-called 'patch clamp' method which allows records to be made from individual ion channels (Neher and Sakmann, 1976a; Neher *et al.*, 1978, Hamill *et al.*, 1981). This technique involves placing an extracellular recording electrode over a small patch of membrane that should ideally contain only one receptor or channel. To achieve maximum resolution the electrode must form an extremely high resistance junction with the membrane. Cultured cells are generally free from overlying connective tissue and their surfaces are often more regular than many cells *in situ*. They also usually have lower densities of ion channels and receptors than their counterparts *in vivo*. For these reasons, cells in culture are ideally suited for use with patch clamp methods; in fact, most of the current work involving this technique is being done on cultured cells.

Present Interests and Future Directions

Currently most of the pharmacological studies on nerves and muscles in culture is aimed at elucidation of the mechanisms of action of specific agents on receptors or ion channels. Much useful information has been obtained and neurophysiologists and pharmacologists in general should consider whether their experiments might be better carried out on cultured preparations.

In addition to studies on nerve and muscle receptors, the culture system is also being used to study mechanisms whereby cells release chemical mediators. Dorsal root ganglion cells have been proposed as a model of nerve terminals because their ionic channels appear to be similar and they are also affected by drugs known to modulate neurotransmitter release. Patch clamp techniques have also been used to investigate changes in the membrane capacitance of cultured chromaffin cells from the adrenal medulla; such changes are believed to result from fustion of storage granule membranes with the external cell membrane (Neher and Marty, 1982).

Further extensions of culture techniques to the study of endocrine cells have also been made recently. Pancreatic islet (Pace *et al.*, 1977), pituitary (Sand *et al.*, 1980; Hagiwara and Ohmori, 1982, 1983), parathyroid (Sand *et al.*, 1981b) and thyroid cells (Sand *et al.*, 1981a; Sinback and Coon, 1982) have been grown in culture and have proved suitable for electrophysiological analysis. It is to be expected that more detailed studies on the biophysical mechanisms of hormone release will be made in the future using cells in culture.

Most of the pharmacological work on cultured nerve and muscle that has been carried out during the last ten years has concerned drugs that interact with specific receptor sites. However, tissue culture methods are also suitable for investigation of side-effects and drug toxicity. Such studies have been largely omitted from the present text because they form a collection of unrelated reports, but obviously this use of tissue culture may be expected to increase with the continuing concern over the safety of drugs and the desire to minimize the use of animals. A general review of the use of culture methods in toxicology is given by Stammati *et al.* (1981).

2 NERVE CELLS IN PRIMARY CULTURE

The first cultures of nervous tissue were of fragments of medullary tube from frog embryos (Harrison, 1907, 1910) and, as described in Chapter 1, these were the first true tissue cultures (see Figure 1.1). Subsequently, Lewis and Lewis (1912) grew sympathetic nerves associated with chick embryo intestine, and Ingebrigtsen (1913) demonstrated that explants of mammalian brain could grow extensive networks of axons (Figure 2.1). Ingebrigtsen extended Harrison's observations by showing that if the connection between the outgrowths and the cell body was cut, the original axon rapidly degenerated but new outgrowths were formed. Since these early reports, a great number of studies have been made on cultured nervous tissue, although most concerned the growth and morphology of nerves in culture. This work has been reviewed by Murray (1965b), Lumsden (1968) and by Hösli and Hösli (1978).

The first reports of the use of electrophysiological techniques with cultured nervous tissue were by Crain (1954, 1956) and by Hild *et al.* (1958). The latter authors made intracellular recordings, although only from astroglial cells which had migrated from explants cultured from kitten brain. Crain's study on explants of spinal ganglia from 7-8-day chick embryos was more successful in that resting membrane potentials in the range –50 to –65 mV and overshooting action potentials were recorded. Fast rising action potentials were subsequently recorded from nerve cell bodies in explants of cerebellum (Hild and Tasaki, 1962) and spinal cord (Hösli *et al.*, 1971). The techniques for successful electrophysiological recording from nerves in dissociated cell cultures developed more slowly. Reliable intracellular measurements were made first from dorsal root ganglion neurones obtained from chick embryos (Scott *et al.*, 1969; Varon and Raiborn, 1971), and then from spinal cord nerves (Fischbach, 1970; Lawson and Biscoe, 1973; Peacock *et al.*, 1973b) and from brain cells (Lawson and Biscoe, 1973; Nelson and Peacock, 1973).

Early pharmacological studies on cultured nerve cells included many experiments on morphological changes brought about by drugs, which were often administered in massive doses (see Murray, 1965b), and studies on drug-induced alterations of the electrical activity of explants (reviewed by Crain, 1976). These experiments have more recently

Figure 2.1: Nerve Cell Outgrowths from Explant Cultures
(a) 4-day spinal ganglion culture from 7-month rabbit. (b) 2-day culture of cerebral cortex from 6-day chick embryo. From Ingebrigtsen (1913), by copyright permission of the Rockefeller University Press.

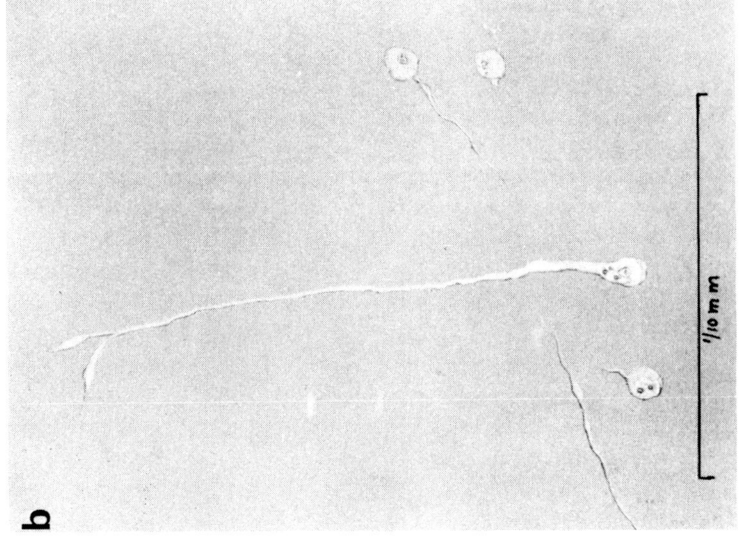

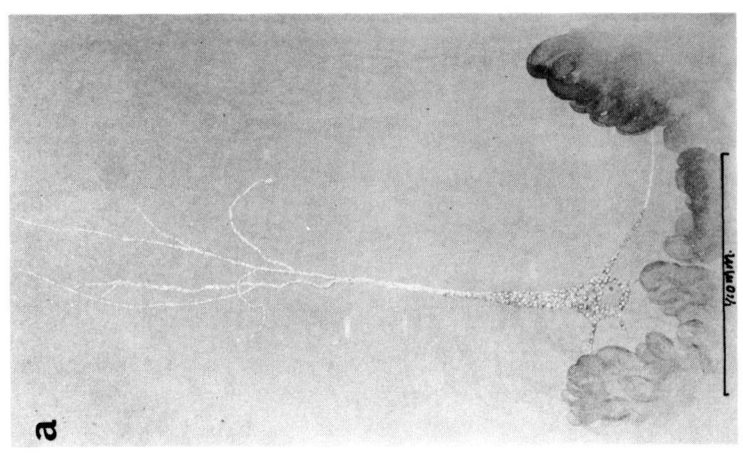

evolved into the widespread use of nerves in tissue culture to examine the actions of possible neurotransmitters on individual cells. This chapter reviews the types of cultures that are commonly used, the electrical properties of nerves in culture, and the pharmacological characterization of the most important neurotransmitter mechanisms. Both central and peripheral nerves are included in this chapter, together with adrenal chromaffin cells as a specialized form of sympathetic neurone. Glial and other accessory cells are, however, not included. Synaptic interactions between cultured nerve cells are considered in Chapter 7.

Types of Culture and Methodology

The tremendous complexity of the central nervous system and the relative inaccessibility of its constituent cells makes it difficult to study experimentally. Of the various strategies used to provide a simple model of the nervous system, tissue culture is perhaps the most promising. The potential to isolate particular types of neurones, then to examine their membrane properties with the most rigorous biophysical techniques, and also to examine their interactions under controlled conditions, is extremely exciting. Against this must be held the disadvantages resulting from the abnormal environment in culture and from disruption of normal cellular connections.

Both explant and cell culture techniques have been used for studies on nerve cells. Explants have been extensively utilized by Crain and his co-workers and this work has been thoroughly reviewed by Crain (1976). Although it is impressive that little pieces of brain can maintain their structural integrity and sustain a wide range of complicated electrical activities when grown as explant cultures, these do not provide ready accessibility to individual cells and it is impossible to see their connections. Hence, much of the potential advantage of a culture system is lost in exchange for having a miniature system that mimics more or less faithfully the complexities of the original. At the opposite end of the spectrum, nerves in dissociated cell cultures are easy to see and to approach with microelectrodes. Drugs can be applied to selected regions of the cell and recordings of synaptic potentials can be readily obtained. However, although some isolated nerves rapidly form functional synapses in cell cultures, there is no guarantee that these interactions are physiologically relevant (see Chapter 7).

Perhaps a useful compromise between the classical explant method and the completely dissociated cell culture is provided by the roller-tube culture method, which was used successfully by Costero and Pomerat (1951) for adult human cerebral cortex and cerebellum. In this technique explants are grown on coverslips in test tubes which are partly filled with medium and slowly rotated (about 8-10 revolutions per hour) so that the cells are intermittently covered by medium. The explants are gradually transformed into monolayers while maintaining many of the original cell-to-cell contacts. The roller-tube method has been used successfully to provide cultures of cerebellum, hypothalamus and hippocampus for electrophysiological studies (reviewed by Gähwiler, 1981).

A large number of different types of nerve cells from a variety of species have now been grown in culture. Details of methods used for preparing explant cultures have been given by Crain (1976) and Gähwiler (1981), and the morphology of some of these cultures is illustrated in Figure 2.2. There are greater variations in conditions used for cell cultures, largely depending on whether mechanical or enzymatic dissociation techniques are used. Methods for virtually every major class of neurone have been published. References to those which have resulted in cultures suitable for electrophysiological experiments are given in Table 2.1 and there is some further discussion of cell culture techniques by Fischbach and Nelson (1977), Nelson *et al.* (1981), Ransom and Barker (1981) and Scott (1982).

Although Varon and Raiborn (1969) described a method for fractionation of cells into relatively homogeneous populations, there has not been a great deal of progress in obtaining 'pure' cultures which have only one type of nerve cell. The most successful technique is to use a source of material that has only a few types of cells and, preferably, a homogeneous population of neurones. For example, relatively pure cultures of peripheral sympathetic nerves can be prepared from sympathetic ganglia. After dissociation, the cells are briefly plated on a plastic dish and the contaminating fibroblasts attach preferentially; the nerve cells in the medium are then replated on a surface which allows them to attach (see, for example, Wakade and Wakade, 1982). (Attempts to provide pure populations of spinal motor neurones are discussed in Chapter 7, p. 183.)

One of the difficulties in growing nerve cells on their own is that such cells appear to need support from other cell types. The nutritional requirements of nerves in culture are far from being understood, but recent work on defined media (i.e. media in which all ingredients are known and to which no mystery ingredient such as serum or embryo

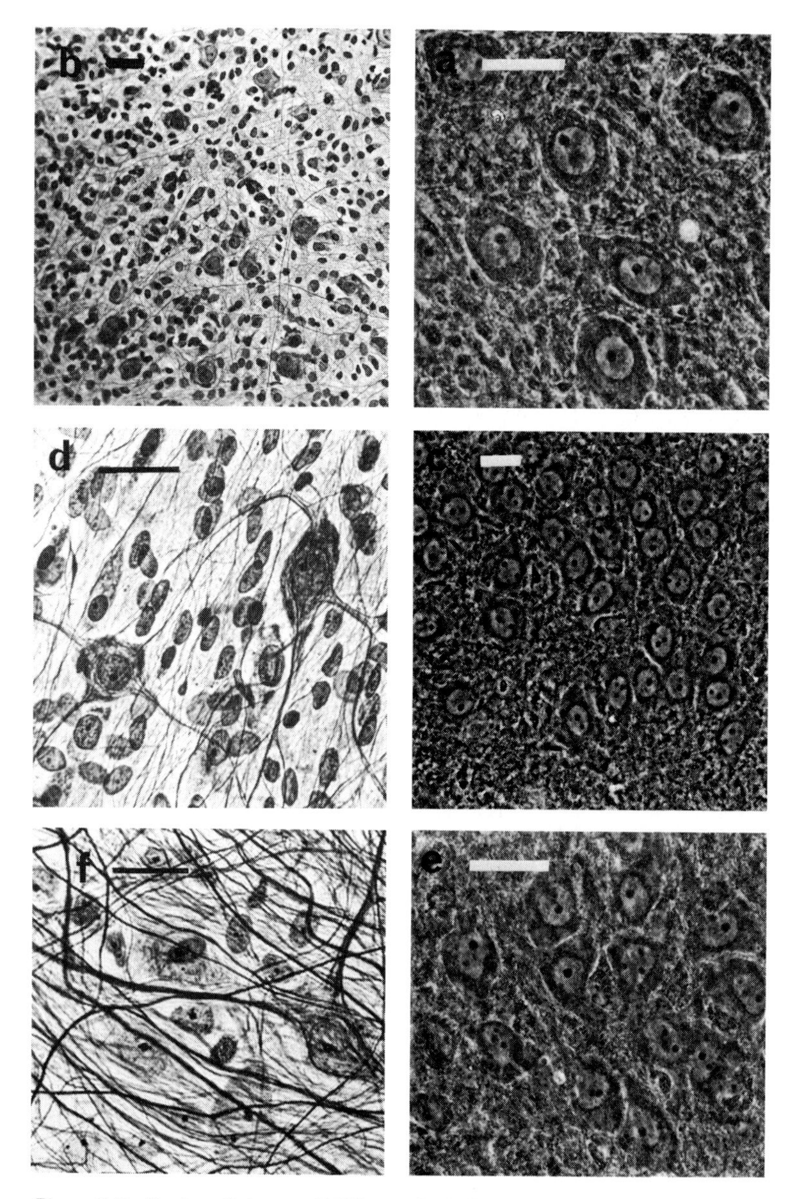

Figure 2.2: Explant Cultures of Different Brain Regions
Left column: phase contrast micrographs. Right column: fixed and silver stained cultures. (a) and (b), cerebellar cells; (c) and (d), hippocampal cells; (e) and (f), hypothalamic cells. Scale markers = 50 μm. Photographs supplied by Dr B. Gähwiler.

Table 2.1: References to Methods for Preparation of Dissociated Nerve Cell Cultures Suitable for Electrophysiology

Nerve cell type	Tissue source	Reference
Brain	fetal rat	Godfrey *et al.* (1975)
Cerebellum	newborn rat	Lasher and Zagon (1972)
	fetal mouse	Nelson and Peacock (1973)
Cerebral cortex	chick embryo	Varon and Raiborn (1969)
	fetal rat	Yavin and Yavin (1974)
		Dichter (1978)
	fetal mouse	Swaiman *et al.* (1982)
Hippocampus	fetal mouse	Peacock *et al.* (1979)
Hypothalamus	newborn rat	Wilkinson *et al.* (1974)
Corpus striatum	newborn rat	Messer (1981)
Lower brain stem	fetal rat	Saji and Miura (1982)
Spinal cord	chick embryo	Fischbach (1972)
	fetal mouse	Ransom and Holz (1977)
Spinal cord motor nerves	chick embryo	Berg and Fischbach (1978)
		Masuko *et al.* (1979)
Peripheral sensory nerves	chick embryo dorsal root ganglia	Scott *et al.* (1969)
		Varon and Raiborn (1971)
	fetal mouse dorsal root ganglia	Peacock *et al.* (1973b)
	adult guinea pig dorsal root ganglia	Fukada and Kameyama (1979)
	newborn rat nodose ganglia	Baccaglini and Cooper (1982a)
Sympathetic	chick embryo paravertebral ganglia	Varon and Raiborn (1972)
	newborn rat superior cervical ganglia	Mains and Patterson (1973)
	adult rat superior cervical ganglia	Wakshull *et al.* (1979a)
Enteric	rat, rabbit and guinea pig gut	Jessen *et al.* (1978, 1983)
Adrenal medullary chromaffin cells	gerbil	Biales *et al.* (1976)
	rat	Brandt *et al.* (1976)
Parasympathetic	chick embryo ciliary ganglia	Nishi and Berg (1977)
		Tuttle *et al.* (1980)

extract is added) indicates that some nerves may be able to be grown selectively (Bottenstein and Sato, 1979; Skaper *et al.*, 1979; Bottenstein *et al.*, 1980; Romijn *et al.*, 1982).

A further difficulty in obtaining cultures of selected types of nerves is the lack of nerve-specific markers to allow unambiguous identification of cell types. With the development of monoclonal antibodies directed against components unique to different nerve types (e.g. McKay and Hockfield, 1982), it should be possible to detect neurones in culture and to follow their purification.

Electrical Properties and Drugs Acting on Ion Channels

Nerves function by conducting electrical signals, therefore their physiology can be described largely by considerations of their electrical properties. The number and type of ion channels in the cell membrane determine the cell's transmembrane potential at rest and the resistance it has during any electrical signal. The lipids in the membrane act as capacitors that are charged and discharged in response to transient signals, and this affects the time course of the cell's response. Any discussion on how accurate a model cultured nerves are for normal function must consider the fundamental electrical properties of the cells in culture. The passive membrane properties of cultured neurones are described in the next section, which is followed by an account of the active electrical properties of such cells, i.e. action potential generation and conduction.

Nerves in culture are often small (10-15 μm maximum diameter) and are therefore hard to penetrate successfully with microelectrodes. The chance of significant damage caused by insertion of a microelectrode has been commented on by many authors and has probably led to the variation in values reported for resting membrane potentials. Although Crain (1956) found that some chick embryo spinal ganglion cells had resting potentials of –50 to –65 mV, only about 16 per cent of the penetrations were successful. Hild and Tasaki (1962) measured membrane potentials of –50 mV or less from cerebellar neurones in explant cultures, but the membrane potentials could not be maintained and, after removal of the recording electrode, the cells died.

The passive membrane properties, viz. resistance, membrane time constant and capacitance, are determined by measuring the voltage response to a small pulse of current injected into the cell (see Figure 2.3). Ideally, separate current-passing and voltage-sensing electrodes should be used, but this is often difficult with small cells. Single electrodes connected to a bridge circuit can be used to inject current and to record membrane potential. However, this method can be unreliable if the stimulation artefact is not completely nullified or balanced out. This can be difficult to ensure as the electrode's characteristics can be altered on insertion into the cell. Also the electrode must behave

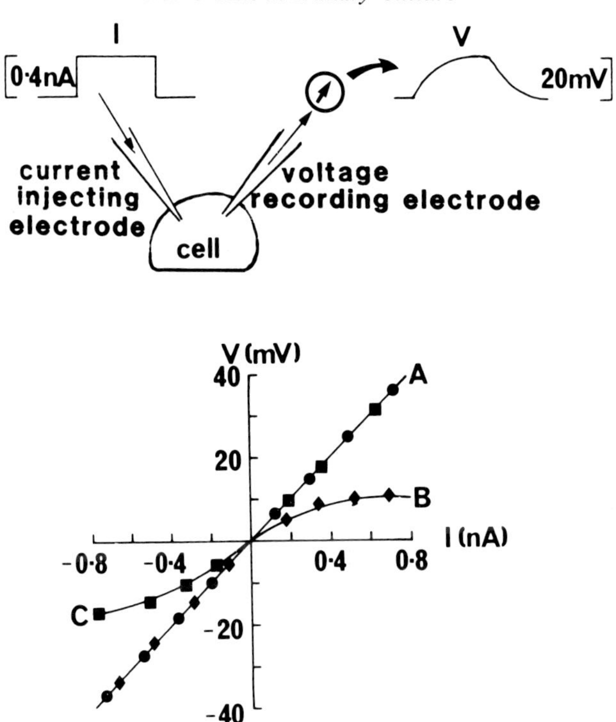

Figure 2.3: Determination of Current-Voltage Relationships
Top: Diagrammatic representation of the experimental arrangement with two intracellular electrodes. Bottom: Hypothetical responses from cells showing completely passive behaviour (●, A), delayed rectification (♦, B) and anomalous rectification (■, C).

as a linear ohmic resistor and such behaviour is not always found with the high resistance electrodes necessary for recording from small cells.

Passive Membrane Properties

The chance of cell damage and the uncertainties that accompany use of high resistance electrodes with high tip potentials make it hard to estimate membrane potentials reliably and dangerous to make extensive comparisons between values obtained in different studies. In general, however, it appears that cultured nerve cells have resting potentials of –40 to –70 mV (Table 2.2). Usually, higher values have been reported in studies on larger cells which have diameters of 20-30 μm. Because of the difficulties involved in making reliable measurements on small cells, few authors have attempted to make developmental studies in culture.

Table 2.2: Passive Membrane Properties of Nerve Cells in Primary Cultures

Tissue type	Species	Membrane potential (−mV)	Input resistance (MΩ)	Time constant (ms)	Specific membrane resistance (Ωcm²)	Specific membrane capacitance (μF/cm²)	Reference
Primary cultures							
Cerebellum	rat	25 – 50	0.9	–	–	–	Lawson and Biscoe (1973)
	mouse	37 ± 2	20 – 60	~4	~2500	1 – 2	Nelson and Peacock (1973)
Cerebral cortex	rat	65 ± 1	50 ± 4	2 – 12	–	–	Dichter (1978)
Hippocampus	mouse	38 ± 0.8	30 – 40	~1.5	~1500	~1	Peacock (1979)
Spinal cord	rat	25 – 50	2.1 ± 1.3	0.8	280	9	Lawson and Biscoe (1973)
	mouse	40	–	~4	3260	1.2	Peacock *et al.* (1973b)
	mouse	51 ± 7	5.6	6.5	–	–	Ransom *et al.* (1977c)
Dorsal root ganglia	rat	25 – 50	2.9 ± 2.2	2.45	580	1.4	Lawson and Biscoe (1973)
	mouse	51	–	~2	1560	1.4	Peacock *et al.* (1973b)
	mouse	50 ± 5	43.5	3	–	–	Ransom *et al.* (1977c)
	human	56 ± 0.5	19.7 ± 0.7	2.5 ± 0.1	984 ± 38	3.2 ± 0.14	Scott *et al.* (1979)
	guinea pig	49 ± 4	26 ± 2	1.6 ± 0.2	–	–	Fukada and Kameyama (1980)
Nodose ganglia	guinea pig	41 ± 3	30 ± 2.5	1.8 ± 0.2	–	–	Fukada and Kameyama (1980)
Sympathetic ganglia	chick embryo	45 ± 2	54 ± 8	3.8 ± 0.4	436 ± 66	11 ± 1.3	Chalazonitis *et al.* (1974)
	guinea pig	50 ± 3	32 ± 4	2.0 ± 0.3	–	–	Fukada and Kameyama (1980)
	bullfrog	46 ± 1.3	27 ± 2	5.0 ± 0.5	1665 ± 181	3.2 ± 0.3	Gruol *et al.* (1981)
Chromaffin cells	rat	50 – 60	430 ± 28	6 – 10	3500	2.3	Brandt *et al.* (1976)
Ciliary ganglia	chick embryo	52 ± 1	138 ± 10	7.3 ± 0.5	–	–	Tuttle *et al.* (1980)
Cell lines							
M and N32	mouse	42 ± 2	30 ± 4	6.4 ± 1.4	5325 ± 923	4.4 ± 0.8	Nelson *et al.* (1971b)
N18	mouse	~40	–	~15	~2500	–	Nelson (1973)
SK-N-SH (undifferentiated)	human	10 – 50	20 – 30	<15	–	–	Kuramoto *et al.* (1981)
SK-N-SH (differentiated)	human	40 – 80	50 – 400	10 – 50	–	–	Kuramoto *et al.* (1981)
Non-cultured cells							
Motor cortex	cat	50 – 70	5 – 15	7 – 12	1000 – 6000	1.5 – 5.0	Cited by Hubbard *et al.* (1969)
Hippocampus	cat	52 ± 2	13	9.9	–	–	Cited by Hubbard *et al.* (1969)
Spinal motor neurone	cat	~70	~1.2	2 – 5	600	5	Cited by Hubbard *et al.* (1969)
Sympathetic ganglia	guinea pig	50 – 70	55 ± 5	9	1000	10	Cited by Chalazonitis *et al.* (1974)
Sympathetic ganglia	toad		16 – 22	2.1 – 4.6	2200 – 4000	1.1	Cited by Hubbard *et al.* (1969)
Ciliary ganglia	chick	50 – 70	28 ± 2	1.6 ± 0.2	–	–	Cited by Tuttle *et al.* (1980)

Note:
Where possible, all values are expressed as means ± standard error of the mean.

Lawson and Biscoe (1973) found that the resting potentials of dorsal root ganglion, but not spinal cord, cells increased with time in culture. However, the highest potentials were still rather low and it is possible that the change reflected an increase in size of the dorsal root ganglion neurones. Moreover Peacock *et al.* (1973b) found no change in electrical properties of fetal mouse spinal cord and dorsal root ganglion cells in cultures aged 16-94 days. There do not appear to be any published reports of the use of voltage sensitive dyes with nerve cells in primary cultures, although such compounds have been used with some success with neuroblastoma cells (see Chapter 3). Hence there is little information about the development of ionic channels that control resting potential in cultured nerves.

As would be expected, the membrane potential of nerve cells in culture is sensitive to changes in the external concentration of K^+ ions. Hösli *et al.* (1972) first demonstrated this with spinal nerves in explant cultures, and similar sensitivities have been found in cell cultures of superior cervical ganglion, dorsal root ganglion and spinal cord nerves.

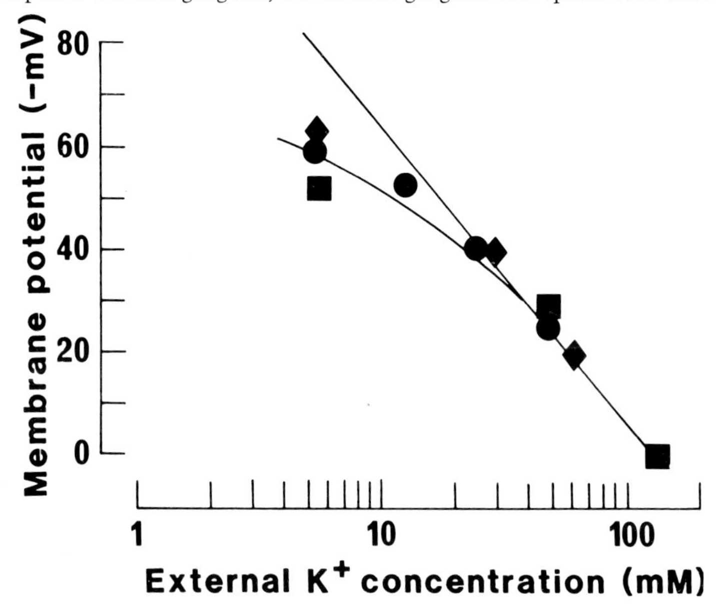

Figure 2.4: Relationship of Membrane Potential to the External Concentration of K^+

●, spinal neurones; ♦, dorsal root ganglion cells; ■, superior cervical ganglion cells in culture. Straight line indicates the slope predicted by the Nernst equation. Graph derived from data of Hösli *et al.* (1972); O'Lague *et al.* (1978b); Choi and Fischbach (1981).

As shown in Figure 2.4, the variation in membrane potential with K^+ concentration is close to that predicted by the Nernst equation, i.e. a 58 mV decrease in membrane potential for a tenfold increase in external K^+. In contrast to most excitable cells in conventional preparations, some neurones in culture have resting potassium conductances that can be blocked by tetraethylammonium and 4-aminopyridine. This has been found for both spinal cord and dorsal root ganglion cells, with the latter being more affected (Heyer and MacDonald, 1982a, b). Additionally, spinal cord, dorsal root ganglion (Heyer and MacDonald, 1982a) and adrenal chromaffin cells (Brandt *et al.*, 1976) appear to have significant resting permeability to Na^+ ions. For example, the average membrane potential of chromaffin cells increased by about 10-20 mV on removal of external Na^+.

Despite the difficulties associated with using one electrode for both current injection and voltage recording, much useful information about the membrane properties of cultured nerve cells has been obtained with single electrode recording, some of which has been satisfactorily cross-checked by two electrode measurements. Some values for input resistance, time constant, specific membrane resistance and specific membrane capacitance of cultured nerve cells and of neurones *in vivo* are given in Table 2.2. There is broad agreement that the input resistance (or total cell resistance) of cultured neurones is 20-60 MΩ. Exceptions are the very low values found for immature cerebellar, spinal cord and dorsal root ganglion cells cultured from fetal or newborn rats (Lawson and Biscoe, 1973), and the higher values for rat adrenal chromaffin cells (Brandt *et al.*, 1976) and chick embryo ciliary ganglion nerves (Tuttle *et al.*, 1980). Since input resistance is higher in small cells, it is likely that the higher values obtained in chromaffin and ciliary ganglion cells reflect their smaller size. However, this explanation does not hold for the cells studied by Lawson and Biscoe (1973) for these were also small. Since many of their recordings were made with two microelectrodes on small cells, it is possible that leakage around the electrodes lowered the apparent input resistance. Alternatively, these results may reveal an early stage in the maturation of nerve cells. Generally, cultured neurones have higher input resistances than the corresponding cells *in vivo* (Table 2.2). This is presumably because the cultured cells are smaller.

The membrane time constant is determined from the time course of the change in the membrane potential in response to a current pulse. If the cell is behaving as a simple circuit comprising a resistor and a capacitor, this time course is exponential and the time constant is the time taken for the membrane potential to fall to $^1/e$ of the steady state value at the end of the current pulse. The time constant also gives an indication of the membrane capacitance, being equal to the product of specific membrane resistance and specific membrane capacitance. Mostly, membrane time constants of cultured nerve cells are in the range 2-8 ms (Table 2.2) which is similar to values reported for cells *in vivo*. Values for specific membrane resistances and capacitances are more variable, probably because of difficulties in calculating the surface area of the cell and determining the influence, if any, of nerve cell processes.

In extensively branched cells the time course of a change in membrane potential in response to a current pulse will not be a simple exponential because of the time taken for the current to spread throughout the processes. Cultured spinal cord nerves have many extensive neurites, whereas dorsal root ganglion cells have generally only one or two large neurites (Figure 2.5). Dorsal root cells in culture respond

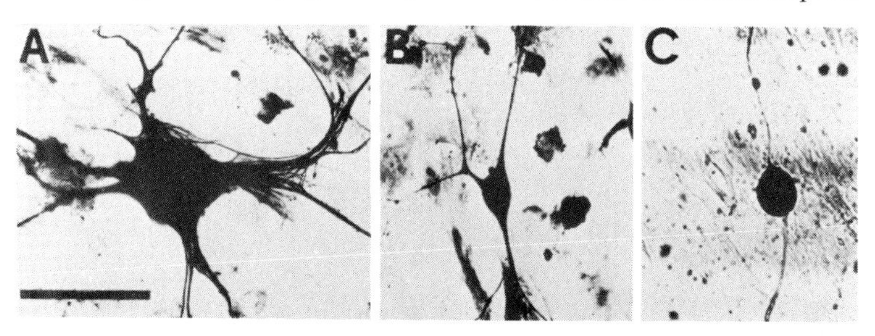

Figure 2.5:
Silver impregnated large (A) and small (B) spinal cord cells compared with sensory ganglion cell (C) grown under the same conditions in another culture. Bar = 50 μm. From Fischbach and Dichter (1974), with permission.

with a simple monotonic function to current pulses, but spinal cord cells reveal at least two processes with different time courses (Ransom *et al.*, 1977c). The electrotonic length of the cultured spinal cell processes was estimated to be 0.9 times the length constant of the cell. This is about half the value obtained from mature spinal motor neurones, presumably reflecting the smaller size and less extensive branching of the processes of the cultured cells compared to the cells *in vivo*.

Action Potentials and Drugs Acting on Ion Channels

The most important functional property of nerve cells *in vivo* is their ability to generate and conduct action potentials. Obviously, this property must be found in cultured neurones if they are to be useful as a model of the nervous system. In fact, most nerve cells in culture appear to be able to generate action potentials at least near the cell body which is the usual recording site. Often many cells, especially those in explant cultures, are spontaneously active (Figure 2.6). Action potentials have been demonstrated with both extracellular and intracellular recording techniques. The success rate with intracellular electrodes depends on the extent of membrane damage caused by the electrode penetration and it is frequently necessary to hyperpolarize

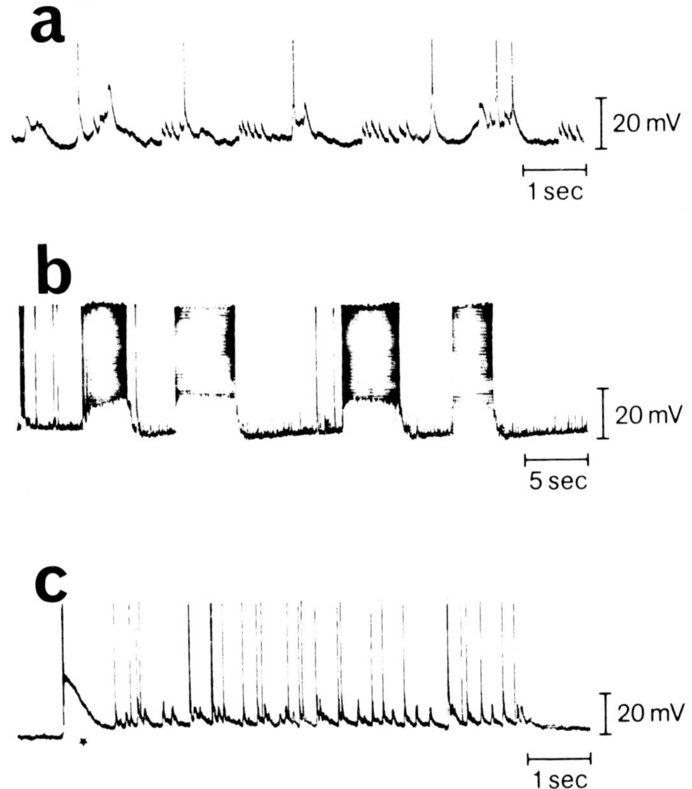

Figure 2.6: Spontaneous Electrical Activity Recorded Intracellularly from Large Neurones in Explant Cultures of Different Brain Areas
(a) Part of a spike train from a cerebellar Purkinje cell displaying random activity with no obvious firing pattern. Note the occurrence of action potentials and both excitatory and inhibitory postsynaptic potentials. Resting membrane potential –64 mV. (b) Phasic activity of a hypothalamic cell from the supra-optic nucleus area. Resting membrane potential –58 mV. (c) Long-lasting depolarizations (*) recorded from a hippocampal pyramidal cell. Resting membrane potential –75 mV. From Gähwiler (1979), with permission.

the membrane potential to a level above the threshold for generation of action potentials.

Crain (1956) was the first to demonstrate that cultured nerve cells could generate action potentials. Using explant cultures of dorsal root ganglia from chick embryos, he found action potentials with very fast rising phases and amplitudes of 80-95 mV. Similar findings were made on isolated neurones in dissociated cell cultures of chick embryo (Scott *et al.*, 1969; Varon and Raiborn, 1971) and mammalian dorsal root ganglia (Peacock *et al.*, 1973b), cells cultured from spinal cord (Fischbach, 1970; Peacock *et al.*, 1973b) and brain tissue (Hild and Tasaki, 1962; Nelson and Peacock, 1973; Godfrey *et al.*, 1975). Most of the recorded action potentials were large enough to cause a temporary reversal of the membrane potential, i.e. they had a positive overshoot, and they were relatively brief events, lasting for only a few milliseconds. Some of the quantitative values for action potential parameters in different cell types are given in Table 2.3.

Most studies on action potentials of cultured nerve cells have dealt with the ionic basis of the action potential; these findings are described below under the separate types of ion channels. A few, mostly earlier studies, were concerned with the conduction of action potentials along the processes of cultured neurones. Using extracellular electrodes, Hild and Tasaki (1962) stimulated processes of cerebellar nerves in explant cultures and found that they could record the evoked action potentials in the cell body. More detailed studies were performed on dissociated cultures of dorsal root ganglia from chick embryos and fetal mice (Scott *et al.*, 1969; Varon and Raiborn 1971; Okun, 1972). Since the length of individual processes in these cell cultures could be measured, the conduction velocity of the action potentials could be calculated. The values obtained (0.05-0.58 m/s) correspond with what would be expected for non-myelinated axons that are only a few microns in diameter. Similar conduction velocities were found in cultured spinal neurones (Fischbach and Dichter, 1974). Although many cells in cultures of dorsal root ganglia have a Ca^{2+} component in the action potentials recorded in the cell body (see below), the conduction of action potentials along processes appears to be entirely dependent on spikes caused by movement of Na^+ ions (Dichter and Fischbach, 1977).

Na^+ Channels. Most evidence indicates that the major portion of inward current during action potential spikes in cultured nerve cells is carried by Na^+ ions. Na^+ dependence of action potentials is usually

Table 2.3: Characteristics of Action Potentials of Nerve Cells in Primary Cultures

Tissue type	Species	Amplitude (mV)	Overshoot (mV)	$\dot{V}$ max. (V/s)	Duration (ms)	Tetrodotoxin resistance	Ca^{2+} Spikes	After hyperpolarization	Reference
Cerebellum	cat or rat	50 – 70	10 – 30	n.d.	1.5 – 3.0	n.d.	n.d.	n.d.	Hild and Tasaki (1962)
Cerebral cortex	rat	78 ± 2	15 ± 2	259	1.1 ± 0.1	no	no	in some cells	Dichter (1978)
Hippocampus	mouse	n.d.	n.d.	139 ± 13	0.9 ± 0.1	n.d.	n.d.	yes	Peacock (1979)
Spinal cord	chick embryo	~60	~9	132 ± 17	1.3 ± 0.1	no	no	yes	Fischbach and Dichter (1974)
	mouse	n.d.	n.d.	102 ± 9	n.d.	n.d.	n.d.	no	Peacock *et al.* (1973b)
	mouse	~70	~10	244 ± 36	0.6 ± 0.2	no	yes	no	Heyer and MacDonald (1982a)
Dorsal root ganglia	chick embryo	80 – 95	~30	300 – 500	2 – 4	n.d.	probably	yes	Crain (1956)
	chick embryo	70 – 90	20 – 30	150 – 350	1.5 – 4	n.d.	n.d.	in some cells	Scott *et al.* (1969)
	chick embryo	~85	36 ± 1	174 ± 10	2.8 ± 0.2	yes	yes	yes	Dichter and Fischbach (1977)
	mouse	n.d.	n.d.	144 ± 16	n.d.	n.d.	n.d.	yes	Peacock *et al.* (1973b)
	mouse	~75	~25	105 ± 11	2.0 ± 1	yes	yes	yes	Heyer and MacDonald (1982a)
	guinea pig	n.d.	n.d.	121 ± 11	n.d.	yes	yes	n.d.	Fukada and Kameyama (1980)
	human	~80	24 ± 1.4	n.d.	2.3 ± 0.1	n.d.	n.d.	yes	Scott *et al.* (1979)
Nodose ganglia	guinea pig	n.d.	~30	108 ± 2	n.d.	yes	yes	n.d.	Fukada and Kameyama (1980)
Sympathetic ganglia	chick embryo	85 ± 2	19 ± 3	60 ± 5	4 ± 0.4	no	no	yes	Chalazonitis *et al.* (1974)
	rat	<100	~20	~230	~1.5	no	yes	yes	O'Lague *et al.* (1978b)
	guinea pig	n.d.	~30	62 ± 8	n.d.	no	yes	yes	Fukada and Kameyama (1980)
Chromaffin cells	gerbil	50 – 70	n.d.	100 – 220	~2	no	n.d.	yes	Biales *et al.* (1976)
	rat	50 – 60	10 – 15	50 – 70	~5	no	yes	yes	Brandt *et al.* (1976)
Ciliary ganglia	chick embryo	76 ± 1.4	~20	n.d.	2.3 ± 0.2	n.d.	n.d.	yes	Tuttle *et al.* (1980)

Notes:
Where possible, values are given as means ± standard error of the mean.
n.d. = not determined.

demonstrated by replacement of the Na^+ in the external medium by an impermeant ion or by sucrose, or by the presence of a blocking effect of tetrodotoxin (Figure 2.7) which specifically blocks the Na^+ channels involved in action potentials. Complete blockade of action potentials by low concentrations of tetrodotoxin has been found in chick embryo

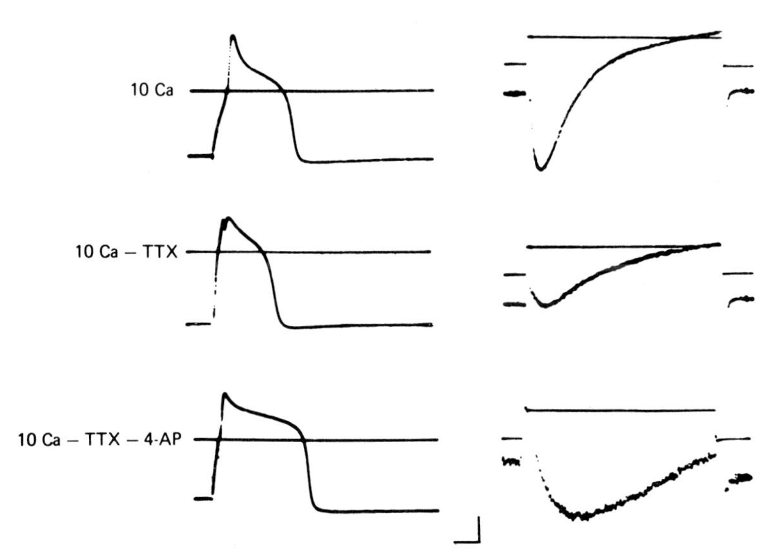

Figure 2.7: Action Potentials (left) and Membrane Currents (right) Recorded from the Same Cells in Cultures of Chick Embryo Dorsal Root Ganglia. Immediately after the spikes were recorded (in the indicated bathing medium), the cells were voltage clamped and the membrane currents recorded during a command voltage step of 60 mV (from –50 to +10mV). Top panels: 10 mM Ca^{2+}. Middle panels: 10 mM Ca^{2+} plus 10^{-7} tetrodotoxin (TTx) g/ml. Lower panels: 10 mM Ca^{2+}, 10^{-7} TTx g/ml and 1 mM 4-aminopyridine (4-AP). Horizontal lines in the action potential panels indicate 0 mV. In the voltage clamp panels the upper traces represent membrane potential and the lower membrane current. Calibrations: 20 mV, 5 ms (left top and middle); 20 mV, 10 ms (left lower); 10 nA, 50 mV, 2 ms (right top and middle); 5 nA, 50 mV, 2 ms (right lower). From Dunlap and Fischbach (1981), with permission.

sympathetic (Chalazonitis *et al.*, 1974) and spinal cord nerves (Fischbach and Dichter, 1974), sympathetic neurones from superior cervical ganglia of neonatal rats (Obata, 1974), fetal mouse spinal cord cells (Ransom and Holz, 1977; though see also Heyer *et al.*, 1981), and nerves from the cerebral cortex of fetal rats (Dichter, 1978).

In some cultured nerves, a portion of the action potential is resistant to tetrodotoxin, although often this depolarization is reduced by removal of external Na^+ ions (Figure 2.7). In these cells, complete blockade of depolarization usually requires removal of Ca^{2+} ions or the presence of a Ca^{2+} channel blocker such as Co^{2+} or Mn^{2+}. Cells found to have tetrodotoxin-resistant Na^+ channels and separate Ca^{2+} channels include adrenal chromaffin cells (Biales *et al.*, 1976; Brandt *et al.*, 1976), dorsal root ganglion nerves (Dichter and Fischbach, 1977; Ransom and Holz, 1977; Matsuda *et al.*, 1978; Heyer and MacDonald, 1982a; Fukada and Kameyama, 1980), neurones from trigeminal and nodose ganglia from guinea pigs (Fukada and Kameyama, 1980) and cells from embryonic chicken cerebral cortex (Mori *et al.*, 1982).

The relative contribution of tetrodotoxin-sensitive and insensitive channels in nerve cells cultured from different autonomic ganglia of guinea pigs was assessed by Fukada and Kameyama (1980) by measuring the maximum rate of depolarization ($\dot{V}_{max}$) in the presence and absence of tetrodotoxin and with Ca^{2+} channels blocked by Co^{2+}. As shown in Table 2.4, there are large differences in the amount of

Table 2.4 Relative Amplitudes of Na^+ and Ca^{2+} Action Potentials in Peripheral Nerves Cultured From Adult Guinea Pigs

Tissue	Na^+ spike $\dot{V}_{max}$ (V/s)	Tetrodotoxin-resistant Na^+ spike $\dot{V}_{max}$ (V/s)	Ca^{2+} spike $\dot{V}_{max}$ (V/s)
Trigeminal ganglia	96 ± 6	18.6 ± 1.7	5.3 ± 0.7
Nodose ganglia	108 ± 2	6.1 ± 1.5	2.9 ± 0.4
Sympathetic ganglia	62 ± 8	0	3.0 ± 0.5
Dorsal root ganglia	121 ± 11	55 ± 5.5	8.7 ± 1.0

Note:
All spikes were measured after hyperpolarization to membrane potentials of −80 to −140 mV in order to remove inactivation. Na^+ spikes were measured in the presence of 1 mM Co^{2+} and 10 mM tetraethylammonium (to block Ca^{2+} and K^+ channels, respectively). Tetrodotoxin-resistant Na^+ spikes were measured in Co^{2+} and tetraethylammonium solution plus 3 μM tetrodotoxin. Ca^{2+} spikes were measured in tetraethylammonium solution containing no Na^+ (replaced by Tris). Values are means ± standard error of the mean. From Fukada and Kameyama (1980).

tetrodotoxin-resistant but Na^+ dependent depolarizations, although the fast Na^+ channel density appeared similar in the different cells. The largest tetrodotoxin-resistant Na^+ spike was found in dorsal root ganglion cells, which also had the largest Ca^{2+} dependent spike (see p. 26).

Na^+ channels could be characterized further by ion flux measurements, but despite the popularity and success of these techniques with neuroblastoma cells (see Chapter 3) there have been few experiments on primary cultures of neurones, even though relatively pure neural cultures can be prepared. Beale *et al.* (1980) studied cultures of rat cerebellum and showed that $^{22}Na^+$ uptake could be stimulated by veratridine and that this effect could be blocked by low concentrations of tetrodotoxin. The amount of Na^+ uptake increased with age in culture, suggesting that some maturation was taking place. A time-dependent increase in the number of binding sites for the Na^+ channel label, scorpion toxin was also found in cultures of fetal mouse brain (Berwald-Netter *et al.*, 1981). Another ion flux study was performed by Pado *et al.* (1980) on cultures prepared from chick embryo retina and telencephalon. Veratridine and scorpion venom stimulated Na^+ accumulation and this was blocked by tetrodotoxin. However, there have apparently been no studies on the ion fluxes through tetrodotoxin-resistant Na^+ channels, and only a preliminary attempt at studying Ca^{2+} channels (Percy *et al.*, 1981).

The use of labelled toxins which bind to sites on the Na^+ channel provides a more direct means of assessing the numbers of channels. Catterall (1981c) found that ^{125}I-scorpion toxin bound specifically to spinal cord neurones in culture. From analysis of autoradiographs it was apparent that the distribution of toxin binding was not uniform. Often, neurones had regions at the origin of one long neurite which had about seven times more labelling than elsewhere on the same cell. Possibly this region corresponds to an area with a lower threshold for initiation of action potentials. Electrophysiological evidence for the presence of such a region in cultured nerve cells was presented by Fischbach and Dichter (1974).

Although it is technically possible, there have been few attempts to study properties of action potential channels in cultured neurones with voltage clamp methods. Dunlap and Fischbach (1981) voltage clamped the cell bodies of cultured dorsal root ganglion nerves. In control solutions, changing the membrane potential from -50 mV to $+10$ mV activated a fast, transient inward current. Since most of this inward current was eliminated by 0.1 μg/ml tetrodotoxin, it probably is carried by Na^+ ions (Figure 2.7). More recently, Na^+ channel properties of cultured adrenal chromaffin cells have been determined using single channel recording techniques (Fenwick *et al.*, 1982b). Since this method uses extracellular electrodes it may be easier to use on small cultured nerve cells. Fenwick *et al.* (1982b) found that the elementary Na^+ current amplitude was about 1 pA and the average channel open time was close to 1 ms.

There have been few investigations of the actions of drugs on the functioning of the Na^+ channels of nerves in culture. As many centrally-active drugs may modify the behaviour of voltage sensitive ionic channels, it can be predicted that this area will receive more attention in the near future. One example of such a study was described by Barker *et al.* (1978a, b, 1980) who reported that Leu-enkephalin could modify the excitability of cultured spinal cord nerves. In about one-third of the cells tested, enkephalin increased the threshold for the generation of action potentials. As this effect was blocked by the specific antagonist naloxone, it was taken to be mediated by some type of opiate receptor, although the underlying mechanism is not known.

Ca^{2+} Channels. As mentioned in the preceding section, some cultured neurones have action potentials that are not wholly Na^+-dependent. Most evidence indicates that the ion responsible for these action potentials is Ca^{2+} and that it is flowing through specific channels, distinct from Na^+ channels.

In retrospect, the presence of Ca^{2+} channels can be suspected after close examination of records of the first intracellularly recorded action potentials (Crain, 1956). The action potentials of some of these chick embryo dorsal root nerves have clearly discernible breaks in the falling phase so that there is almost a plateau before complete repolarization. This suggests that there may be an additional inward current that has slower kinetics than the rapidly activating and inactivating Na^+ current. Peacock *et al.* (1973b) and Dichter and Fischbach (1977) confirmed that action potentials recorded from dorsal root ganglion nerves in cell culture had inflections on their falling phase. In contrast, action potentials of cultured spinal cord nerves have a smooth, monophasic repolarization. Dorsal root ganglion cells can generate active responses in the absence of external Na^+ or in the presence of tetrodotoxin (Figure 2.7). These responses are proportional to the concentration of Ca^{2+} in the bathing solution and blocked by Co^{2+} (Matsuda *et al.*, 1976). In dorsal root ganglion nerves cultured from chick embryo, the maximum rate of depolarization in the absence of Na^+ is only about 20 V/s, compared with the normal value of about 170 V/s (Dichter and Fischbach, 1977). It should be noted that nerves such as spinal cord cells may have functional Ca^{2+} channels, although their contribution under normal circumstances may be too small to be detected (Heyer *et al.*, 1981) (see also Table 2.3).

Although the properties of the Ca^{2+} currents of dorsal root ganglion neurones in primary culture have not been studied in detail, there are broad similarities with the properties of Ca^{2+} currents of N1E-115 neuroblastoma cells (Moolenaar and Spector, 1979a,b; see Chapter 3, p. 65). The current activates about −50 mV, reaches a peak near 0 mV and has a reversal potential in the region of +50 to +80 mV

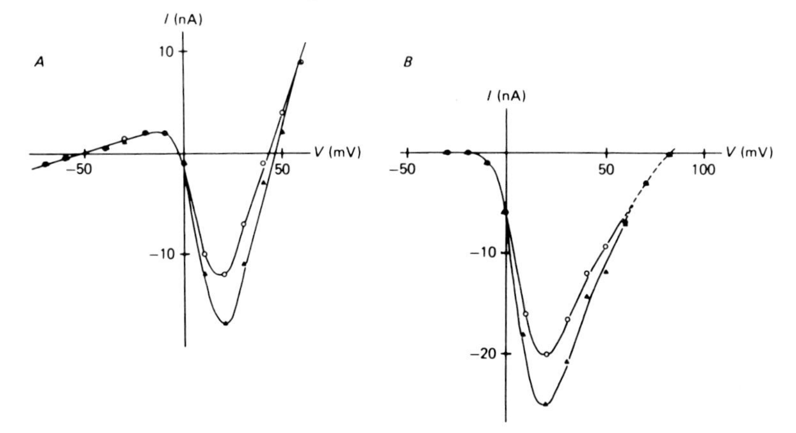

Figure 2.8: Effect of Noradrenaline on the Voltage-Ca^{2+} Current Relationship in a Cultured Dorsal Root Ganglion Cell
Maximum inward currents plotted against voltage for a cell bathed in 10 mM Ca-TTX-TEA (tetraethylammonium) in the presence (○) and absence (▲) of 10^{-4}M noradrenaline. The curve in A was not corrected for leakage conductance. The result of subtracting the leakage conductance is shown in B. Time-invariant outward leakage currents that remained in the presence of 10 mM Co^{2+}-TTX-TEA were averaged in three cells. This 'average leakage curve', which was concave upward, was subtracted from the data in A. The null potential was approximated by extrapolation. From Dunlap and Fischbach (1981), with permission.

(Figure 2.8) (Dunlap and Fischbach, 1981). Single channel Ca^{2+} currents have been recorded with patch clamp techniques from dorsal root ganglion cells (Brown *et al.*, 1982) and from cultured adrenal chromaffin cells (Fenwick *et al.*, 1982b).

In a study on explant cultures of dorsal root ganglia from fetal mice, Matsuda *et al.* (1978) found that not all nerves had Ca^{2+}-dependent responses and the authors postulated that Ca^{2+} channels were lost during maturation. In neurones cultured from *Xenopus* embryos, a very marked developmental transition does occur (Spitzer and Lamborghini, 1976). After a few hours in culture, action potentials are about 100 ms long, and are insensitive to low Na^{+} or tetrodotoxin although blocked by Co^{2+} or La^{3+}. By 24 h in culture the action potentials are shorter (about 10 ms) and have a fast rising phase, which is blockable by tetrodotoxin or low Na^+, and a plateau which can be reduced by Co^{2+} or La^{3+}. After 3-4 days in culture, the nerves have action potentials which are only a few ms long and which are completely dependent on Na^+. At this stage Co^{2+} and La^{3+} have no effect on the action potentials. Similar developmental sequences have been described for neuroblastoma cells and for cardiac and skeletal muscle in culture. However, Dichter and Fischbach (1977) argue that most adult dorsal root ganglion cells probably do have functional Ca^{2+} channels, although no distinct plateau phase may be visible, and that the presence of Ca^{2+}-dependent action potentials in cultured nerves is not an artefact or a reflection of the immaturity of the cells.

Release of neurotransmitters is strongly dependent on Ca^{2+} but the properties of the Ca^{2+} channels involved are hard to study directly because of the small size of most nerve terminals. One possible use of cultured nerve cells is to provide an accessible model for neural Ca^{2+} channels, so that the actions of neurotransmitters or drugs that may act on nerve endings by affecting Ca^{2+} fluxes can be studied. Dunlap and Fischbach (1978) demonstrated that the duration of the plateau phase of the action potential of cultured dorsal root ganglion cells could be shortened by up to 60 per cent by local application of GABA, noradrenaline or 5-hydroxytryptamine, though not by acetylcholine or glycine. Glutamate gave a small, inconsistent response. As GABA, noradrenaline and 5-hydroxytryptamine had little affect on the amplitude of the initial spike, they were not likely to be acting on Na^+ channels, and because similar effects were obtained in cells whose K^+ channels were presumed to be blocked by Ba^+, it was suggested that the neurotransmitters acted on Ca^{2+} channels. Subsequently, Dunlap and Fischbach (1981) demonstrated by voltage clamp experiments that noradrenaline, GABA and 5-hydroxytryptamine decreased the size of the inward Ca^{2+} current (Figure 2.8). It is not yet known whether this effect results from a decrease in the number of functioning channels or a reduction in the single channel conductance. However, these neurotransmitters do seem to act directly on the Ca^{2+} channels as they are not producing other obvious changes in membrane properties such as conductance or potential. Activation of naloxone-sensitive opiate receptors by enkephalins can also decrease the duration of action potentials of cultured dorsal root ganglion cells (Mudge *et al.*, 1979; Werz and MacDonald, 1982a). Phenobarbitone and pentobarbitone have both been shown to decrease Ca^{2+}-dependent action potentials in similar cultures (Heyer and MacDonald, 1982b). As pentobarbitone was about five times more active than phenobarbitone, it was suggested that such a difference may in part underlie the differences in the pharmacological profile of the sedative and the anticonvulsant barbiturates.

K^+ Channels. Determination of input resistance at different membrane potentials provides evidence for the presence or absence of voltage sensitive K^+ channels. If the membrane is completely passive, the current-voltage relationship will be linear (Figure 2.3A). If there are channels which open on depolarization and carry outward current, the measured input resistance in the depolarizing direction will be less than expected (Figure 2.3B). This phenomenon is often called delayed or outward rectification and is usually associated with the opening of depolarization-activated K^+ channels. Another type of behaviour is when input resistance falls during hyperpolarization (Figure 2.3C) and this has been called anomalous or inward rectification and is probably due to opening of the so-called 'inward rectifier' K^+ channels.

Most studies on the current-voltage relationship of cultured nerve cells have revealed the presence of delayed rectification. This is found in spinal cord and dorsal root ganglion cells from fetal mice (Ransom *et al.*, 1977c) and from chick embryos (Fischbach and Dichter, 1974; Dichter and Fischbach, 1977), brain cells from fetal rats (Godfrey *et al.*, 1975), superior cervical ganglion nerves from newborn rats (O'Lague *et al.*, 1978b) and rat adrenal chromaffin cells (Brandt *et al.*, 1976) (Figure 2.9). The majority of cells behaved linearly in the hyperpolarizing direction, although changes similar to anomalous rectification were seen in some mouse spinal cord and dorsal root ganglion cells (Peacock *et al.*, 1973b; Ransom *et al.*, 1977c), some rat brain cells (Godfrey *et al.*, 1975) and chick spinal cord neurones (Fischbach and Dichter, 1974; Brookes, 1978). Large hyperpolarizations were usually necessary to reveal this type of rectification.

There was no delayed rectification in response to depolarizing pulses in the cells examined by Lawson and Biscoe (1973). In fact, input resistance increased with depolarization and decreased with hyperpolarization. It is possible that this behaviour reflects the functional immaturity of these cells as action potentials could not be stimulated by depolarization in any of the cells.

The delayed rectification in sympathetic nerves was reduced by the K^+ channel blocker tetraethylammonium (O'Lague *et al.*, 1978b) and also by addition of Co^{2+} (Figure 2.9). Assuming that Co^{2+} is acting in its normal manner to block Ca^{2+} channels, this result indicates a

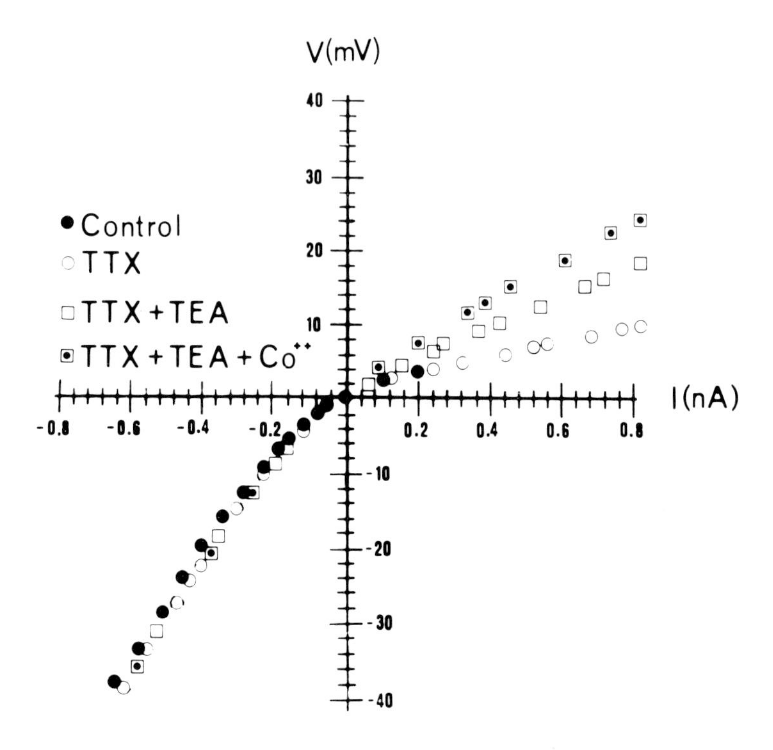

Figure 2.9: Current-Voltage Relationship of a Rat Sympathetic Neurone
The cell was impaled with two microelectrodes, one for passing current (I; duration of pulse 80 ms) and one for measuring the voltage response (V). In control solution (●), there is evidence of pronounced delayed rectification although the measurable range in the depolarizing direction is limited because of action potential generation. Delayed rectification is more obvious when action potentials are blocked with tetrodotoxin (TTx, 3 μM, ○). Addition of the potassium channel blocker, tetraethylammonium (TEA, 5 mM, □) reduces delayed rectification, and this is further reduced by addition of the calcium channel blocker, Co^{2+} (2.5 mM,▣). Drug effects were reversible. From O'Lague *et al.* (1978b), with permission.

contribution of Ca^{2+} channels to the membrane conductance and possibly the presence of a Ca^{2+}-activated K^+ conductance. Studies on action potentials (discussed below) support the suggestion that these cells have K^+ channels which can be activated by Ca^{2+} ions.

Evidence for the involvement of K^+ channels during action potentials is obtained from findings that the K^+ channel blocking compounds tetraethylammonium and 4-aminopyridine (Figure 5.4) can prolong the action potentials of cultured sympathetic (O'Lague *et al.*, 1978b) and sensory nerves (Dunlap and Fischbach, 1981) (Figure 2.7).

In many nerve cells in culture the action potential is followed by a variable period of hyperpolarization (see Table 2.3). This is more pronounced in dorsal root ganglion cells than in spinal cord cells (Peacock *et al.*, 1973b; Ransom *et al.*, 1977c). In sympathetic (O'Lague *et al.*, 1978b), dorsal root ganglion (Ransom *et al.*, 1977c; Dunlap and Fischbach, 1981) and adrenal chromaffin cells (Biales *et al.*, 1976) the direction of the after-potential reverses when the membrane potential is altered to about –85 mV, i.e. close to the equilibrium potential for K^+, suggesting that the hyperpolarizing after-potential is a consequence of the opening of K^+ channels. On repetitive stimulation of dorsal root ganglion or spinal cord neurones in culture, a post-tetanic hyperpolarization occurs (Ransom *et al.*, 1975, 1977c). In dorsal root ganglion cells this also involves an increase in K^+ permeability but in spinal cord cells an electrogenic Na^+ pump appears to be the major mechanism.

Long-lasting after-hyperpolarizations found in older cultures of sympathetic nerves (O'Lague *et al.*, 1978b), in cultured nerves from dorsal root ganglia (Dichter and Fischbach, 1977; Dunlap and Fischbach, 1981) and ciliary ganglia (Bader *et al.*, 1982) may result from a Ca^{2+}-activated increase in K^+ permeability. In most dorsal root ganglion cells, blockade of inward Ca^{2+} currents by Co^{2+} substitution led to a decrease in the late outward K^+ current, although Co^{2+} did not affect the outward current of some cells (Dunlap and Fischbach, 1981). More direct evidence for a Ca^{2+}-activated K^+ conductance was provided for cultured adrenal chromaffin cells (Marty, 1981) and for sympathetic ganglia (Adams *et al.*, 1982). The channels had a short open time but a large single channel conductance. This K^+ current in sympathetic nerves was very sensitive to tetraethylammonium but not to 4-aminopyridine (Adams *et al.*, 1982), although a similar current in parasympathetic cells in culture was found to be insensitive to tetraethylammonium but sensitive to Cs^+ (Bader *et al.*, 1982).

Receptor Properties

The ability to study the actions of potential neurotransmitter compounds on identified cells has been a great stimulus to use isolated nerves in tissue culture. The prospect of detailed investigations into the mechanisms of centrally-acting drugs has also attracted many workers to use cultured nerve cells. However, there is always the worry that properties expressed in culture may not reflect accurately the behaviour of cells *in vivo*. Evidence from other systems, such as skeletal muscle, is encouraging for it indicates that the molecular properties of receptors of cells in culture are very similar to those of cells *in situ*. With the increasing capability to identify the source and type of neurones in culture, studies on cultured nerves can be expected to make an even greater contribution to the understanding of the physiological and pharmacological properties of receptor mechanisms in the nervous system. The many different types of receptor that have been found on various cultured neurones are dealt with individually in the following sections.

Acetylcholine Receptors

The presence of receptors for acetylcholine in cultures of nervous tissue was implied by the finding that tubocurarine enhanced the rhythmic electrical activity of explants of spinal cord and cerebral cortex (Crain and Bornstein, 1964). However, a direct demonstration of acetylcholine responses in cultured nerve cells was only provided ten years later: Chalazonitis *et al.* (1974) found that about 20 per cent of cells in cultures of dissociated sympathetic ganglia from chick embryos were depolarized by iontophoretic application of acetylcholine.

Cholinoceptors on Peripheral Nerves. In cultured sympathetic cells the average sensitivity to acetylcholine reported by different workers is quite variable. In cell cultures of chick embryo sympathetic ganglia, sensitivity was low (< 10 mV/nC, i.e. the peak depolarization induced by a given electric charge passed through the acetylcholine-containing electrode), but other workers have found higher sensitivities on sympathetic nerves cultured from the superior cervical ganglia of neonatal rats: up to 300 mV/nC (Obata, 1974) and 500-2500 mV/nC (O'Lague *et al.*, 1978c). The depolarization induced by acetylcholine in superior cervical ganglion cells in culture is accompanied by an increase in membrane conductance and the response has an equilibrium potential of around -30 mV (Ko *et al.*, 1976a). This response appears to be mediated by nicotinic cholinoceptors as hexamethonium, mecamylamine and tubocurarine could abolish it. Alpha-bungarotoxin had no effect on responses of cells in cultures of superior cervical ganglia but this toxin selectively blocks at the nicotinic receptors of skeletal muscle. Difficulties associated with its use as a receptor label with cultured nerve cells are discussed at the end of this section (p. 35).

Muscarinic receptors have also been found on some sympathetic nerves in culture. Iontophoresis of acetylcholine could give monophasic or biphasic depolarizations, depending on the length of drug application (Freschi and Shain, 1980). The second phase could last up to 1 min. and was blocked by atropine. The muscarinic response was a slow inward current associated with a decrease in input resistance (Freschi, 1982). As expected, tubocurarine reduced the size of the initial fast response but not the slow phase. Similar slow responses to muscarinic agonists were found with some nerves in explant cultures of bullfrog sympathetic ganglia (Gruol *et al.*, 1981). Exposure of cultured chick embryo sympathetic ganglia to the muscarinic agonists methacholine and arecoline stimulated the release of noradrenaline (Greene and Rein, 1978). This effect was very sensitive to atropine (the concentration for 50 per cent inhibition, IC_{50}, was about 10^{-8} M), whereas stimulation by nicotine was resistant to atropine (IC_{50} about 3×10^{-5} M) but readily antagonized by the ganglion blocker mecamylamine.

Adrenal chromaffin cells in tissue culture also have functional acetycholine receptors (Biales *et al.*, 1976; Brandt *et al.*, 1976; Fenwick *et al.*, 1982a; Holz *et al.*, 1982; Kidokoro *et al.*, 1982). Sensitivity to iontophoretically-applied acetylcholine was 200-500 mV/nC (Brandt *et al.*, 1976) and the concentration of acetylcholine required to produce half the maximal depolarization was about 10^{-5} M (Kidokoro *et al.*, 1982). From analysis of the voltage noise induced by acetylcholine, the average open time (τ) of the receptor-associated channels was calculated to be about 40 ms (measured at a membrane potential of about −75 mV) and the single channel conductance (γ) was estimated to be about 23 pS (Kidokoro *et al.*, 1982). A more direct estimate of the properties of acetylcholine-activated ion channels has been obtained with patch clamp techniques (Fenwick *et al.*, 1982a). Single channel conductance was found to be about 44 pS and open time was about 27 ms at −80 mV and at room temperature. These values agree reasonably well with estimates of channel properties on intact ganglion preparations (Rang, 1981). However, there is disagreement about the pharmacological characteristics of the cholinoceptors on cultured chromaffin cells. Despite using the same species and similar culture methods, Brandt *et al.* (1976) found that the stimulatory action of acetylcholine on the frequency of spontaneous action potentials was blocked by 10^{-7} M atropine, but not by hexamethonium or tubocurarine, whereas Kidokoro *et al.* (1982) found that depolarization responses to acetylcholine were antagonized by hexamethonium but only by very high concentrations of atropine. Additionally, Kidokoro *et al.* (1982) reported that nicotine mimicked the effects of acetylcholine but muscarine and pilocarpine did not have any effect.

Similarly, Kumakura *et al.* (1980) and Holz *et al.* (1982) have reported that the carbachol- or nicotine-induced catecholamine release from cultured adrenal cells was blocked by nicotinic antagonists but not by atropine.

There is no obvious reason for the discrepancy between the results of Brandt *et al.* (1976) and Kidokoro *et al.* (1982), but is is possible that activation of muscarinic receptors may affect the threshold for action potential firing without producing depolarization. If that happened, Kidokoro *et al.* (1982) would not have measured any effect. Nevertheless, depolarization induced by stimulation of nicotinic receptors would be expected to influence the frequency of action potentials but this was not found by Brandt *et al.* (1976). Muscarinic receptors on cultured bovine chromaffin cells have been shown to be involved in the incorporation of ^{32}P into phosphatidic acid and phosphatidylinositol but not to be involved in $^{45}Ca^{2+}$ uptake or noradrenaline release, which were stimulated by activation of nicotinic receptors (Fisher *et al.*, 1981).

As would be predicted from the physiology of intact preparations, parasympathetic nerves in tissue culture also have receptors for acetylcholine. Ravdin and Berg (1979) found that cells from chick ciliary ganglia had an average sensitivity to iontophoretically applied acetylcholine of around 75 mV/nC. Responses were blocked by tubocurarine, although a high concentration was apparently required. In a similar study (Bader *et al.*, 1982), tubocurarine and hexamethonium blocked acetylcholine responses but atropine at 1 μM had no effect.

Sensory nerves cultured from dorsal root ganglia are not sensitive to acetylcholine (Obata, 1974), although some of the cells in such cultures retain properties characteristic of pain fibres, responding to bradykinin, capsaicin and prostaglandin E_2 (Baccaglini and Hogan, 1983). Other presumed sensory nerves cultured from nodose ganglia do respond to acetylcholine (Baccaglini and Cooper, 1982a, b). The concentration needed to produce half the maximal depolarization response was estimated to be about 3×10^{-5} M, and the responses were sensitive to hexamethonium but not to atropine.

Cholinoceptors on Central Neurones. About 30 per cent of cells tested in cultures of dissociated spinal cord from fetal mice were depolarized by acetylcholine (Ransom *et al.*, 1977a); however, there was no pharmacological characterization of the receptors. The acetylcholine responses had slow time courses and appeared to be associated with an increase in membrane resistance (Barker and Ransom, 1978b).

Subsequently it has been shown that muscarine depolarizes cultured spinal neurones, an action blocked by atropine (Nowak and MacDonald, 1983). The mechanism underlying this effect appears to be a decrease in a voltage-dependent K^+ current. Direct evidence for the presence of muscarinic cholinoceptors in cultures of spinal cord cells has been provided by the demonstration of specific binding of the muscarinic antagonist ^{3}H-quinuclidinyl benzilate (QNB) (Brookes and Burt, 1980). Over a three-week culture period the amount of QNB binding increased about fourfold to 340 pM/g protein.

Specific binding of QNB has also been demonstrated on membranes prepared from cultures of cerebral cortex (Dudai and Yavin, 1978) and cerebellum (Burgoyne and Pearce, 1982). The apparent dissociation constant for QNB was 0.5×10^{-9} M and the specific binding was displaced by muscarinic drugs more readily than by nicotinic drugs. The number of binding sites was found to increase about five times during the first three weeks in culture. Evidence for functional muscarinic cholinoceptors on some cultured cerebral cortex cells was obtained by Robbins *et al.* (1982) who found that acetylcholine stimulated somatostatin release by a receptor which was blocked by atropine but not by hexamethonium. Specific QNB binding has also been demonstrated on cultured glial cells (Repke and Maderspach, 1982).

Acetylcholine receptors are present on some cells in cultures of chick embryo retinas because carbachol induces an increased uptake of $^{22}Na^+$ (Betz, 1981). This response was partly blocked by hyoscine and partly by tubocurarine, suggesting that both muscarinic and nicotinic cholinoceptors were involved.

α-Bungarotoxin Binding Sites. The snake venom toxin, α-bungarotoxin is a specific antagonist at nicotinic receptors of skeletal muscle (see Chapter 4) but it does not block the nicotinic cholinoceptors of autonomic ganglia. However, in 1973 Greene and his colleagues reported that ^{125}I-α-bungarotoxin bound specifically to sympathetic neurones in cell cultures of chick embryo paravertebral ganglia. The density of the sites was 100-200 /μm^2 and the binding was inhibited by preincubation with nicotinic agonists and antagonists. From this work, the authors concluded that α-bungarotoxin could be used as a probe for acetylcholine receptors of cultured nerves. Similar binding studies were subsequently performed on nerves cultured from chick embryo retinas (Vogel *et al.*, 1976). However, there was no evidence that the α-bungarotoxin binding sites were on physiologically functional acetylcholine receptors. It has since been demonstrated that α-bungarotoxin does not inhibit the acetylcholine responses of cultured sympathetic (Carbonetto *et al*., 1978; Dvorak *et al*., 1978; Kouvelas *et al*., 1978) or parasympathetic nerves (Ravdin and Berg, 1979) or of retinal cells (Betz, 1981). Some other toxins have been isolated along with α-bungarotoxin and these have been reported to block responses of ciliary ganglion cells in culture (Ravdin and Berg, 1979). If these toxins were to become more readily available, more useful experiments on labelling neural acetylcholine receptors could be performed. In view of the very large number of snake toxins that have been isolated and characterized (see Karlsson, 1979; Dufton and Hider, 1983), it is perhaps surprising that so few toxins have been tried as labels for cholinoceptors on nerves.

The significance of the α-bungarotoxin binding site is unknown at present. They perhaps represent acetylcholine receptors which have not been fully assembled and hence have no ion channel. Certainly the binding properties are similar to those of cholinoceptors. The numbers of the α-bungarotoxin sites in ciliary ganglion cultures is reduced by pretreating with carbachol (Messing, 1982). However, in cultures of nerves from chick retina, carbachol pretreatment decreases QNB but not α-bungarotoxin binding (Betz, 1982).

Monoamine Receptors

There have been few investigations of the actions of monoamine transmitters on nerves in culture and no detailed mechanistic studies have been published. Most experiments on nerves of peripheral origin have yielded negative results. For example, nerves in cultures of dorsal root ganglia do not alter their membrane potential in response to noradrenaline (Obata, 1974). Sympathetic nerves in cultures of superior cervical ganglia do not respond to iontophoretic application of adrenaline, noradrenaline or dopamine (Obata, 1974; O'Lague *et al*., 1978c), although very high concentrations of adrenaline and noradrenaline could cause some hyperpolarizations of cells in explants of adult bullfrog sympathetic ganglia (Gruol *et al*., 1981). About half the cells in cultures of nodose ganglia respond with depolarization to 5-hydroxytryptamine but not to adrenaline (Baccaglini and Cooper, 1982b).

In contast to these rather negative reports on cultured peripheral nerves, most investigations of actions of monoamine transmitters on central nerves in culture have revealed some activity. In dissociated cell cultures of fetal rat brain, two out of eight cells tested were found to hyperpolarize in response to noradrenaline (Godfrey *et al*., 1975). In contrast, depolarization of some cells in cultures of dissociated neonatal mouse brain by bath application of noradrenaline, dopamine, and 5-hydroxytryptamine was reported by Bonkowski and Dryden (1976), although iontophoretic application of noradrenaline caused hyperpolarization, while 5-hydroxytryptamine could be both depolarizing and hyperpolarizing (Bonkowski and Dryden, 1977). Evidence for β-adrenoceptors in re-aggregate cultures of fetal mouse brain was provided by Wehner *et al.* (1982). Isoprenaline increased the levels of cyclic AMP, with half the maximal effect occurring about 10^{-8} M. Adrenaline was more potent than noradrenaline at increasing cyclic AMP levels, and these effects were blocked by propranolol, with 50 per cent inhibition at about 5×10^{-7} M. The presence of β-adrenoceptors was also indicated by the specific binding of the β-receptor antagonist, ^{125}I-iodohydroxybenzylpindolol, which bound to a single saturable site with an apparent dissociation constant of 100-300 pM. Specific binding was displaced by β-adrenoceptor agonists.

There have also been studies using cultures made from specific regions of the brain. In explants of neonatal rat hypothalamus, histamine could both increase or decrease the spontaneous activity recorded with extracellular electrodes (Geller, 1981). The specific H_2-receptor agonist dimaprit only inhibited action potential firing and its effects could be blocked by the H_2-receptor antagonists metiamide and cimetidine, but not by an H_1-receptor antagonist, promethazine. Attempts to associate the excitatory action of histamine with activation of H_1-receptors were not very successful. A specific H_1-agonist (2-(2 pyridyl)-ethylamine) did not produce consistent results, although there was an indication that its excitatory actions were decreased by promethazine but not metiamide.

Cell cultures of the lower brain stem of fetal rats were used to test the effects of a wide range of possible neurotransmitters (Saji and Miura, 1982). The drugs were applied by iontophoresis and the frequency of firing of spontaneous action potentials was recorded with extracellular electrodes. Noradrenaline, dopamine and 5-hydroxytryptamine inhibited the activity in most cells. The mechanisms underlying the actions of these compounds on action potential firing were not determined but dopamine (and adenosine) have been shown to increase the cyclic AMP levels of chick embryo retina cells in culture (De Mello *et al.*, 1982).

Some direct effects of noradrenaline and 5-hydroxytryptamine on the action potential characteristics of cultured dorsal root ganglion neurones have been studied in detail (Dunlap and Fischbach, 1978, 1981). Although neither noradrenaline nor 5-hydroxytryptamine had any effect on the membrane potential or membrane conductance of these cells, they both shortened the duration of the action potential by decreasing the inward Ca^{2+} current (Figure 2.8). This effect was concentration-dependent, the maximal shortening being about 60 per cent and the half maximal concentrations for both compounds being about 6×10^{-7}M. Noradrenaline's action would appear to be on α-adrenoceptors as it could be mimicked by phenylephrine and dopamine but not by isoprenaline, and it could be blocked by phentolamine but not propranolol. The type of receptor activated by 5-hydroxytryptamine has not been characterized although it was not affected by 10^{-4} M methysergide. The mechanism for the coupling of the α-adrenoceptor and the 5-hydroxytryptamine receptor to the Ca^{2+} channel is unknown but the culture system would appear to be an ideal model for further investigations on chemical modulation of voltage-sensitive ion channels.

Amino Acid Neurotransmitters

There is increasing evidence that some amino acids act as central neurotransmitters. L-glutamate and L-aspartate are thought to be excitatory transmitters, whereas glycine, γ-aminobutyric acid (GABA) and β-alanine may be inhibitory. Much of the pharmacological work with nerves in tissue culture has concentrated on examining the effects of these compounds. Mostly, primary cultures of spinal cord have been used as peripheral nerves are generally insensitive and there does not appear to be a neural cell line which is sensitive to the amino acid transmitter candidates.

Glutamate and Aspartate. Both glutamate and aspartate depolarize nerves (Figure 2.10) in spinal cord explant (Hösli *et al.*, 1971, 1976) and dissociated cell cultures (Ransom *et al.*, 1975, 1977a; MacDonald and Wojtowicz, 1982). Similar effects were found on both human and rat spinal cord in explant cultures (Hösli *et al.*, 1973). However, there are probably separate receptors for each amino acid as some spinal cord cells respond to glutamate but not aspartate (Ransom *et al.*, 1977a). Generally, aspartate is less potent than glutamate. Local application of glutamate to regions containing pre-synaptic terminals in cultured spinal cord cells caused an increase in the frequency of excitatory and inhibitory post-synaptic potentials (Ransom *et al.*, 1977a; Brookes, 1978) so that it would appear that glutamate can act pre-synaptically to modulate transmitter release. These synaptic areas tended to correspond to regions of the cell with higher than average

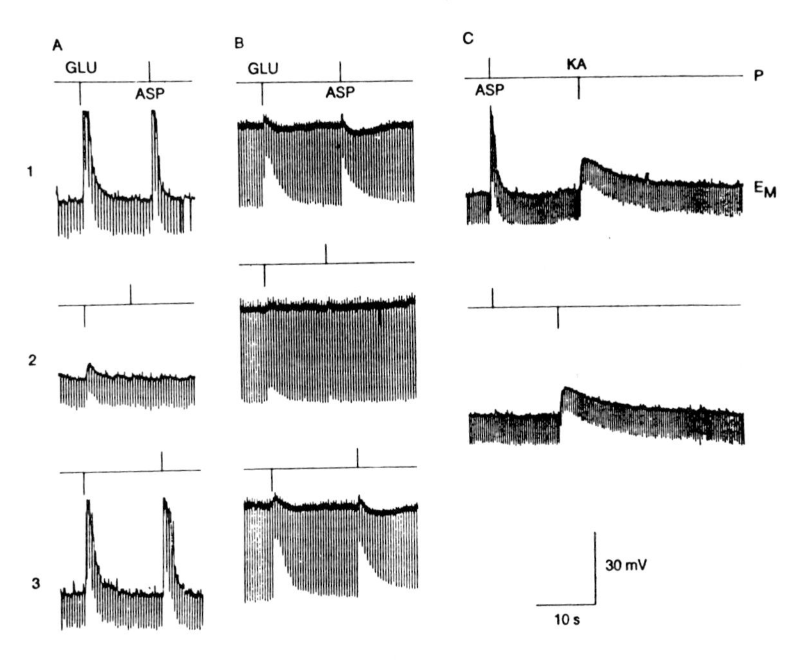

Figure 2.10: The Blockade by DL-α-aminoadipate of both the Increase and Decrease of Membrane Conductance (G_m) in a Cultured Brain Neurone with Application of Glutamate (GLU) and Aspartate (ASP) but the Sparing of the Response to Kainate (KA)
(A) Applications of glutamate (100 μM; 90 μM free glutamate) indicated by downward deflections and aspartate (500 μM; 300 μM free aspartate) by upward deflections (P). Membrane potential was −70 mV. During co-application of DL-α-aminoadipate (2 mM), control depolarizations associated with decreased G_m (A1) were greatly diminished (A2) but showed full reversal (A3). (B) Depolarization of this neurone to −20 mV converts this response to an increase in G_m (B1). Note that hyperpolarizing voltage transients are larger than in (A) because current pulses were increased from 0.15 nA to 0.06 nA. DL-α-aminoadipate (2 mM) blocks these responses (B2) which was followed by recovery (B3). (C) Recording from the same neurone at −70 mV. In this case the glutamate pipette was replaced by one containing kainate (100 μM). Note that the control measure response to aspartate (C1) was again blocked by the same concentration of DL-α-aminoadipate but that to kainate was spared (C2). Recovery was complete but not shown. From MacDonald and Wojtowicz (1982), by permission of the National Research Council of Canada.

sensitivity to glutamate, suggesting that there may be a concentration of glutamate receptors at synapses.

Some brain cells are also sensitive to glutamate (Geller and Woodward, 1974; Godfrey *et al.*, 1975; Pearce and Dutton, 1982) and all cells in cultures of rat brain stem were found to be stimulated by glutamate (Saji and Miura, 1982). In contrast, cells in cultures of superior cervical (Obata, 1974; Wakshull *et al.*, 1979a), dorsal root (Obata, 1974; Ransom *et al.*, 1977a) and nodose ganglia (Baccaglini and Cooper, 1982b) are not affected by glutamate.

The depolarization of spinal cord cells by glutamate and aspartate is dependent on the presence of Na^+ in the external medium (Hösli *et al.*, 1973, 1976) but glutamate responses are unaffected by changes in the concentration of Cl^- (Barker and Ransom, 1978a). The null potential for the glutamate response was estimated to be around -20 mV (Ransom *et al.*, 1977a). A null potential for glutamate of 0 mV was determined by voltage clamp techniques (Barker *et al.*, 1977) and a value of -8 mV was recently determined by MacDonald and Wojtowicz (1982). Earlier work by these authors had revealed that glutamate may be able to activate more than one conductance mechanism in some cells (MacDonald and Wojtowicz, 1980; Wojtowicz *et al.*, 1981); this obviously complicates the determination of a single reversal potential. In addition, pre-synaptic stimulation by glutamate may lead to transmitter release and the added post-synaptic responses may alter the apparent reversal potential (Brookes, 1978).

Although glutamate and aspartate can be readily shown to depolarize spinal cord cells in tissue culture, the fundamental properties of the receptors have not yet been characterized. The limiting slope of the log response against log concentration graph for glutamate is close to one, suggesting that one molecule of glutamate is needed to activate each receptor (Barker and Ransom, 1978a). In studies on voltage clamped spinal cord cells, L-aspartic acid was found to decrease the slope conductance between holding potentials of -70 to -30 mV (Figure 2.11) (MacDonald *et al.*, 1982). The authors postulated that such an effect could result from activation of a voltage-dependent increase in the conductance to Na^+ and/or Ca^{2+} or a decrease in K^+ conductance. Since the Ca^{2+} channel blocker verapamil had no effect and since the change in input resistance produced by aspartic acid was blocked by Na^+ free solutions, it was suggested that aspartic acid activates a voltage-dependent Na^+ conductance.

In a related study, MacDonald and Wojtowicz (1982) found that cultured spinal nerves were depolarized by glutamate and aspartate and by several analogues (ibotenate, kainate, homocysteate and N-methyl-D-aspartate). However, the accompanying conductance changes were more variable as both increases and decreases were observed. At membrane potentials more negative than -30 mV, glutamate and aspartate

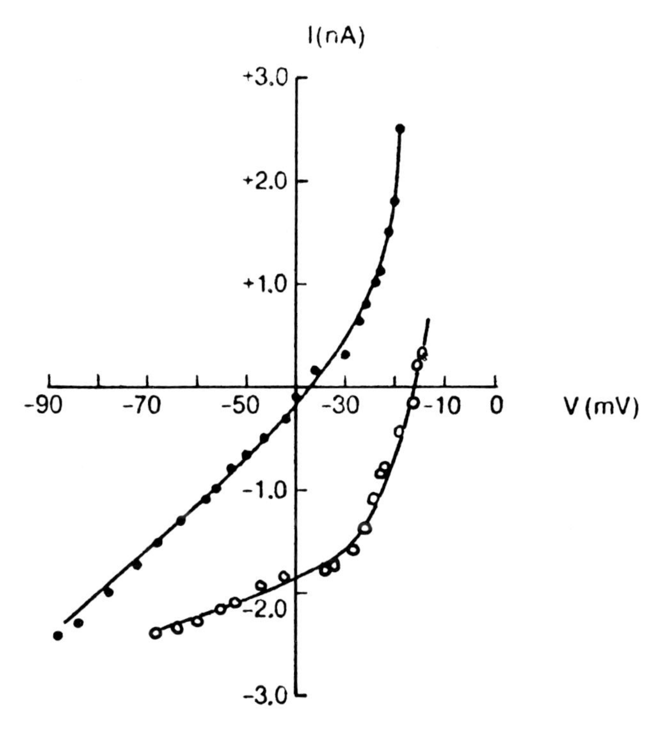

Figure 2.11: The Current-Voltage Relationship for a Cultured Spinal Neurone under Voltage Clamp, and the Effect of L-aspartic Acid
Resting membrane potential was –38 mV and both depolarizing and hyper-polarizing command steps were employed to construct the curve. The control curve (●) was performed in the bathing solution supplemented with tetrodotoxin (2 μM) and repeated during a constant microperfusion with L-aspartic acid (500 μM; ○). A net inward current (negative by convention) was evoked by L-aspartic acid. The slope of this steady-state relationship was reduced in the range from –70 to –30 mV. From MacDonald *et al.* (1982), with permission.

decreased membrane conductance in a concentration-dependent manner. The reversal potential for this effect was about –85 mV, and it was suggested that the response could be mediated by a decrease in K^+ permeability. It is not understood why the excitatory amino acids can produce two different effects on membrane conductance. Different receptors may be involved but there are few antagonists that could be used in attempts to distinguish them. D, L-α-aminoadipate was found to inhibit the responses to all the excitatory amino acids except kainate (Figure 2.10), but it was not specific in blocking the conductance increase or decrease (MacDonald and Wojtowicz, 1982).

Despite the lack of specific glutamate or aspartate antagonists, the responses of spinal cord cells in tissue culture to these amino acids have been shown to be sensitive to certain drugs. For example, pentobarbitone decreased depolarizations produced by glutamate (Ransom and Barker, 1975). This inhibition is probably a combination of a direct blocking action and an indirect shunting effect resulting from activation of a membrane conductance by pentobarbitone (Barker and Ransom, 1978b). The effect of pentobarbitone on membrane conductance can be reduced by the GABA antagonist picrotoxin (see next section), but this drug has no effect on the direct depressant action of pentobarbitone on glutamate responses.

The responses to glutamate are also decreased by Leu-enkephalin (Barker *et al.*, 1978a, b). The depression was rapid in onset and well sustained, but readily reversed. Increasing doses of enkephalin increased the blockade of glutamate responses up to a maximal level of about 50 per cent. On voltage clamped cells, enkephalin caused a slowing of the time course of the glutamate-induced current and the blocking effects were not obviously voltage-dependent, although the effect of enkephalin increased at higher concentrations of glutamate. Barker *et al.* (1978a) suggested that enkephalin acted in a non-competitive manner rather than by decreasing the affinity of the receptor for glutamate or by altering the reversal potential.

Prolonged application of glutamate to spinal cord cells in culture did not produce any desensitization (Hösli *et al.*, 1976; Ransom *et al.*, 1975, 1977a) but prolonged responses were followed by a long after-hyperpolarization. This hyperpolarization was not associated with changes in input resistance but was blocked by ouabain, suggesting that it was a consequence of activation of a Na^+ pump rather than opening of ionic channels (Ransom *et al.*, 1975).

GABA, Glycine and β-Alanine. The presence of receptors for GABA and glycine in cultures of the central nervous system was suggested initially by findings that the glycine antagonist strychnine (Corner and Crain, 1969; Zipser *et al.*, 1973) and the GABA antagonists bicuculline and picrotoxin (Zipser *et al.*, 1973; Crain and Bornstein, 1974) depressed inhibitory synaptic activity and enhanced excitatory potentials in 'organotypic' cultures of spinal cord, medulla, hippocampus and cerebral cortex (see Chapter 7). The first direct demonstration that GABA and glycine could act on brain cells in tissue culture, came with the findings that application of the compounds could decrease the frequency of spontaneous action potentials recorded with extracellular electrodes from cerebellar explant cultures (Geller and Woodward, 1974). Such GABA responses were blocked by bicuculline and picrotoxin but not by strychnine (Gähwiler, 1975). More recently both GABA and glycine have been found to inhibit action potential firing in cell cultures of fetal rat brain stem (Saji and Miura, 1982). Similarly, GABA and glycine can alter membrane potentials of some nerves in dissociated cultures of whole brain (Godfrey *et al.*, 1975; Bonkowski and Dryden, 1977) and of cerebral cortex (Dichter, 1980). Earlier, Hösli *et al.* (1971) had demonstrated that iontophoretic application of glycine could hyperpolarize nerve cells in spinal cord explants. Spinal cord cells in dissociated cultures are also sensitive to glycine and to GABA and β-alanine (Ransom and Barker, 1975; Ransom *et al.*, 1977a; Barker and Ransom, 1978a). In cultures of the peripheral nervous system, GABA was found to depolarize cells in dissociated cultures of dorsal root and superior cervical ganglia (Obata, 1974) and nodose ganglia (Baccaglini and Cooper, 1982b), and in explants of bullfrog sympathetic ganglia (Gruol *et al.*, 1981). The sensitivity of cells in cultures of chick embryo dorsal root ganglia to GABA decreases with time in culture (Choi and Fischbach, 1981; Dunlap, 1981). In addition, dorsal root ganglion nerves in culture do not respond to glycine or β-alanine (Ransom *et al.*, 1977a; Choi and Fischbach, 1981). GABA can also affect action potential generation (Lawson *et al.*, 1976) and action potential duration (Dunlap and Fischbach, 1978, 1981) in dorsal root ganglion cells without affecting other membrane properties. This has been discussed earlier in the section on Ca^{2+} channels (p. 28). GABA receptors have also been detected by binding studies (Table 2.5).

It is apparent that GABA and glycine can produce either depolarization or hyperpolarization of spinal cord nerves in culture. Both responses involve an increase in the membrane conductance (Hösli *et al.*, 1971; Ransom *et al.*, 1977a) and are dependent on Cl^- ions (Hösli *et al.*, 1973; Ransom *et al.*, 1977a). The null (or reversal) potential for GABA, glycine and β-alanine responses can be shifted by changes in the concentrations of Cl^- both inside and outside the cell (Figure 2.12) (Barker and Ransom, 1978a; Dichter, 1980). Changes in external Na^+ concentrations do not affect GABA responses (Dichter, 1980; Choi and Fischbach, 1981). If electrodes filled with potassium acetate are used to record GABA responses from cultured spinal cord cells, the null potential is about –60 mV but if KCl-filled electrodes are used, the null potential for GABA is about –20 mV, presumably because of leakage of Cl^- from the electrode tip inside the cell. Whether a depolarizing or hyperpolarizing response to GABA is recorded will

Table 2.5: Binding of Labelled GABA to Nerve Cell Membranes Prepared from Cultured and Non-cultured Cells

Tissue	K_B (nM)	B_{max} (fmol/mg protein)	Reference
Non-cultured cells			
Adult rat cerebral cortex	K_{B1} 8	620	Frere *et al.* (1982)
	K_{B2} 390	3,900	
Adult rat spinal cord	K_{B1} 5.6*	250*	Frere *et al.* (1982)
	K_{B2} 340	9,800	
Cultured cells			
Fetal rat cerebral cortex	n.d.	40	DeFeudis *et al.* (1980)
Fetal mouse cerebral cortex	K_{B1} 9	240	Frere *et al.* (1982)
	K_{B2} 510	1,300	
Fetal mouse brain	K_{B1} 9 ± 3	380 ± 35	Ticku *et al.* (1980a)
	K_{B2} 250 ± 30	2,400 ± 125	
Fetal mouse spinal cord	21 ± 3	2,522 ± 690	Ticku *et al.* (1980b)
Fetal mouse spinal cord	K_{B1} 13	120	Frere *et al.* (1982)
	K_{B2} 640	3,200	

Notes:
* = determined with ^{3}H-muscimol.
n.d. = not determined.
K_{B1} = dissociation constant of high affinity binding site.
K_{B2} = dissociation constant of low affinity binding site.

therefore depend on the resting membrane potential of the cell and the electrolyte in the recording electrode. The null potentials for glycine and β-alanine responses are similar to that for GABA responses, indicating a similar ionic basis for the responses (Barker and Ransom, 1978a).

Responses to GABA, glycine and β-alanine are concentration-dependent (Barker and Ransom, 1978a; Dichter, 1980; Choi and Fischbach, 1981). In general, GABA is more potent than glycine which is more active than β-alanine. Barker *et al.* (1982) calculated that the relative amount of charge transferred when spinal cord cells were activated was GABA (1.0):glycine (0.74):β-alanine (0.32). On cells in cultures of cerebral cortex, the threshold for eliciting GABA responses was about 1 μM and the half maximal concentration was about 10 μM, but for glycine the threshold was about 50 μM and the half maximal

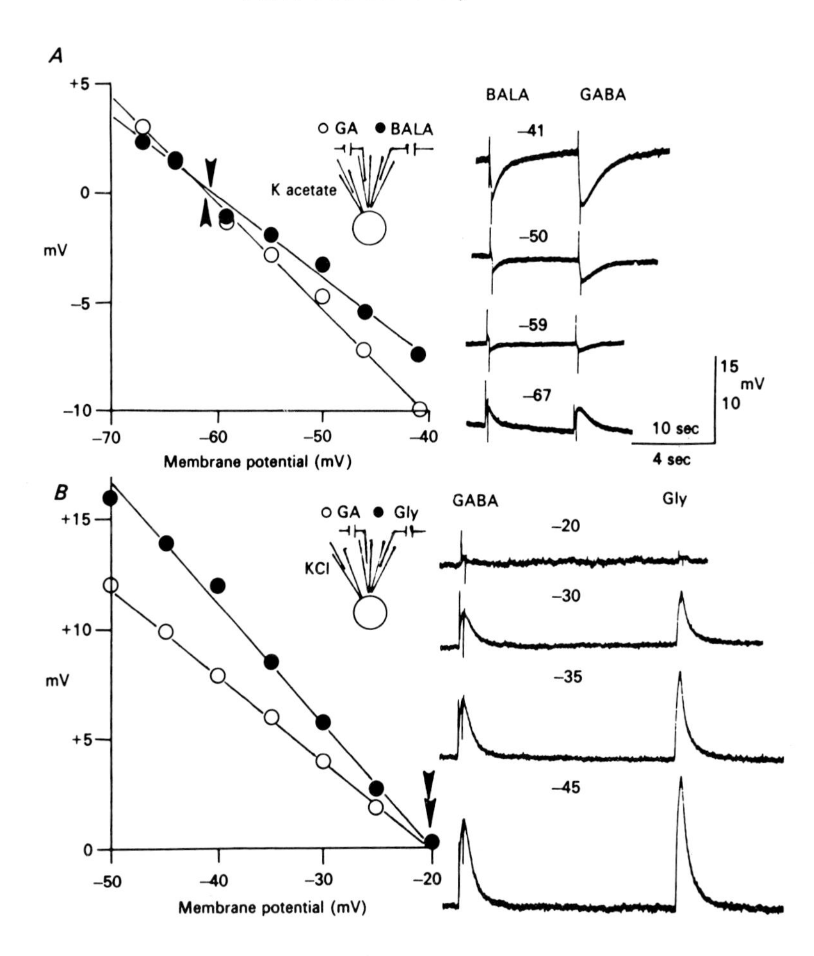

Figure 2.12: Comparison of Reversal Potentials of Cultured Spinal Neurones to GABA, Glycine and β-alanine
Recordings from two different spinal cord neurones. Specimen records on right. A, K acetate recording electrode. Reversal potential of β-alanine (BALA) response similar to that of GABA (–60 mV). β-alanine pulse: 30 nA, 100 ms; GABA pulse: 20 nA, 100 ms. B, KCl recording electrode. Reversal potential of glycine response similar to that of GABA (–20 mV). Glycine pulse: 16 nA, 100 ms; GABA pulse: 24 nA, 100 ms. Lower calibrations apply to B. From Barker and Ransom (1978a), with permission.

concentration about 100 μM (Dichter, 1980). The slope of the log concentration against log effect curve for GABA has been determined to be between 2-3, suggesting that more than one molecule of GABA interacts with each receptor (Barker and Ransom, 1978a; Choi and Fischbach, 1981; Nowak *et al.*, 1982). Dose response curves for glycine (Dichter, 1980) have similar slopes to those for GABA.

Since responses to GABA, glycine and β-alanine seem to be mediated by similar changes in Cl^- permeability, it is feasible that all three amino acids act on the same receptor site. However, there are several pieces of evidence against this. First, some cells respond to GABA but not to glycine, e.g. some cells in spinal cord cultures and dorsal root ganglion cells in general (Ransom *et al.*, 1977a); secondly, GABA and glycine responses can be selectively desensitized (Nelson *et al.*, 1977); and thirdly, as discussed in detail below, certain drugs can block responses to GABA or to glycine selectively. It remains possible that there are distinct recognition sites coupled to the same ionic channel. In fact, there is some evidence for interactions between GABA and glycine responses on cultured spinal cord cells (Barker and McBurney, 1979a).

The increase in membrane conductance produced by GABA on spinal neurones and on cells in cultures of cerebellar brain stem regions wanes several times more slowly than those produced by glycine or β-alanine (Barker and Ransom, 1978a). These differences in time course of the decay of amino acid induced conductance increases are a reflection of the different mean channel open times with the different compounds. From noise analysis experiments in voltage clamped spinal cord cells, the average channel lifetime (τ) for GABA was 27$\pm$ 8 ms compared to 5$\pm$ 1 ms for glycine (Barker and McBurney, 1979a). With β-alanine, channel lifetime was about 6 ms (Barker *et al.*, 1982). The mean single channel conductances (γ) for glycine and β-alanine, however, were almost twice the value for GABA (about 15 pS). Noise spectra for these drugs could be well described by single Lorentzian curves, indicating that one type of process was present (Figure 2.13). Several analogues of GABA open ion channels for times different to that for GABA but all the values for single channel conductance are similar (Table 2.6) (Barker and Mathers, 1981). There was no obvious change in τ and γ values for GABA when holding potential was varied between –40 to –90 mV (Barker *et al.*, 1982). The channels activated by GABA and glycine show similar sensitivity to changes in temperature (Mathers and Barker, 1981). Gamma values increase only slightly with rising temperature but τ decreases rapidly. The Q_{10} for τ (in the range 18-28°) was 2.8 for GABA and 3.0 for glycine. Activation energy

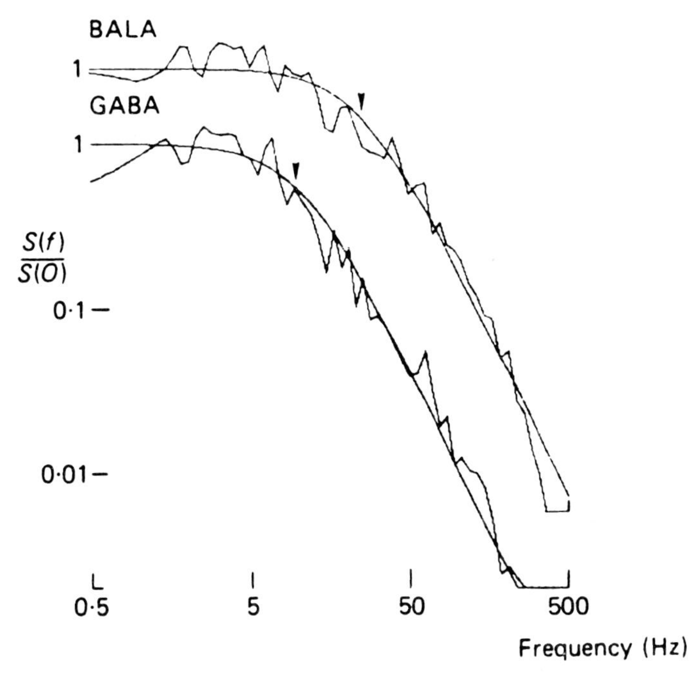

Figure 2.13: Noise Analysis on Cultured Spinal Neurones
Normalized power spectra of membrane current fluctuations produced by pairs of inhibitory amino acids on the same cell (voltage clamped at −70 mV). The half power frequencies (marked by arrow heads) for the spectra obtained from β-alanine (BALA) and GABA fluctuations in this cell are 25.0 and 10.3 Hz, giving estimated channel durations of 6.4 and 15.5 ms, respectively. From Barker *et al.* (1982), with permission.

for channel closing was also similar with both drugs: 18-19 Kcal/mol.

The actions of GABA and an analogue, muscimol have been studied by patch clamp techniques (Mathers *et al.*, 1981; Jackson *et al.*, 1982b). Both drugs caused discrete, all-or-none jumps of inward current when they were applied in a patch electrode to spinal cord cells held at −80 mV. The amplitude of the currents was 1.5-2.0 pA but the drugs produced events of different duration. With muscimol there was evidence of many current jumps of short duration, suggestive of very rapid opening and closing of the channels, perhaps occurring during one activation (Figure 2.14).

Drug Interactions at GABA and Glycine Receptors. Responses of cultured nerves to GABA and to glycine can be antagonized selectively

Table 2.6: Channel Properties of Cultured Spinal Cord Cells Associated with Inhibitory Amino Acid Transmitters and Their Analogues

Drug	Holding potential (−mV)	Channel open time (ms)	Mean channel conductance (pS)	Reference
Glycine	60 – 70	6.2 ± 0.8	32.3 ± 5.0	Barker and Mathers (1981)
GABA	60 – 70	30.4 ± 0.3	15.4 ± 0.4	Barker and Mathers (1981)
GABA*	50 – 100	22.6 ± 1.5	14.4 ± 0.1	Barker *et al.* (1982)
β-alanine	50 – 100	5.6 ± 0.6	22.1 ± 1.9	Barker *et al.* (1982)
(−) Pento-barbitone	50 – 90	153.4 ± 9.0	14.7 ± 3.0	Mathers and Barker (1980)
Muscimol	60 – 70	76.3 ± 4.0	17.9 ± 0.9	Barker and Mathers (1981)
Taurine	60 – 70	2.3 ± 0.2	15.5 ± 1.7	Barker and Mathers (1981)

Notes:
All experiments were performed at about 23°C. KCl-filled electrodes were used so that the reversal potentials were around −20 mV.
Values are means ± standard error of the mean.
* Measured on cultured dorsal root ganglion cells.

by drugs. In explant cultures of spinal cord, glycine produces complex electrical discharges which can be reduced by low concentrations of strychnine but not by bicuculline or picrotoxin (Crain, 1974). Similarly, effects of GABA on spinal cord or cerebral cortex explants are blocked by picrotoxin and bicuculline, but not by strychnine (Crain, 1974). The depressant effect of GABA on the frequency of firing of action potentials in Purkinje cells in cerebellar explants is also antagonized by picrotoxin but not strychnine (Gähwiler, 1975).

In these studies on cells in explants, drugs were added to the perfusing fluid; hence, they cannot reveal at what level the antagonism takes place as each cell probably receives a complex synaptic input. Iontophoretic application of agonists directly to identified cells should allow a more precise examination of the nature of the drug interactions. Although the first such study (Geller and Woodward, 1974) produced rather uncertain results with bicuculline and picrotoxin, later workers using dissociated cell cultures have been more successful. For example, GABA responses on cultured spinal neurones were inhibited by 1-2 μM bicuculline and by 15 μM picrotoxin but not by 2 μM strychnine, which was sufficient to block glycine responses (Obata *et al.*, 1978).

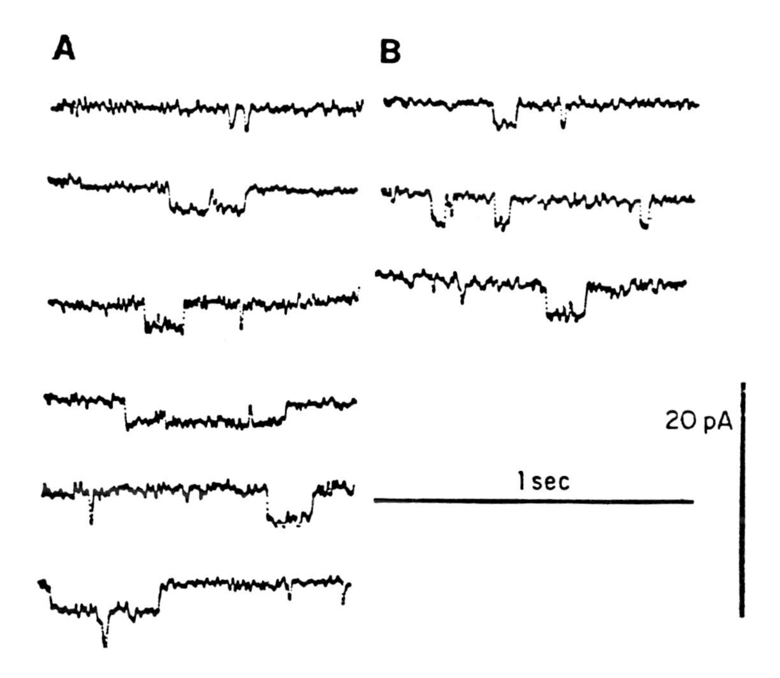

Figure 2.14: Single Channel Currents Recorded from Cultured Spinal Neurones in Response to (A) Muscimol and (B) GABA
Muscimol records typically show a mixture of short and long jumps, the longest event shown (4th sweep from top) is over 400 ms long. From Jackson *et al.* (1982b). Reprinted with permission; copyright 1982, Society for Neuroscience.

Glycine responses were unaffected by bicuculline and picrotoxin. Similar specificity was found by Choi and Fischbach (1981), who also demonstrated that strychnine antagonized responses to β-alanine and taurine as well as to glycine. In cell cultures of fetal rat cerebral cortex, 10 μM bicuculline and picrotoxin also blocked GABA responses (Dichter, 1980). Although 10 μM strychnine could decrease responses to GABA, this concentration was about 20 times that needed to block glycine. The blocking action of bicuculline was found to be quantitatively similar in spinal cord and cerebral cortical neurones in dissociated cell cultures from fetal mice (Nowak *et al.*, 1982). The antagonism was concentration-dependent, with 50 per cent inhibition requiring about 1 μM, and it appeared to be competitive. GABA responses on cultured spinal cord cells were also found to be blocked competitively by sodium penicillin and leptazol (pentylenetetrazol) (MacDonald and Barker, 1977). Neither of these drugs blocked responses to glycine, β-alanine or glutamate. Presumably the convulsant activity of penicillin and leptazol may be related to their blockade of GABA-mediated inhibition in the central nervous system.

Binding studies confirm that bicuculline antagonizes the interaction of GABA and its receptor (DeFeudis *et al.*, 1980; Ticku *et al.*, 1980 a, b; Frere *et al.*, 1982). However, picrotoxin has little effect on GABA binding, suggesting that it is not acting by blocking the GABA recognition site. It may act on the ion channel or on some associated part of the GABA receptor-channel complex (Ticku, unpublished, 1980; see review by Olsen, 1982).

Chick embryo dorsal root ganglion nerves in culture can respond to GABA either with a decrease in the duration of action potentials (Dunlap and Fischbach, 1978, 1981) or, especially in young cultures, with an increase in membrane conductance (Choi and Fischbach, 1981). These two effects seem to be mediated by different GABA receptors as muscimol can increase membrane conductance but has no effect on action potential duration whereas another GABA agonist, baclofen can shorten action potentials but does not affect membrane conductance (Dunlap, 1981). Bicuculline blocked the actions of muscimol but not baclofen; 50 per cent inhibition of the muscimol response occurred with about 0.7 μM bicuculline.

Responses to GABA but not to glycine or β-alanine can be enhanced by drugs such as barbiturates, benzodiazepines and anticonvulsants (see reviews by MacDonald and Barker, 1981; MacDonald and Young, 1981; Olsen, 1982). These effects do not appear to be a consequence of changes in GABA uptake or metabolism but are brought about by direct interactions at the receptor. Studies in tissue culture have allowed investigators to assemble a detailed picture of the cellular pharmacology of such drugs and have provided valuable insight into their mechanisms of action.

Pentobarbitone was the first to be shown to augment GABA responses of spinal nerves in culture (Ransom and Barker, 1975). Although it decreased glutamate responses, pentobarbitone enhanced the responses to GABA (but not to glycine) in about 50 per cent of cells tested. At higher concentrations pentobarbitone also had a direct action to increase membrane conductance. The direct effect of pentobarbitone was slower in onset than the response to GABA, but it had the same null potential as the GABA response and it was blocked by picrotoxin (Barker and Ransom, 1978b). Pentobarbitone could also reverse picrotoxin inhibition of GABA responses. These results have led to the suggestion that pentobarbitone can directly activate the ion channel associated with the GABA receptor. More recently, it has been shown that pentobarbitone opens ion channels with similar conductance to GABA-operated channels but with five times longer open time (Table 2.6; Mathers and Barker, 1980; Mathers *et al.*, 1981).

With lower concentrations of pentobarbitone there is no direct effect on membrane conductance but responses to GABA are augmented. Noise analysis experiments reveal that the single channel conductance of GABA channels is unchanged (about 18 pS) but the mean channel lifetime is greatly prolonged in the presence of pentobarbitone (from about 24 ms to 120 ms) (Barker and McBurney, 1979b). Pentobarbitone was more potent than phenobarbitone at enhancing GABA responses and prolonging their time course (MacDonald and Barker, 1978).

Other drugs with anticonvulsant activity can augment the responses to GABA on spinal cord cells in tissue culture without affecting glycine or β-alanine responses (MacDonald and Barker, 1979; MacDonald and Bergey, 1979; Choi *et al.*, 1981a; MacDonald and Barker, 1982). Active compounds include methylphenobarbitone, diazepam, chlordiazepoxide, flurazepam and valproic acid. Barbiturates can also block glutamate responses but benzodiazepines do not. The augmentation of GABA produced by benzodiazepines has been analysed by Choi *et al.* (1981a) and Study and Barker (1981). Half maximal augmentation of GABA was produced by 17 μM chlordiazepoxide (Choi *et al.*, 1981a). In contrast to the effects of pentobarbitone, diazepam had only a small action on the mean channel open time of GABA-activated channels (Study and Barker, 1981). Since diazepam did not increase single channel conductance or alter the ionic driving force, it was suggested that it somehow increased the frequency of channel opening in the presence of GABA. This could be confirmed by patch clamp experiments.

Direct effects of some benzodiazepines on membrane conductance (MacDonald *et al.*, 1979; MacDonald and Barker, 1982) and benzodiazepine binding to specific receptor sites (McCarthy and Harden, 1981) have been demonstrated on cultured central neurones. These may be endogenous benzodiazepine-like ligands, which could be the purines inosine and hypoxanthine (MacDonald *et al.*, 1979). These may serve to modulate GABA receptor function (see review by Olsen, 1982).

Peptides

There is a rapidly increasing interest in the possibility that many naturally occurring peptides can act as neurotransmitters or as modulators of synaptic function. Among the compounds that have been studied using primary cultures of nerve cells are substance P and some opioid peptides.

In cultures of mouse spinal neurones, iontophoretic or pressure application of substance P to cell bodies was found to produce two distinct effects: an excitatory depolarizing response and an inhibition of glutamate responses (Vincent and Barker, 1979). The ionic basis for the direct effect of substance P was investigated by Nowak and MacDonald (1982). Underlying the depolarization was a decrease in membrane conductance, probably to K^+. Determination of the responses at different membrane potentials indicated that this effect of substance P on membrane conductance was present at the normal resting potential, was increased by depolarization but was not observed at hyperpolarized levels. This is a similar profile to that of the muscarine-sensitive K^+ conductance described earlier (p. 135) (Nowak and MacDonald, 1983).

Endogenous opiate peptides and opiate drugs have been shown to have several actions on nerve cells in culture. Effects on the duration of action potentials have been mentioned earlier (p. 28) in the section on Ca^{2+} channels, and some effects on synaptic transmission are described in Chapter 7. Not all of the reported actions of opiates on cultured nerves are necessarily mediated through specific opiate receptors because they are not all reversed or prevented by opiate antagonists such as naloxone. For example, the enkephalin-induced decrease in glutamate responses (Barker *et al.*, 1978a), the direct depolarization of some spinal cord cells by enkephalin (Barker *et al.*, 1978b), and the blockade by morphine of GABA and glycine responses (Werz and MacDonald, 1982b) have been reported not to be antagonized by naloxone.

However, other actions on cultured nerve cells do seem to involve opiate receptors. These include decreasing the electrical excitability of some spinal cord cells (Barker *et al.*, 1978b), reducing transmitter release from dorsal root ganglion (Crain *et al.*, 1978; MacDonald and Nelson, 1978) and hippocampal cells (Gähwiler, 1980). The depression of transmitter release is presumably a consequence of the shortening of the action potential produced by opiates (Werz and MacDonald, 1982a, c). This action is mediated by a reduction in the Ca^{2+} component of the action potential. Based on the sensitivities to Leu-enkephalin and morphiceptin, Werz and MacDonald (1982c) concluded that some cells had opiate receptors of the μ-subtype whereas others had δ-type receptors, but that both types of receptors could influence Ca^{2+} currents.

3 NEURAL CELL LINES

One promising approach to the study of the properties of nervous tissue is to use cells of neural origin that have adapted to permanent life in tissue culture. By obtaining cultures from individual cells, the resulting clonal cell lines should each comprise a homogeneous population, and the problems associated with the use of complex mixtures of cell types should be avoided. Since cloned cells divide in culture, large quantities of cells can be obtained for quantitative biochemical studies. The problem lies in finding cells that will grow readily in continuous culture yet still express specialized neural properties. Use of transformed cells is the rule, and hence the clonal cells will be genetically abnormal (but see Bulloch *et al.*, 1977, for a different approach to establishing nerve cell lines).

Tumours of the nervous system obviously have the capability to multiply by cell division and most neural cell lines have been derived from tumour material. Although there had been a few previous reports of human neural tumour cells in culture, the first clonal lines were established in 1969 from the mouse neuroblastoma tumour C-1300 (Augusti-Tocco and Sato, 1969; Schubert *et al.*, 1969). This tumour had arisen spontaneously in A/J mice more than 20 years previously and had been maintained by repeated passaging in mice. The precise origin of the tumour is uncertain although it was found near the spinal cord. In animals or in suspension cultures, the cells are undifferentiated and spherical but the cells extend axon-like processes when they are allowed to grow on a compatible surface. It was quickly discovered that clones derived from C-1300 neuroblastoma contained electrically-excitable cells (Nelson *et al.*, 1969, 1971b) and enzymes for synthesis of various neurotransmitters (Augusti-Tocco and Sato, 1969; Amano *et al.*, 1972). Functional receptors were also present (Harris and Dennis, 1970; Nelson *et al.*, 1971a). Hence, these neuroblastoma clones provide a model system for the study of nerve cells.

Since the successful introduction of clonal lines with neural properties from a mouse neuroblastoma, other lines have been established from a variety of sources, including human (Kuramoto *et al.*, 1977, 1981; Schlesinger *et al.*, 1979) and rat (Bottenstein and Sato, 1979) neuroblastomas. Several clones were derived from tumours induced in rat brains by treatment with nitrosethylurea (Schubert *et al.*, 1974b). About 25 per cent of the clones which were characterized by Schubert *et al.* (1974b) were thought to originate from nerve cells because they could generate action potentials. Greene and Tischler (1976) isolated a clone, PC12 from a rat adrenal phaeochromocytoma. These cells synthesized noradrenaline and, in response to nerve growth factor, could extend long fine processes which were highly branched and had numerous varicosities. PC12 cells therefore resemble sympathetic neurones rather than chromaffin cells of the adrenal medulla. However, cells in this line also synthesize and release acetylcholine in addition to noradrenaline (Greene and Rein, 1977a, b). Neural cell lines can also be derived from brain tissue without the use of exogenous chemicals or viral transforming agents, although this procedure is technically more demanding (Bulloch *et al.*, 1977).

In an attempt to increase the range of properties available in cloned nerve cells, several groups of workers have made hybrids of neuroblastoma cells and other cell types (see review by Hamprecht, 1977a). For example, mouse neuroblastoma cells have been fused with mouse (Minna *et al.*, 1971) and human fibroblasts (Peacock *et al.*, 1973b), rat glioma cells (Amano *et al.*, 1974), rat sympathetic ganglion cells (Greene *et al.*, 1975) and Chinese hamster brain cells (MacDermot *et al.*, 1979). The hybrid clones sometimes offer advantages over the parent neuroblastoma cells because neural properties can be expressed more strongly. For example, cells of neuroblastoma clone NG18TG-2 do not form synapses but NG18TG-2 neuroblastoma × C6BU-1 glioma hybrids (line NG108-15 or 108CC15) form functional neuromuscular junctions when added to cultures of skeletal muscle (Nelson *et al.*, 1976; see Chapter 7). In addition, cultures of neuroblastoma hybrids often have a higher proportion of electrically-excitable cells than parent cultures (Amano *et al.*, 1974; Chalazonitis *et al.*, 1975).

Although some cultured neuroblastoma cells spontaneously extend neurites, the proportion is rather low and the rate of differentiation is slow. Hence, a variety of procedures have been tried in attempts to induce differentiation in culture. Differentiation appears to be inversely related to cell division so that slowing the growth rate of cultures by reducing the serum concentration increases the number of differentiated cells (for reviews, see Nelson, 1975; Kimhi, 1981). Use of metabolic inhibitors such as aminopterin (Peacock *et al.*, 1972) also increases the proportion of differentiated cells. Two other treatments that successfully induce differentiation of neuroblastoma cells are inclusion in the culture medium of dibutyryl cyclic AMP (Furmanski *et al.*, 1971) or dimethylsulphoxide (Kimhi *et al.*, 1976). The morphological changes in N1E-115 neuroblastoma cells induced by dimethyl-

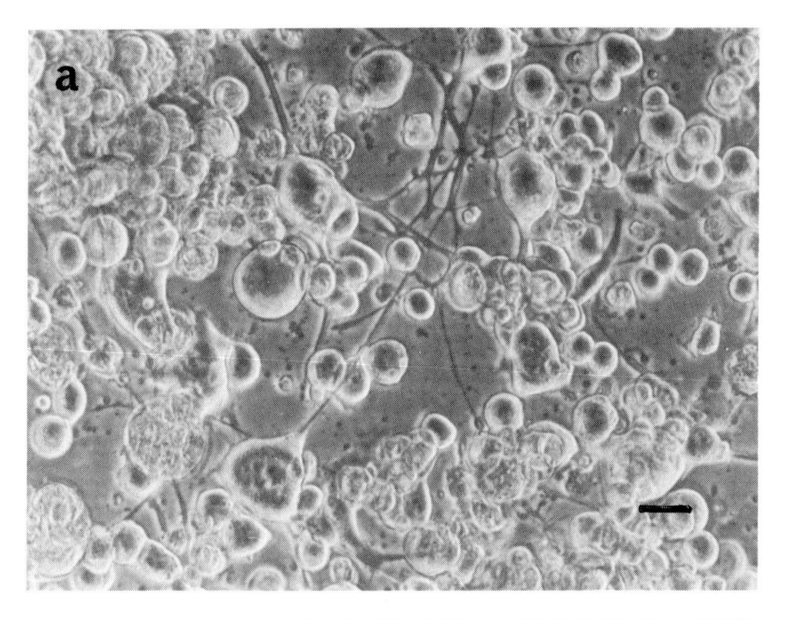

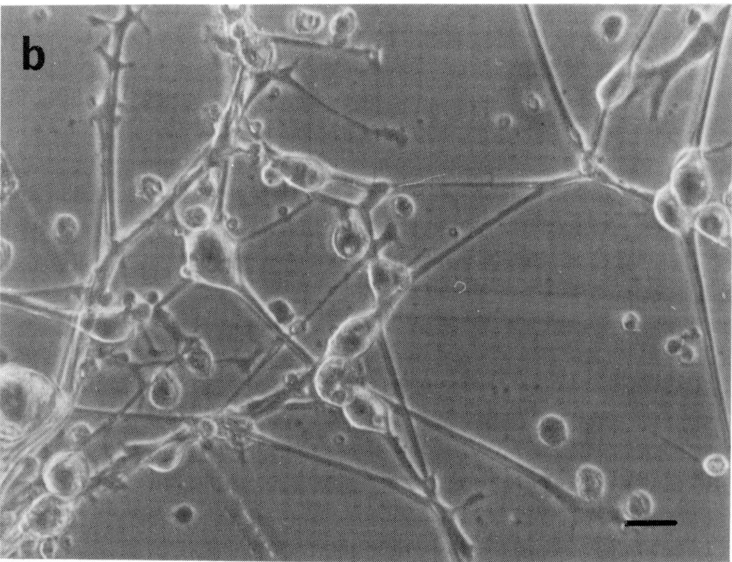

Figure 3.1: Phase Contrast Micrographs of N1E-115 Neuroblastoma Cells
(a) Undifferentiated cells after seven days in Dulbecco's MEM. (b) Differentiated cells in a culture treated with 2 per cent dimethylsulphoxide. Calibration bars = 50 μm. Micrographs supplied by Prof. P.N.R. Usherwood.

sulphoxide are shown in Figure 3.1. Hybrid cells respond well to dibutyryl cyclic AMP but not to other agents (Hamprecht, 1977a). For a fuller discussion of the morphological and biochemical development of neural cell lines in culture, see the reviews by Hamprecht (1977a) and Kimhi (1981).

The remainder of this chapter is concentrated on the development and properties of electrically-excitable membranes and on the functions of specific receptor systems in clonal lines of neural origin. Early work on cloned neural cells was reviewed by Nelson (1973, 1975) and Schubert *et al.* (1973).

Electrical Properties

In the first studies of the electrical properties of cultured neuroblastoma and neuroblastoma hybrid cells, the membranes were classified as active or passive on the basis of their responses to injected pulses of current. The cells were usually hyperpolarized to a standard membrane potential in order to remove any voltage-dependent inactivation of membrane ion channels. Active responses to injected current could be either of two forms: an action potential or a rectification response without an initial action potential spike (see Figure 3.2). In a study on a mixed population of cells derived from C-1300 neuroblastoma, only about 10 per cent of cells tested responded with action potentials and 35 per cent showed passive responses (Nelson *et al.*, 1969). The passive membrane properties of cultured neuroblastoma cells have not been studied extensively. Membrane potentials of up to –80 mV and input resistances of about 50-400 MΩ were recorded in some cells of a human line (Kuramoto *et al.*, 1981). Lower input resistances were characteristic of mouse neuroblastoma cells (Moolenaar and Spector, 1978) and TXC11 hybrid cells (Freschi and Shain, 1982). Values from cloned cells can be compared with those of nerve cells in primary culture (Table 2.2).

In cloned mouse neuroblastoma cultures, the size of the active responses increased when the cultures went from the actively growing stage to the stationary phase and the cells began to differentiate (Peacock *et al.*, 1972). As the neuroblastoma cells differentiated there was an increase in the average membrane potential from about –14 mV to around –40 mV (Peacock *et al.*, 1972). Treatments that induce morphological differentiation of neural cell lines tend to increase their electrophysiological maturation. After growth in dibutyryl cyclic AMP, the proportion of excitable cells in cultures of N1E-115 neuroblastoma rose from about 50 to 90 per cent (Chalazonitis and Greene, 1974). This change was accompanied by a significant increase in membrane potential (from –20 mV to –30 mV) and an increase in the maximum rate of rise of action potentials from 14 V/s to about 80 V/s. Similar changes took place in mouse neuroblastoma clones N-18, NS-20 and N1E-115 after incubation with 2 per cent dimethylsulphoxide (Kimhi *et al.*, 1976). Treatment of PC12 cells (the cell line derived from a rat phaeochromocytoma) with nerve growth factor led to a dramatic increase in electrical excitability: none of the untreated but about 75 per cent of treated cells could generate action potentials (Dichter *et al.*, 1977). Treatment with nerve growth factor did not cause any change in membrane potential. Cells of the human neuroblastoma line SK-N-SH were inexcitable until after morphological differentiation had been induced by growth in dibutyryl cyclic AMP (Kuramoto *et al.*, 1977, 1981). However, as discussed extensively by Schubert *et al.* (1973), there is little clear experimental evidence for a true qualitative change in the membrane properties of neuroblastoma cells as they undergo morphological development. Undifferentiated cells can also show active electrical responses; negative results from such cells could often arise because of the technical difficulties associated with recording from such small cells. Use of extracellular patch clamp electrodes would help resolve this problem.

The changes in the electrical properties of cultured neural cells induced by dibutyryl cyclic AMP or dimethylsulphoxide develop over several days and are stable for at least a few weeks. Spector *et al.* (1975) demonstrated that changes could be induced rapidly after addition of valinomycin (10^{-8}-10^{-6} M) to cultures of N1E-115 mouse neuroblastoma cells. Within 10 min. the membrane potential had increased from –16 to –34 mV and the maximum rate of depolarization during action potentials changed from 20 V/s to about 50 V/s. Valinomycin is a K^+ ionophore but it is not certain whether all the changes are due simply to the increase in potassium conductance after addition of valinomycin, and it is not clear how long lasting these effects would be.

Although all cells in a clone must be descended from the same original cell, not every cell in neuroblastoma cultures develops identical electrical properties (e.g. Nelson *et al.*, 1971b; Chalazonitis and Greene, 1974). This should be remembered when comparing results from methods such as those of electrophysiology that examine individual cells, and from methods such as ion flux measurements that give an average value for a population of cells.

Na^+ Channels

Nelson *et al.* (1971b) found that action potentials in mouse neuroblastoma cells were reduced, but not abolished, by tetrodotoxin, a specific blocker of the fast Na^+ channels responsible for the inward current during nerve action potentials. This was confirmed for clones N1E-115 and N-18, in which tetrodotoxin (or removal of Na^+ ions) could reduce the amplitude of the action potential to about 14 per cent of the peak response (Spector *et al.*, 1973). The size of the tetrodotoxin-insensitive response could be increased by raising the extracellular concentration of Ca^{2+} (Figure 3.2) and it was blocked by Co^{2+}, suggesting the involvement of channels for Ca^{2+} ions (see p. 65).

The cell bodies of neuroblastoma and neuroblastoma hybrid cells can be large (up to 150 μm) and they behave as isopotential spheres (Moolenaar and Spector, 1977; Grinvald *et al.*, 1981). Therefore, they would appear to be eminently suited for study with voltage clamp techniques. However, apart from a detailed series of studies on neuroblastoma clone N1E-115 (Moolenaar and Spector, 1977, 1978, 1979 a, b; Fishman and Spector, 1981; Romey and Lazdunski, 1982), this opportunity has not been taken. Moolenaar and Spector (1977, 1978) demonstrated that the initial fast inward current (Figure 3.3) was carried by Na^+ ions. This current was abolished by tetrodotoxin (0.1 μg/ml; about 3×10^{-7} M) and by removal of Na^+ from the external medium. The reversal potential for the Na^+ current was found to be about +40 mV (Figure 3.3) and the average value for Na^+ conductance was about 85 mS/cm^2, both values being similar to the corresponding values in squid giant axon. Further details of this work can be found in the review by Spector (1981). More recently, properties of individual Na^+ channels of N1E-115 cells have been studied using the patch clamp technique (Quandt and Narahashi, 1982). The single channel current was about 1 pA, corresponding to a single channel conductance of about 10 pS. The average open time was about 2 ms. These channels were blocked by tetrodotoxin in nanomolar concentrations. Batrachotoxin (Figure 3.4) was found to prevent inactivation of Na^+ channels and to cause channels to open spontaneously, although it also decreased single channel conductance. Rather similar results with batrachotoxin were obtained with internally perfused NG108-15 hybrid cells (Huang *et al.*, 1982).

Although these electrophysiological studies relate only to the properties of Na^+ channels in the neuroblastoma cell body, there is evidence that similar channels are present in at least some of the neurites.

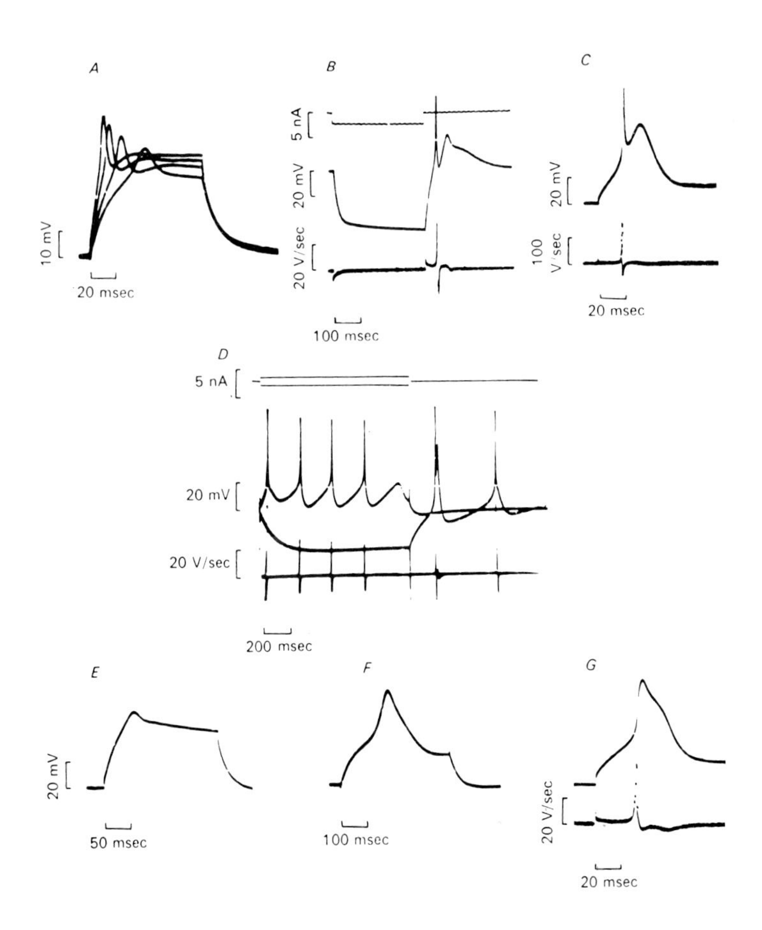

Figure 3.2: Electrical Activity of N1E-115 Neuroblastoma Cells under Constant Current Conditions

(A) Graded active responses during depolarizing current steps applied on the resting potential (–42 mV). (B) Anodal break excitation at the end of a hyperpolarizing step; all-or-none action potential is followed by a depolarizing afterpotential. (C) Action potential elicited by a depolarizing step from a steady hyperpolarized potential level (–80 mV); maximum rate of rise 150 V/s. (D) Example of repetitive firing behaviour; depolarizing and hyperpolarizing current steps were applied on resting potential (–55 mV) resulting in multiple discharges. (E) Voltage response in Na^+ free solution (1.8 mM Ca^{2+}). (F) Slow action potential in Na^+ free solution containing 10 mM Ca^{2+}. (G) Prolonged action potential in normal solution containing 10mM tetraethylammonium (TEA). Membrane potential in E, F and G was adjusted to –75 mV. Temp. 20-23°C. From Moolenaar and Spector (1978), with permission.

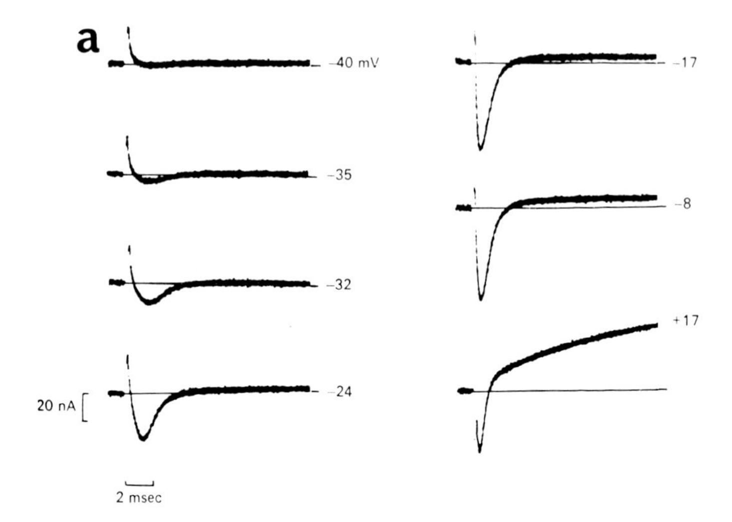
a
-40 mV
-35
-32
-24
20 nA
2 msec
-17
-8
+17

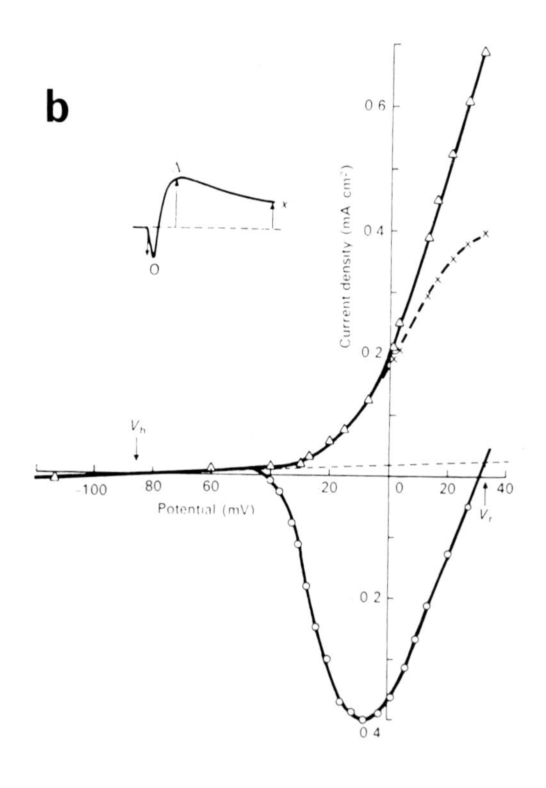
b
0.6
0.4
0.2
0.2
0.4
Current density (mA cm⁻²)
V_h
-100
80
60
20
0
20
40
Potential (mV)
V_r
O

Use of high resolution optical techniques and voltage sensitive dyes revealed that action potentials initiated in the cell body can propagate along the processes to the nerve endings (Grinvald *et al.*, 1981). Conduction velocity was estimated to be 0.2-0.4 ms, which is only slightly slower than that of non-myelinated nerves in the body.

The pharmacological properties of the Na^+ channels of neuroblastoma cells have been extensively investigated using ion flux techniques (see review by Catterall, 1981a). The uptake of $^{22}Na^+$ by neuroblastoma cells can be stimulated by the alkaloid veratridine and this can be blocked by tetrodotoxin (Catterall and Nirenberg, 1973). (Ouabain is present to prevent Na^+/K^+ exchange being stimulated by the increase in intracellular Na^+.) Clones that were known not to have action potentials did not respond to veratridine, and in responsive clones the extent of stimulation varied, ranging from about two times in mouse neuroblastoma × mouse fibroblast hybrids to nearly five times in some mouse neuroblastoma and mouse neuroblastoma × rat glioma hybrid clones (Catterall and Nirenberg, 1973). Presumably these differences are a measure of the different densities of Na^+ channels in the different clones. These types of studies are often carried out on cells that are not maximally differentiated. Although the number of Na^+ channels available to contribute to an action potential will be determined by the membrane potential (which increases with differentiation), it will not matter for ion flux measurements if membrane potential is low since veratridine will open inactivated Na^+ channels. For a full account of ion flux techniques, see Catterall (1981a).

Richelson (1977a) showed that Li^+ ions could pass through the Na^+ channels of mouse neuroblastoma cells. Hence, the properties of these channels can be monitored with a non-radioactive technique (atomic absorption spectrophotometry for Li^+). Similar results were obtained with human neuroblastoma clones, although in these cells only about half of the veratridine-induced uptake could be blocked by

← **Figure 3.3: Action Currents of N1E-115 Neuroblastoma Cells under Voltage Clamp**

Holding potential was −85 mV. (A) Records of the initial inward current in response to the voltage step to the indicated value of membrane potential. (B) Current densities as a function of membrane potential for the same cell as in (A). Currents were measured as indicated in the inset and relative to a small holding current. ○, peak inward current; △, maximum value of the outward current; X, value of the outward current after 400 ms. Dashed line represents an estimate of leakage current. V_h, holding potential; V_r, reversal potential for the inward current. Estimated surface area 2.0×10^{-4} cm^2. From Moolenaar and Spector (1978), with permission.

tetrodotoxin (Schlesinger *et al*., 1979), and with neuroblastoma × glioma hybrids (Reiser *et al*., 1982). These hybrid cells could fire action potentials when Na^+ was substituted by Li^+.

Veratridine is not the only substance that stimulates Na^+ uptake: similar activation of Na^+ channels in neuroblastoma cells was found with batrachotoxin, aconitine and grayanotoxin (Catterall, 1975a, 1977). Their structures are given in Figure 3.4. The effectiveness of these compounds differs both in terms of the maximal degree of stimulation and the effective concentrations. The most active is batrachotoxin which was estimated to activate 95 per cent of the total population of Na^+ channels, with half maximal activation occurring at $2\text{-}7 \times 10^{-7}$ M (Catterall, 1975a, 1977) (Figure 3.5). The corresponding values for the other three drugs are: grayanotoxin 51 per cent and 1.2×10^{-3} M, veratridine 8 per cent and $3\text{-}5 \times 10^{-5}$ M, and aconitine 2 per cent and $4\text{-}8 \times 10^{-6}$ M. Competition experiments demonstrated that these compounds act at a common site which is distinct from the tetrodotoxin binding site. In addition, a polypeptide from the venom of the scorpion *Leiurus quinquestriatus* and toxin ATx II from the sea anemone *Anemonia sulcata* interact co-operatively with the alkaloid toxins to decrease the concentrations necessary for activation of Na^+ uptake and to increase the maximum fraction of Na^+ channels that they can open (Catterall, 1977; Catterall and Beress, 1978). In an electrophysiological study, ATx II was found to block inactivation of Na^+ channels on N-18 neuroblastoma cells and to shift the voltage dependence of the maximal rate of rise of action potentials (Miyake and Shibata, 1981).

Na^+ flux studies have now been carried out on several cell lines of neural origin (Stallcup, 1977, 1979; Huang *et al*., 1979; Schlesinger *et al*., 1979). Although qualitatively similar effects were found in the various clones with veratridine and tetrodotoxin, there were quantitative differences between the concentrations that were effective (Figure 3.5). Tetrodotoxin was more potent in mouse neuroblastoma cells (half maximal inhibition of veratridine-induced Na^+ uptake at 8-18 nM) than in human or rat neuroblastoma cells (half maximal inhibition at 50 nM-1.5 μM) (Stallcup, 1977). Tetrodotoxin was also less active in PC12 cells (50 per cent inhibition at 0.5 μM) (Stallcup, 1979). Moreover, there were differences in the nature of the co-operative interactions between veratridine and the venom of the scorpion *Tityus serrulatus* in different clones, suggesting that there are subtle differences between their Na^+ channels (Stallcup, 1977). The basis for this apparent spectrum of fast Na^+ channels remains to be discovered.

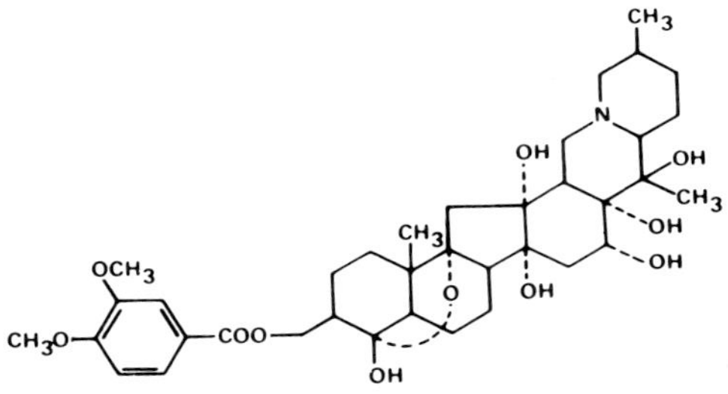

Veratridine

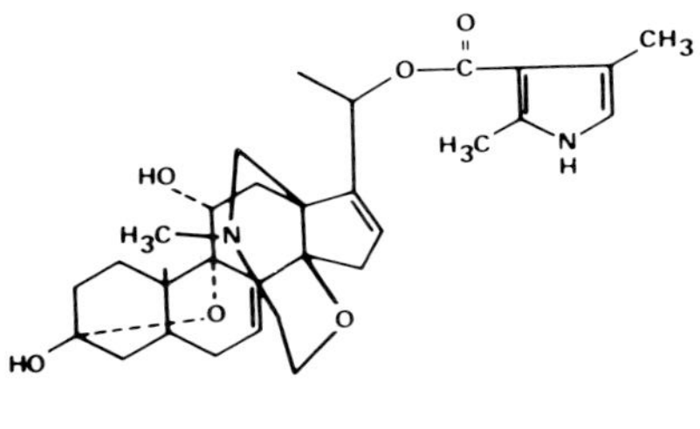

Batrachotoxin

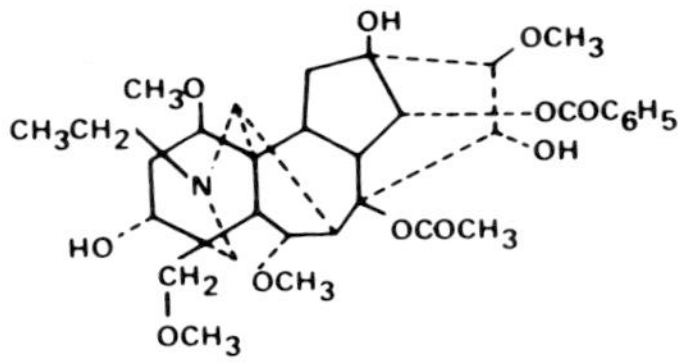

Aconitine

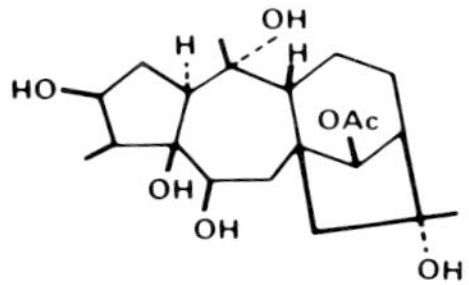

Grayanotoxin

Figure 3.4: Structures of Neurotoxins that Activate the Action Potential Na^+ Channel
Veratridine, a plant alkaloid, and batrachotoxin, a frog toxin, are both seroids containing an unusual hemiketal bridge (3α, 9α and 4α, 9α, respectively) and a tertiary nitrogen in a ring attached to the steroid D ring. Aconitine and the non-nitrogenous diterpenoid grayanotoxin share a fused 5-, 7- and 6-membered ring system and bear little structural similarity to veratridine and batrachotoxin. From Catterall (1975a), with permission.

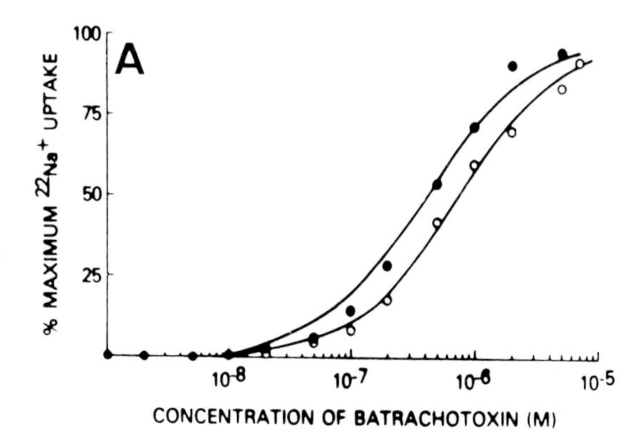

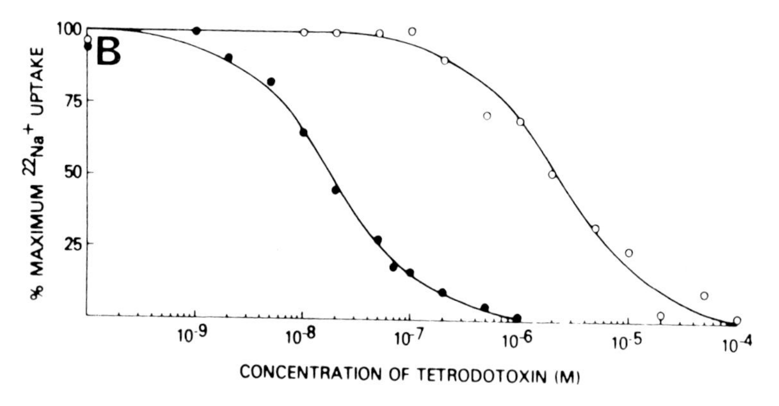

Figure 3.5: Stimulation of $^{22}Na^+$ Uptake by Two Neuroblastoma Clones by Batrachotoxin and its Inhibition by Tetrodotoxin
(A) Effect of batrachotoxin concentration on $^{22}Na^+$ uptake in C9 (○) and N18 (●) clones. Cells were incubated at 37°C with batrachotoxin in Na^+-free high K^+ medium for 60 min. and then Na^+ uptake was measured for 30 s in high choline assay medium containing Na^+. Maximum Na^+ uptake was similar in the two clones. (B) Inhibition of batrachotoxin-induced Na^+ uptake by tetrodotoxin in C9 (○) and N18 (●) clones. 10^{-6} M batrachotoxin was used. The concentration of tetrodotoxin for 50 per cent inhibition was about 18 nM with N18 cells and 1.5 μM with C9 cells. From Huang *et al*. (1979), by copyright permission of the Rockefeller University Press.

The actions of four antidysrhythmic drugs (lignocaine, quinidine, procainamide and phenytoin) on $^{22}Na^+$ uptake by N-18 neuroblastoma cells have been examined (Catterall, 1981b). All of these drugs inhibited the Na^+ influx stimulated by veratridine or batrachotoxin. The blockade was not voltage-dependent since the effects in cultures depolarized by high K^+ were similar to those in control cultures whose average membrane potential was about −40 mV. None of the drugs interfered with the binding of ^3H-labelled saxitoxin (which acts like tetrodotoxin), indicating that these antidysrhythmics were not acting at the tetrodotoxin/saxitoxin site. The interactions with veratridine and batrachotoxin were more complicated, suggesting that the antidysrhythmic drugs were competitive antagonists of the more powerful activator, batrachotoxin, but had mixed competitive and non-competitive interactions with the weaker activator, veratridine. Catterall (1981b) concluded that the four antidysrhythmic compounds may interfere with Na^+ channel function by allosteric inhibition of the binding site for batrachotoxin and veratridine. It should be remembered that some of these drugs have other actions that might also be important for their clinical effectiveness: quinidine blocks K^+ channels (Fishman and Spector, 1981) and phenytoin can block Ca^{2+} channels in mouse neuroblastoma cells (Tuttle and Richelson, 1979).

Ca^{2+} Channels

As mentioned earlier, not all of the action potential of neuroblastoma cells is sensitive to blockade by tetrodotoxin (Nelson *et al.*, 1971b; Spector *et al.*, 1973), and the tetrodotoxin-resistant portion of the spike can be eliminated by Co^{2+} which is known to block Ca^{2+} channels (Spector *et al.*, 1973). In a developmental study of mouse neuroblastoma clone N-18, the contribution of Ca^{2+} channels to action potentials was found to diminish with morphological differentiation of the cells (Miyake, 1978). In fully differentiated cultures, some cells still had biphasic action potentials made up of an initial fast spike that could be abolished by tetrodotoxin and a slower component whose amplitude was proportional to the concentration of Ca^{2+} in the medium. The action potentials of other cells consisted only of a fast, tetrodotoxin-sensitive spike. It is not known whether this difference between cells reflects two different differentiated states or whether there is a tendency for nerves to mature by the conversion of Ca^+ channels to Na^+. Unpublished results cited by Moolenaar and Spector (1978) apparently show no evidence for the latter possibility.

The properties of the Ca^{2+} current in N1E-115 mouse neuroblastoma cells have been investigated with voltage clamp techniques (Moolenaar and Spector, 1977, 1978, 1979a, b; Romey and Lazdunski, 1982). Since this current is much smaller than the other membrane currents, it is easier to study in the presence of solutions containing no Na^+ (to eliminate Na^+ currents) and tetraethylammonium (to block K^+ currents) and containing elevated Ca^{2+} concentrations (see Figure 3.2). On depolarization from a holding potential of –80 mV, the Ca^{2+} current, which is inward in direction, was found to activate at about –55 mV (Figure 3.6). The reversal potential was estimated to be about +50 mV. Ba^{2+} and Sr^{2+} ions could substitute for Ca^{2+} but the current was blocked by Mg^{2+}, Mn^{2+}, Co^{2+} and La^{2+} (in order of increasing potency). However, verapamil and D600 did not affect the Ca^{2+} current (Moolenaar and Spector, 1979a, b), although Ca^{2+} spikes in NG108-15 neuroblastoma hybrid cells have been shown to be sensitive to D600 (Atlas and Adler 1981) and D600 reduces the increase in cyclic GMP content of N1E-115 cells produced by high K^+ (Study *et al.*, 1978). More recently, verapamil and D600 have been reported to be able to block the Ca^{2+} current of N1E-115 cells, although high concentrations (about 10^{-4} M) are required (Romey and Lazdunski, 1982).

The influx of Ca^{2+} ions seems to activate a very slow outward current that is probably carried by K^+ (see p. 68). From their experiments Moolenaar and Spector (1979b) were uncertain whether stimulation of this late outward current was by Ca^{2+} flowing through a second population of channels. A subsequent study (Fishman and Spector, 1981) provided evidence that there may be two types of Ca^{2+} channel, one which inactivates with a time constant of about 30 ms and another with an inactivation time constant of about 2 s.

Calcium-dependent action potentials have also been recorded in several neuroblastoma × glioma hybrid lines, including NG108-5, NG108-15 and NG108-25 (Reiser *et al.*, 1977; Atlas and Adler, 1981). The ability of some α-adrenoceptor antagonists to block Ca^{2+} channels in these cells was highlighted by Atlas and Adler (1981). Although Ca^{2+}-dependent action potentials were not seen in tetrodotoxin-treated PC12 cells (Dichter *et al.*, 1977), functional voltage-dependent Ca^{2+} channels do seem to be present in these cells as judged by the increase in $^{45}Ca^{2+}$ uptake caused by depolarization (Stallcup, 1979). This apparent discrepancy probably results from the low density of the Ca^{2+} current in normal physiological solutions. Ion flux studies in PC12 cells (Stallcup, 1979) also provide evidence that there are separate Na^+ and Ca^{2+} channels. Calcium influx stimulated by high K^+ was not blocked by tetrodotoxin but was almost abolished by 1 mM Mn^{2+} or Co^{2+}. Evidence for the proposal that different clones may have different densities of Ca^{2+} channels was provided by Freedman *et al.* (1983) who compared depolarization-induced uptake of $^{45}Ca^{2+}$ in N4TG1 neuroblastoma and the hybrids, NG108-15 and NCB-20. NCB-20 cells had the largest uptake, and these cells were used to quantify the

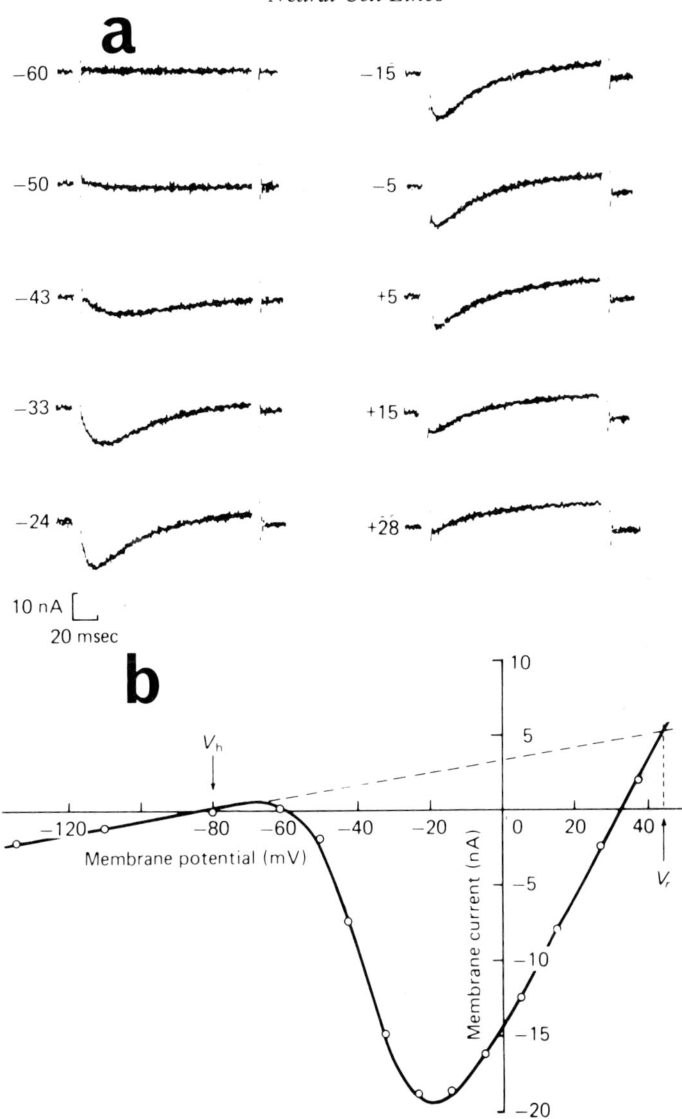

Figure 3.6: Inward Ca^{2+} Currents (I_{Ca}) of a N1E-115 Neuroblastoma Cell in Na^+-free Solution Containing 20 mM Ca^{2+} and 25 mM Tetraethylammonium
Holding potential (V_h) was −80 mV. (a) Records of the current flow as a result of voltage steps to the command voltages indicated. (b) Peak values of I_{Ca} as a function of membrane potential for the same cell as in (a). Dashed line represents an estimate of leakage current as measured by hyperpolarizing voltage steps. V_r, reversal potential for I_{Ca}. From Moolenaar and Spector (1979b), with permission.

potency of ten organic Ca^{2+} channel blocking drugs. IC_{50}s ranged from 0.58 nM for nisoldipine to 4.5 μM for diltiazem.

Lazdunski and colleagues have used N1E-115 neuroblastoma cells to study the specificity of action of several compounds on ion channels. Veratridine and batrachotoxin were both capable of blocking Ca^{2+} currents in addition to their actions on Na^+ channels; scorpion toxin, ATx II and aconitine only affected Na^+ channels (Romey and Lazdunski, 1982). Phencyclidine, which is a potent blocker of K^+ channels, also can reduce Na^+ currents, although its effects on Ca^{2+} channels were not reported (Tourneur *et al.*, 1982).

K^+ Channels

In some of the early studies of the electrophysiological properties of neuroblastoma cells in culture (Nelson *et al.*, 1969, 1971b) there was evidence for the presence of rectification during depolarization responses to injected current. This can be interpreted as reflecting the opening of K^+ channels that normally function during the falling phase of the action potential. Subsequently, it was shown that addition of the K^+ channel blocker tetraethylammonium increased the duration of action potentials in N1E-115 mouse neuroblastoma cells (Moolenaar and Spector, 1978) (Figure 3.2). In voltage clamp experiments the delayed outward current of these cells was blocked by tetraethylammonium (15 mM). The reversal potential for this current was about −70 mV, which is close to the theoretical equilibrium potential for K^+.

In addition to the voltage-dependent K^+ channels that can be blocked by tetraethylammonium and 4-aminopyridine (Fishman and Spector, 1981), some neuroblastoma cells have a K^+ current that is activated by the influx of Ca^{2+} ions (Moolenaar and Spector, 1979a, b). These K^+ channels are responsible for the delayed afterhyperpolarizations that follow action potentials; this phenomenon may be important in the control of repetitive activity of nerve cells. This slow K^+ current is not blocked by tetraethylammonium (Moolenaar and Spector, 1979a, b), but is blocked by quinine or quinidine (Fishman and Spector, 1981) and by apamin (Hugues *et al.*, 1982). Additionally, some neuroblastoma cells have a Ca^{2+}-activated conductance that appears to be caused by non-selective channels for cations (Yellen, 1982). Channels with similar properties had previously been found in cultured cardiac cells (see Chapter 5, p. 145).

Ion permeation through K^+ channels can be studied by measuring the fluxes of radioactive K^+ or Rb^+ (see, for example, Arner and Stallcup, 1981). Despite the availability of clones lacking Na^+ channels but having K^+ channels, comparatively little work has been done on the pharmacology of the K^+ channels of clonal nerve cells. Perhaps this reflects the lack of highly potent and specific drugs which interact with K^+ channels; this situation may be rectified with discovery of a new type of scorpion toxin that has such properties (Carbone *et al*., 1982).

Receptor Properties

Clonal lines of nerve cells provide a powerful and convenient system for studying the functions of specific receptors. First, individual cells can be studied with electrophysiological techniques in order to detect responses to particular neurotransmitters and to elucidate the membrane conductance changes that occur on receptor activation. Secondly, ligand binding techniques can be used as the cells should be part of a relatively homogeneous population and hence interpretation of binding data should be reasonably straightforward. Thirdly, ion fluxes through receptor-operated channels can be measured. Finally, the availability of large quantities of cells facilitates measurement of biochemical consequences of receptor activation, such as changes in concentrations of cyclic nucleotides. Thus, there are at least four distinct experimental approaches to the study of receptors in these cells. It should be possible to correlate results from the different types of experiments and in this way obtain a detailed understanding of the mechanism of action of important neurotransmitters and how they are influenced by centrally-acting drugs. Unfortunately, there has been little combined use of these techniques and consequently there are many difficulties in attempting to relate results from one method obtained under one set of conditions to those made with a different method and perhaps a different set of conditions. Thus, despite the large volume of published results, comparatively little has been made of an excellent opportunity to provide a more fundamental knowledge of the workings of the many receptors found on mammalian nervous tissue.

Neuroblastoma and other neural cell lines have been reported to have a great many functional receptors. These include receptors for acetylcholine, noradrenaline, dopamine, 5-hydroxytryptamine, prostaglandins and opiates. Although the physiological significance of some of these is obscure, they do provide potential models for the investigation of the pharmacology of nerve cells. In the following sections, each major receptor system is considered in turn.

Acetylcholine Receptors

The presence of acetylcholine receptors in neuroblastoma cells was first demonstrated with intracellular recording techniques and iontophoretic application of acetylcholine (Harris and Dennis, 1970; Nelson *et al.*, 1971a; Peacock *et al.*, 1973a). There were differences in the results of these studies, both in the types of responses found and in the percentage of responding cells. This was perhaps a consequence of the different degrees of maturation in the different cultures.

Harris and Dennis (1970), working with the mouse neuroblastoma clones derived by Schubert *et al.* (1969), found that all cells tested were depolarized by acetylcholine. Although the overall sensitivity was very low (about 1 mV/nC), the acetylcholine-induced depolarization could be sufficient to generate action potentials. The cell bodies and the ends of the processes were sensitive, but there was no response when acetylcholine was applied to some of the processes. In contrast, Nelson *et al.* (1971a) found that less than 60 per cent of cells in mouse neuroblastoma cultures (lines N-32 and M) responded to acetylcholine. There were three types of responses: a rapid transient depolarization, a secondary longer and slower depolarization and, in some cells, a hyperpolarizing response that normally followed the initial rapid depolarization. In both these studies (Harris and Dennis, 1970; Nelson *et al.*, 1971a), there was no evidence for synaptic transmission between cells.

In a more extensive study of the chemosensitivity of mouse neuroblastoma clone N-18, Peacock and Nelson (1973) again found that not all cells responded to acetylcholine and that there was more than one type of response (Figure 3.7). Less than 10 per cent of cells were depolarized by acetylcholine, about 20 per cent were hyperpolarized and about 4 per cent had biphasic responses. In some cells the hyperpolarization response reversed direction at membrane potentials of about −80 mV, suggesting that this response might be due to an increased membrane permeability to K^+. The depolarization responses got smaller, without actually reversing, as the resting membrane potential was reduced towards 0 mV; these responses could involve an increase in permeability to both Na^+ and K^+. The two types of response appeared to be due to two different types of cholinoceptor rather than to the same receptor coupled to different ion channels because the nicotinic antagonist tubocurarine more effectively blocked the depolarization responses whereas the muscarinic antagonist atropine selectively blocked the hyperpolarizations (Figure 3.7). In some cells sensitivity was restricted to the cell body. However, in a different clone

(N-2A), there seemed to be spatial separation of receptor types, with depolarization responses being found on the cell body, hyperpolarization responses from the processes and biphasic responses at the junctions between cell body and processes (Peacock *et al.*, 1973a). In another examination of the acetylcholine responses of N-18

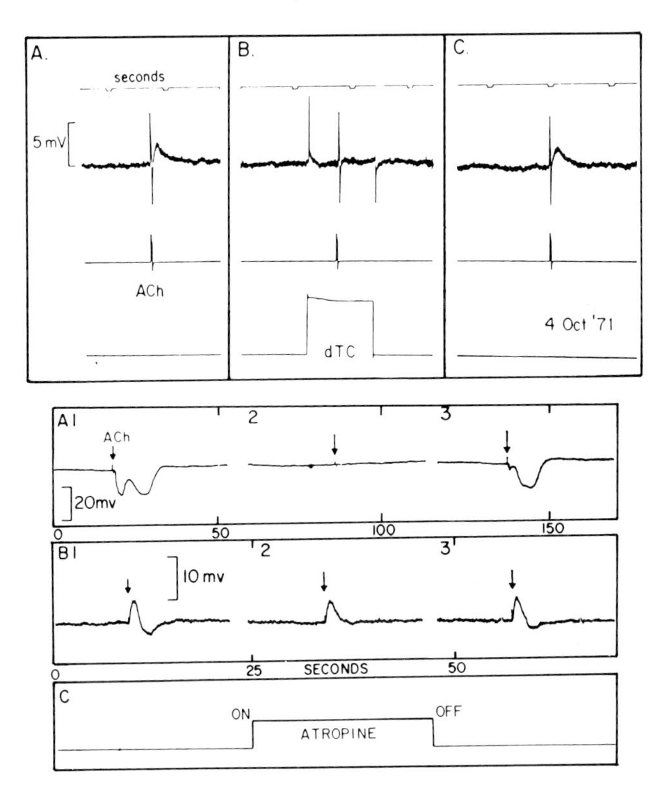

Figure 3.7: Responses of Mouse Neuroblastoma Cells to Iontophoretic Application of Acetylcholine

(A) Monophasic depolarization produced by acetylcholine (ACh). (B) ACh response is blocked by iontophoretic application of tubocurarine (dTC). (C) Full recovery of ACh sensitivity several seconds after the tubocurarine pulse. (A1) Hyperpolarization produced by ACh. (A2) Iontophoretic application of atropine blocks the ACh response. (A3) Partial recovery of the ACh response. (B1) Biphasic depolarization-hyperpolarization response to ACh. (B2) Selective blockade of the hyperpolarization component by atropine. (B3) Partial recovery of the hyperpolarization response to ACh several seconds after the end of the atropine pulse. From Peacock and Nelson (1973), with permission; copyright John Wiley & Sons, Inc.

neuroblastoma cells, Koike and Miyake (1977) found that activation of both nicotinic and muscarinic cholinoceptors resulted in depolarization and they did not report hyperpolarization responses reported by Peacock and Nelson (1973). The reason for this difference is not known. Concanavalin A was found to block the depolarizations induced by acetylcholine.

In all of these studies, it was not possible to predict from the morphology of the cells or from their electrical properties whether they would respond to acetylcholine. Needless to say, the factors controlling the variation in the expression of acetylcholine receptors in presumably homogeneous populations of cloned cells are not yet known.

Other neural cell lines which have acetylcholine receptors as determined by electrophysiological methods include mouse neuroblastoma clone N1E-115 (unpublished results, cited by Matsuzawa and Nirenberg, 1975; Wastek *et al.*, 1981) and some mouse neuroblastoma X mouse fibroblast clones (Chalazonitis *et al.*, 1977), which are reported to give only hyperpolarizing responses; PC12 rat phaeochromocytoma-derived cells (Dichter *et al.*, 1977), the mouse neuroblastoma clones N4TG-1 and N18TG-2 (Chalazonitis *et al.*, 1977), the neuroblastoma X sympathetic ganglion cell hybrid clone NX-31 (Chalazonitis *et al.*, 1975), the neuroblastoma X glioma hybrid line NG108-15 (Christian *et al.*, 1978a), all of which appear to have only depolarizing responses; and some neuroblastoma X fibroblast hybrid clones (Peacock *et al.*, 1973a) and a human cell line TE671 (Syapin *et al.*, 1982) which have both depolarization and hyperpolarization responses to acetylcholine.

Nicotinic Cholinoceptors. Activation of nicotinic receptors typically results in a rapidly developing and short-lived depolarization, whereas muscarinic receptors are usually involved in slower, more prolonged changes in membrane potential that can be either hyperpolarizing or depolarizing. Nicotinic cholinoceptors are usually distinguished from muscarinic cholinoceptors on the basis of their selective responses to appropriate cholinoceptor agonists and antagonists. Not all studies of cholinoceptor properties in neural cell lines have included a rigorous classification of the receptor type. Blockade of the response by tubocurarine is frequently regarded as being sufficient to identify a receptor as being nicotinic, and sensitivity to atropine is equated with the presence of muscarinic receptors. In some cases only the drug predicted to be selective is tested, and in others the concentrations of the antagonists used are too high for them to be acting specifically. For these reasons, claims for the presence of a particular class of cholinoceptor have to be treated with caution.

The most studied nicotinic receptor system in a cloned nerve cell line is that of the rat phaeochromocytoma-derived line PC12. Treatment of PC12 cultures with nerve growth factor to increase their morphological differentiation also greatly enhanced their sensitivity to acetylcholine (Dichter *et al.*, 1977). Most cells became responsive to brief iontophoretic pulses of acetylcholine and sensitivities of 100 mV/nC could be measured. Tubocurarine (10^{-5} M) abolished the depolarizations induced by acetylcholine. In some cells, both in nerve growth factor-treated and untreated cultures, there was a very slow depolarization in response to long applications of acetylcholine. Similar slow responses were found in NX-31 hybrid cells (Chalazonitis *et al.*, 1975). There was some evidence that these responses were insensitive to tubocurarine, and the mechanism underlying them is not known.

Activation of cholinoceptors of PC12 cells by carbachol causes an increase in the uptake of $^{22}Na^+$ and $^{45}Ca^{2+}$ (Patrick and Stallcup, 1977; Stallcup, 1979). Some of the Ca^{2+} uptake is associated with the opening of depolarization-activated Ca^{2+} channels which can be blocked by Mn^{2+}, but some of the Ca^{2+} flows through receptor-operated ion channels (Stallcup, 1979). The increased ion flux induced by carbachol can be abolished by the nicotinic antagonists tubocurarine and hexamethonium, but not by the snake venom toxin α-bungarotoxin (Patrick and Stallcup, 1977; Stallcup, 1979). Antibodies directed against the nicotinic receptor of eel electroplaques also block the effects of carbachol (Stallcup, 1979). The muscarinic antagonist QNB (in high concentrations) and local anaesthetics (procaine and QX222) also block carbachol-induced Na^+ uptake in PC12 cells (Patrick and Stallcup, 1977; Stallcup and Patrick, 1980). In addition substance P and histrionicotoxin could reduce Na^+ influx. Some of this effect is probably due to direct block of the receptor-associated ion channel as it can be reversed by increasing the concentration of Na^+, but Stallcup and Patrick (1980) also proposed that receptor desensitization was enhanced by these compounds.

Although atropine and QNB antagonized the effects of agonists, the concentrations are at least 100 times higher than would be necessary to block muscarinic receptors. It has been reported, however, that PC12 cells have specific binding sites for muscarinic drugs (Jumblatt and Tischler, 1982), although it is not known whether these are physiologically functional. The ability of mecamylamine and hexamethonium to block the receptors on PC12 cells may indicate that these nicotinic receptors are similar to ganglionic cholinoceptors rather than to those at the neuromuscular junction. Alpha-bungarotoxin did not block responses to carbachol or acetylcholine. Similar results with α-bungarotoxin and with erabutoxins were obtained on N1E-115 neuroblastoma cells (Kato and Narahashi, 1982a). This inability of α-neurotoxins to inhibit acetylcholine responses could be predicted because these toxins are not antagonists at ganglionic nicotinic receptors (Bursztajn and Gershon, 1977). However, α-bungarotoxin does bind to specific sites on PC12 cells and this binding can be blocked by cholinergic drugs (Patrick and Stallcup, 1977). This 'α-bungarotoxin-receptor' does not appear to be coupled to any known function in the cells. Similar specific binding of snake toxins to non-functional receptors has been noted in other preparations (see Chapter 2, p. 35 for further discussion). Therefore, the presence of specific high affinity binding sites for drugs should not, on its own, be taken as evidence for the presence of functional receptors. For example, Simantov and Sachs (1973) described neuroblastoma clones in which levels of acetylcholine receptors and acetylcholinesterase were separately controlled; however, the receptor was characterized only by binding of an α-neurotoxin.

The agonist-induced depolarization of PC12 cells can lead to release of noradrenaline (Greene and Rein, 1977a) and dopamine (Ritchie, 1979) from the cells. This release has several characteristics of normal neurotransmitter secretion, being Ca^{2+}-dependent and blocked by Mg^{2+}. Transmitter release is stimulated by the nicotinic agonist DMPP (dimethylphenylpiperazinium) (Greene and Rein, 1977a), but not by the muscarinic agonists methacholine (Greene and Rein, 1977a) or pilocarpine (Ritchie, 1979). There is good agreement between the effective concentrations of agonists and antagonists necessary to affect ion flux and to influence transmitter release. Similar results have been obtained with primary cultures of adrenal chromaffin cells (Amy and Kirshner, 1982; Holz *et al.*, 1982).

Muscarinic Cholinoceptors. The first evidence for the existence of muscarinic receptors on cloned neural cells was the demonstration that the hyperpolarization response of N-18 neuroblastoma cells to acetylcholine could be blocked better with atropine than with tubocurarine (Peacock and Nelson, 1973) (Figure 3.7). In hybrid lines from neuroblastoma × L cells, the concentration of atropine that caused 50 per cent inhibition of the hyperpolarization response was about 10^{-8} M whereas tubocurarine was ineffective, even at 10^{-5} M (Chalazonitis *et al.*, 1977). In both these studies, the hyperpolarizing response was found to reverse about −80 mV, which is close to the predicted K^+ equilibrium potential. However, little attention has been paid subsequently to the conductance changes mediated by the muscarinic receptors on neuroblastoma cells. Cells of the neuroblastoma clone N1E-115 were also reported to hyperpolarize in response to carbachol (Matsuzawa and Nirenberg, 1975), although no results were presented. More recently, carbachol (10^{-3} M) has been demonstrated to cause about 10 mV hyperpolarization of N1E-115 cells (Wastek *et al.*, 1981). Atropine (10^{-6} M), but not tubocurarine (10^{-5} M), blocked the response. In addition to conventional intracellular recording techniques, membrane potential was also estimated from the distribution of radiolabelled tetraphenylphosphonium, a lipophilic substance whose distribution is voltage-dependent.

Neuroblastoma × glioma cells (clone NG108-5) depolarize when acetylcholine is applied (Hamprecht, 1974). This response can be blocked by tubocurarine and α-bungarotoxin as well as atropine. In some cells of this clone, acetylcholine-induced depolarizations are followed by hyperpolarizations. Tubocurarine appears to block the depolarization response preferentially, whereas atropine can selectively eliminate the hyperpolarization (Hamprecht, 1974).

From the relative lack of data concerning the electrophysiological effects of muscarinic activation of cloned cells, it can be seen that there is a need for more detailed investigations of this system. Our understanding of the properties of these receptors and their associated ion channels will remain incomplete until data from noise analysis or patch clamp studies become available.

The main reason for the paucity of electrophysiological information about muscarinic receptor function is the popularity of biochemical studies on cyclic nucleotide levels and their modification by various drugs. Matsuzawa and Nirenberg (1975) demonstrated that carbachol could increase the intracellular concentration of cyclic GMP of N-18 and N1E-115 neuroblastoma cells by 40 and 210 times, respectively. The effect of carbachol is rapid in onset and quickly wanes unless phosphodiesterase activity is inhibited. Carbachol also increased cyclic GMP levels in a neuroblastoma × L cell hybrid clone by six times but had no effect on a rat glioma clone (Matsuzawa and Nirenberg, 1975). The maximal effect of carbachol was obtained at 10^{-3} M, with the half maximal concentration being 10^{-4} M. Atropine blocked the carbachol-induced increase in cyclic GMP levels with a half maximal concentration of 10^{-7} M. Tubocurarine, even at 10^{-4} M, had no effect on the response to carbachol. On its own, carbachol had comparatively little effect on cyclic AMP concentrations but it markedly reduced the increase in cyclic AMP in response to prostaglandin E_1. Similarly, acetylcholine was found to block the increase in cyclic AMP produced by prostaglandin E_1 in NG108-15 and NG108-25 neuroblastoma × glioma lines (Traber *et al.*, 1975a). This effect of acetylcholine was antagonised by atropine (IC_{50} of about 10^{-8} M) but not by α-bungarotoxin or tubocurarine. The cells used in this study were not differentiated and their selective blockade by atropine contrasts with earlier studies on differentiated cells in which the depolarization response to acetylcholine was blocked by both muscarinic and nicotinic drugs (Hamprecht, 1974). The events underlying this apparent change in sensitivity during differentiation have not been clarified.

Muscarinic regulation of cyclic nucleotide levels has been demonstrated in a number of other clonal lines of neural origin. It should be noted that the early reports of extremely high sensitivity of the hybrid clone 108CC15 (or NG108-15) to carbachol (with a maximal increase in cyclic GMP levels claimed to happen at 10^{-10}M) are unreliable (Gullis *et al.*, 1975; but see Hamprecht, 1977b; Gullis, 1977). The pharmacological properties of muscarinic cholinoceptors have subsequently been studied in mouse neuroblastoma clones NS-20 (Blume *et al.*, 1977) and N1E-115 (Richelson, 1977b; Richelson *et al.*, 1978), and in two subclones of the neuroblastoma × sympathetic ganglion cell hybrid line NX-31 (Blosser *et al.*, 1978; Myers *et al.*, 1978). When tested against the prostaglandin- or adenosine-induced increase in cyclic AMP (Blume *et al.*, 1977), acetylcholine was the most potent agonist (the half maximal effective concentration, EC_{50} was about 10^{-6} M), followed by carbachol and muscarine (EC_{50}s about 10^{-5} M), and by methacholine (EC_{50} about 10^{-4} M). Curiously, two other muscarinic agonists, pilocarpine and tetramethylammonium were ineffective if the cultures had been pretreated with a phosphodiesterase inhibitor; a similar effect with weak muscarinic agonists (pilocarpine, tetramethylammonium and arecoline) was noted by Strange (1978). The effects of carbachol on NS20 cells were blocked by the muscarinic antagonists, atropine (IC_{50} 10^{-9} M), isopropamide (IC_{50} 10^{-9} M) and QNB (IC_{50} 10^{-10}M) (Blume *et al.*, 1977). Hence, these receptors have properties consistent with those of muscarinic receptors. However, decamethonium and suxamethonium, both normally thought of as nicotinic agonists, also reduced the rise in cyclic AMP caused by prostaglandin E_1. Since QNB (10^{-10}M) blocked suxamethonium and decamethonium, they are presumably acting on muscarinic receptors (Blume *et al.*, 1977).

A series of antipsychotic drugs, including chlorpromazine and haloperidol, was found to block competitively the actions of carbachol on cyclic GMP levels of N1E-115 cells (Richelson, 1977b). The order of potency of these drugs against carbachol was similar to that determined by binding studies to central nervous system tissue. Similar results were reported recently with a number of antidepressants (El-Fakahany and Richelson, 1983). Several local anaesthetics were also shown to block the effect of carbachol on cyclic GMP levels (Richelson *et al.*, 1978). Unexpectedly, this antagonism was apparently competitive. However, results from binding studies of the interactions of tetracaine and cocaine with labelled hyoscine indicate that these local anaesthetics bind non-competitively to the muscarinic receptors of N1E-115 cells (Burgermeister *et al.*, 1978).

Although stimulation of muscarinic receptors evidently leads to an increase in cyclic GMP and also to a decrease in the stimulatory effect of various compounds on cyclic AMP, the underlying mechanism has to be determined. Since receptor stimulation can have the same biochemical effects even in cells that have opposite membrane potential responses (e.g. on N1E-115 cells which hyperpolarize and on NG108-15 cells which depolarize in response to muscarinic stimulation), the question is raised of whether there is one common mechanism or whether biochemical and electrophysiological effects depend on different populations of receptors.

Muscarinic receptor stimulation increases levels of cyclic GMP even in the presence of phosphodiesterase inhibitors. Therefore it is not acting by preventing the destruction of cyclic GMP, rather it must be affecting cyclic GMP synthesis by influencing the activity of guanylate cyclase. Bartfai *et al.* (1978) confirmed that carbachol increased the guanylate cyclase activity of N1E-115 cells, and that this was abolished by 10^{-6} M atropine. The carbachol effect was dependent on extracellular Ca^{2+} ions, and the Ca^{2+} ionophore A23187 could also increase guanylate cyclase activity. Since most of the activity of this enzyme was thought to be soluble in the cell cytoplasm, Bartfai *et al.* (1978) suggested that carbachol may increase Ca^{2+} entry and that Ca^{2+} then acts to stimulate the activity of guanylate cyclase. Further evidence for this hypothesis was presented by Study *et al.* (1978). Carbachol, veratridine and high concentrations of K^+ were all shown to increase the cyclic GMP content of N1E-115 cells and all these treatments required extracellular Ca^{2+}. Veratridine was blocked by tetrodotoxin, suggesting that activation of Na^+ channels could lead to depolarization which would in turn activate voltage-sensitive Ca^{2+} channels; the effect of high K^+, which would depolarize the membrane potential directly, was not blocked by tetrodotoxin but by the Ca^{2+} channel blockers D600 and Co^{2+}. Since the effects of carbachol were antagonized by atropine, they presumably involved muscarinic receptors. Since these effects were not blocked by tetrodotoxin but were antagonized by D600 and, to a lesser extent, by Co^{2+}, carbachol's action probably involves Ca^{2+} channels rather than Na^+ channels.

There could be separate Ca^{2+} channels opened by muscarinic receptor stimulation and by depolarization. Both are blocked by D600 and by phenytoin (Study, 1980) but the voltage-sensitive channels are more sensitive to Co^{2+} (Study *et al.*, 1978). Since carbachol produces hyperpolarization of N1E-115 cells (Wastek *et al.*, 1981), it would not be expected to activate depolarization-sensitive channels as occurs in, for example, PC12 cells (Ritchie, 1979; Stallcup, 1979). It is not clear whether the postulated Ca^{2+} channels associated with these muscarinic receptors are the same channels that mediate the hyperpolarization that follows stimulation of muscarinic receptors. In two subclones of the neuroblastoma × sympathetic ganglion cell hybrid line NX-31, acetylcholine mediates biochemical changes without affecting resting membrane potential (Myers *et al.*, 1977, 1978), suggesting that, in some cells at least, there can be separate receptors.

The muscarinic receptors of clones N1E-115 and NG108-15 have also been characterized by study of the binding of labelled QNB and hyoscine (Burgermeister *et al.*, 1978). Both clones have high and low affinity binding sites. QNB bound to the high affinity sites with an apparent dissociation constant of 6×10^{-11}M in N1E-115 cells and 1×10^{-10}M in NG108-15 cells. The density of the high affinity sites was relatively low: approximately 15,000-25,000 sites per cell, and there were roughly twice as many low affinity sites. Competition binding studies with muscarinic agonists and antagonists revealed good agreement between the concentrations needed to increase cyclic GMP or block the prostaglandin-induced increase in cyclic AMP and those displacing labelled QNB. Interestingly, analysis of antagonist binding showed no evidence for interactions between receptors whereas binding of agonists apparently involves negative co-operativity (Burgermeister *et al.*, 1978). Similar results have been found with muscarinic receptors of cultured heart muscle (Galper and Smith, 1978).

The muscarinic receptors of neuroblastoma clones undergo short-term desensitization induced by high concentrations of specific agonists, and are also susceptible to longer-term modulation by prolonged exposure to agonists. Desensitization in N1E-115 cells is specific to muscarinic receptors (i.e. there is no cross-desensitization with other types of receptors) and is prevented by pre-incubation with atropine (Richelson, 1978a). Since desensitization by carbachol is not affected by removal of Ca^{2+} from the external medium, which abolishes the carbachol-induced increase in cyclic GMP, desensitization cannot be a consequence of raised intracellular levels of cyclic GMP.

Incubation of both N1E-115 neuroblastoma and NG108-15 neuroblastoma × glioma hybrid cells with muscarinic agonists for several hours leads to a reduction in receptor numbers, as judged by binding of labelled QNB (Klein *et al.*, 1979; Taylor *et al.*, 1979; Shifrin and Klein, 1980). The affinity of receptors for QNB was not markedly changed by this treatment, and antagonists themselves did not alter receptor numbers. The loss of receptor sites was quite rapid, with a new steady state being reached in 4-9 hours. The number of receptors recovered more gradually on removal of the agonist. Recovery was dependent on protein synthesis, being prevented by cycloheximide.

Monoamine Receptors

In addition to cholinoceptors, many clonal nerve cell lines have other types of specific receptors, including those for the monoamine transmitters. These receptors have been demonstrated to mediate both electrophysiological and biochemical changes, but they have not been as fully characterized as the acetylcholine receptors.

Adrenoceptors. N-18 neuroblastoma cells did not respond to iontophoretic application of noradrenaline (Peacock and Nelson, 1973) but in neuroblastoma × glioma hybrid clones, noradrenaline could cause sufficient depolarization to trigger action potentials (Traber *et al.*, 1975b). The ability of noradrenaline to depolarize neuroblastoma × glioma cells (clone NG108-15) was also found by Myers and Livengood (1975) and Christian *et al.* (1978a). None of these studies contained any characterization of the adrenoceptor subtype that mediated the depolarization response. In TCX11 cells (a subclone of NX-31, the neuroblastoma × sympathetic ganglion hybrid line) noradrenaline also causes depolarization (Myers *et al.*, 1977) and α-adrenoceptors would appear to be involved because the response can be blocked by the α-antagonist phentolamine and is not mimicked by β-receptor agonist isoprenaline.

In neuroblastoma × glioma cells, noradrenaline also blocked the increase in cyclic AMP induced by prostaglandin E_1 (Traber *et al.*, 1975b). From the order of potency of agonists (noradrenaline $>>$ phenylephrine $>$ isoprenaline) and antagonists (phentolamine $>$ propranolol), it can be taken that this effect is mediated by α-adrenoceptors. Although noradrenaline acts like acetylcholine to reduce the response to prostaglandins, it is probably acting directly to depress adenylate cyclase activity rather than via cyclic GMP (Sabol and Nirenberg, 1979a). Long-term incubation with noradrenaline leads to a compensatory increase in adenylate cyclase activity (Sabol and Nirenberg, 1979b). On the basis of their pharmacological data, Sabol and Nirenberg (1979a) postulated that noradrenaline acted on α_2-adrenoceptors on NG108-15 cells. The binding of clonidine and related drugs to these cells was found to be influenced by the presence of ATP, cyclic AMP and GTP (Atlas and Sabol, 1981); perhaps when adenylate cyclase is functional, addition of agonist converts the receptors from a high to a low affinity state.

Some neuroblastoma lines appear to have β-adrenoceptors rather than α-adrenoceptors. Although an initial study showed no effect of isoprenaline on the cyclic AMP content of four mouse neuroblastoma clones, N4-TG1, S-20, N-10 and N-18 (Gilman and Nirenberg, 1971), it was later found that the neuroblastoma line N4-TG1 and some neuroblastoma × glioma hybrid lines increased their cyclic AMP levels in response to isoprenaline and noradrenaline (Gilman and Minna, 1973). Because the response was not blocked by the β_1-selective antagonist practolol but was decreased by propranolol, it was concluded that β_2-adrenoceptors were involved. In homogenates of cells from neuroblastoma clone NBA_2, noradrenaline increased the activity of adenylate cyclase and this effect was blocked more readily by propranolol than by phentolamine (Prasad and Gilmer, 1974). The sensitivity to noradrenaline increased when the cells differentiated, but the effectiveness of isoprenaline decreased, suggesting that the adrenoceptors may change from β- to α-type. However, this has not been confirmed. Additionally, it has not been shown that these receptors are functional in intact cells. Whitsett *et al.* (1981) obtained evidence from binding studies that homogenates of some human neuroblastoma clones had β_1-adrenoceptors. Some lines appeared to have specific binding sites but they did not change their adenylate cyclase activity when exposed to catecholamines, indicating that the receptors were not functionally coupled to the enzyme.

Dopamine Receptors. The demonstration of receptors specific for dopamine is complicated because dopamine can act on adrenoceptors and on 5HT receptors. It has been postulated (Blosser *et al.*, 1978; Myers *et al.*, 1978) that catecholamine receptors may exist in undifferentiated cells in an immature form which can be activated by both dopamine and noradrenaline; with differentiation, some of these receptors become more specific for dopamine. This interesting suggestion deserves further investigation. Although neuroblastoma clones provide a ready model for receptor development in the central nervous system, little work has been done on this topic.

Evidence for specific dopamine receptors was provided by Peacock and Nelson (1973) who showed that about 25 per cent of N-18 neuroblastoma cells hyperpolarized in response to dopamine but did not respond to noradrenaline. Similarly, there was some evidence that dopamine and noradrenaline acted on distinct receptors to increase adenylate cyclase activity in homogenates of NBA_2 cells (Prasad and Gilmer, 1974).

In a hybrid line (TCX11) formed from neuroblastoma and sympathetic ganglion cells, dopamine produced depolarization (Figure 3.8) in about 90 per cent of differentiated cells (Myers *et al.*, 1977). The reversal potential for this response was estimated to be about −15 mV and the response was thought to involve an increase in membrane permeability to both Na^+ and K^+. Two other dopamine receptor agonists, apomorphine and piribedil (ET495), also caused depolarization. The specificity of the response to dopamine was not absolutely clear as it was blocked by some dopamine receptor antagonists (trifluoperazine, promethazine, chlorpromazine and bulbocapnine) but also by phentolamine. Unexpectedly, tubocurarine, but not α-bungarotoxin or hexamethonium, was found to be very effective at blocking dopamine responses, although these cells did not respond to acetylcholine itself. Dopamine responses have also been studied in N1E-115 neuroblastoma cells (Kato and Narahashi, 1982b). Under voltage clamp, dopamine induced an inward current that became outward at holding potentials more positive than +14 mV. The dopamine responses were blocked much more effectively by haloperidol than by phentolamine.

5-Hydroxytryptamine Receptors. Peacock and Nelson (1973) tested N-18 neuroblastoma cells for responses to 5-hydroxytryptamine (5HT, serotonin). Although some of the cells were sensitive to acetylcholine and dopamine, 5HT gave no effect. However, 5HT has been found to depolarize other cloned nerve cell lines (Figure 3.8), including TCX11 (a neuroblastoma × sympathetic ganglion cell hybrid clone) (Myers *et al.*, 1977; Freschi and Shain, 1982), NG108-15 neuroblastoma × glioma hybrid cells (Christian *et al.*, 1978a; MacDermot *et al.*, 1979; Kondo *et al.*, 1981), N18TG2 and N1E-115 mouse neuroblastoma

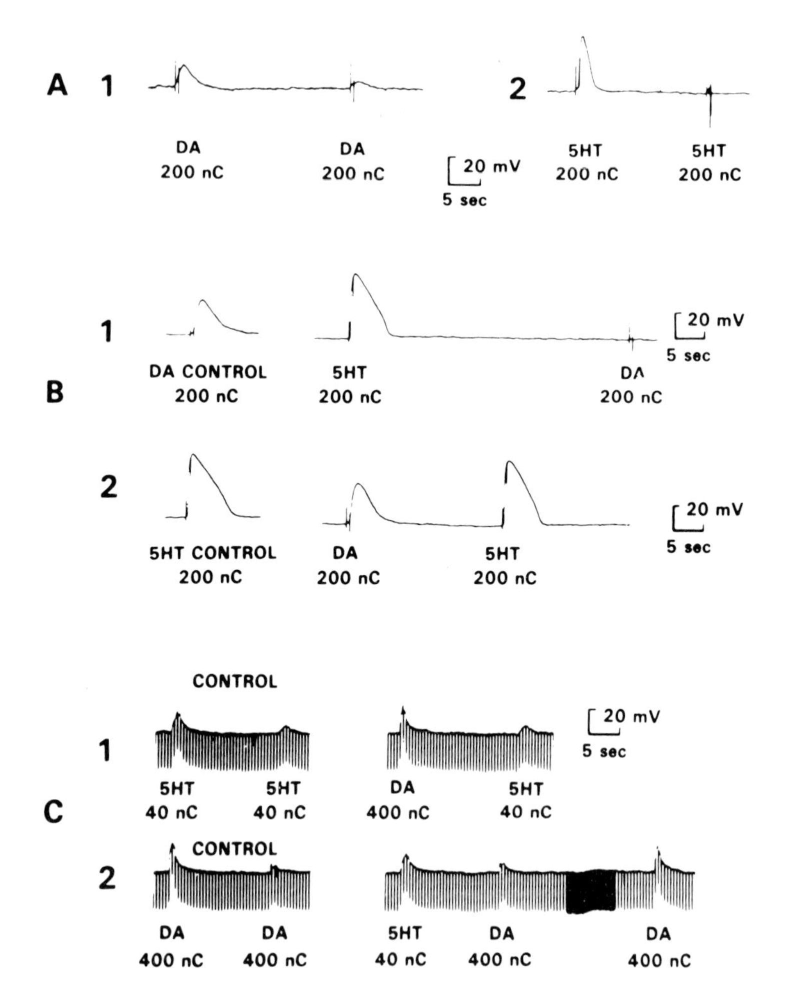

Figure 3.8: Depolarization Responses of TCX11 Cells to Dopamine (DA) and 5-Hydroxytryptamine (5HT), and Cross-desensitization
Sequential doses of DA (A1) and 5HT (A2) cause reduction in the response amplitude. When large equal doses of 5HT and DA are used, 5HT desensitizes the cell to the subsequent DA dose (B1), but DA does not appear to affect the response of the cell to the subsequent 5HT dose (B2). When the 5HT dose is lowered to the amount that causes the response equal to that caused by DA, each amine reduces the responsiveness of the cell to the subsequent dose of the other amine (C). A, B, and C are from different cells. From Freschi and Shain (1982), with permission. Copyright 1982, Society for Neuroscience.

lines (MacDermot *et al.*, 1979; Guharay and Usherwood, 1981). In NG108-15 cells, the maximum sensitivity to iontophoretic application of 5HT was about 100 mV/nC, which was about a fifth of that to acetylcholine (Christian *et al.*, 1978a). The 5HT-induced depolarization of N1E-115 cells was blocked by Co^{2+}, suggesting that Ca^{2+} channels were involved in the response (Guharay and Usherwood, 1981). The depolarization response to 5HT appears to involve activation of channels with simple open-closed transitions (Figure 3.9).

As mentioned earlier, TCX11 cells also respond to dopamine but the receptors involved may be those stimulated by 5HT. Although Myers *et al.* (1977) found that there was no cross-desensitization between dopamine and 5HT responses, Freschi and Shain (1982) demonstrated that equi-active amounts of dopamine and 5HT did produce cross-desensitization (Figure 3.8). When both drugs were tested on the same cell, they had identical reversal potentials of about 0 mV. There was also cross-desensitization between 5HT and dopamine responses in NG108-15 cells (although not between 5HT and acetylcholine responses) (Christian *et al.*, 1978a). Since relatively large amounts of dopamine were required to produce effects on these cells, it is probable that dopamine was acting non-specifically on 5HT receptors.

Although 5HT could depolarize NG108-15 cells, it had little effect on the activity of their adenylate cyclase (MacDermot *et al.*, 1979). Similar results were found with N18TG2 and N1E-115 neuroblastoma cultures. 5HT did increase the adenylate cyclase activity of homogenates of NCB-20 cells, the hybrid line from neuroblastoma × Chinese hamster brain cells, and these effects could be blocked by the 5HT antagonists cyproheptadine and mianserin. In NCB-20 cells 5HT also caused depolarization and stimulated the release of acetylcholine (MacDermot *et al.*, 1979). The lack of a consistent relationship between depolarization responses and effects on adenylate cyclase activity suggest that there may be separate 5HT receptors linked to ion channels and to adenylate cyclase.

Histamine Receptors. In a series of reports, Richelson (1978b, c; Taylor and Richelson, 1979) showed that histamine increased the cyclic GMP content of N1E-115 neuroblastoma cells. Histamine was active in the range 10^{-6}-10^{-5} M, and the maximum effect was about a 25 times increase in cyclic GMP levels, although this was short lasting. The response was dependent on extracellular Ca^{2+}. The histamine receptors were of the H_1 type as they were stimulated by 2-methylhistamine, a specific agonist at H_1 receptors, but not by 4-methylhistamine, which

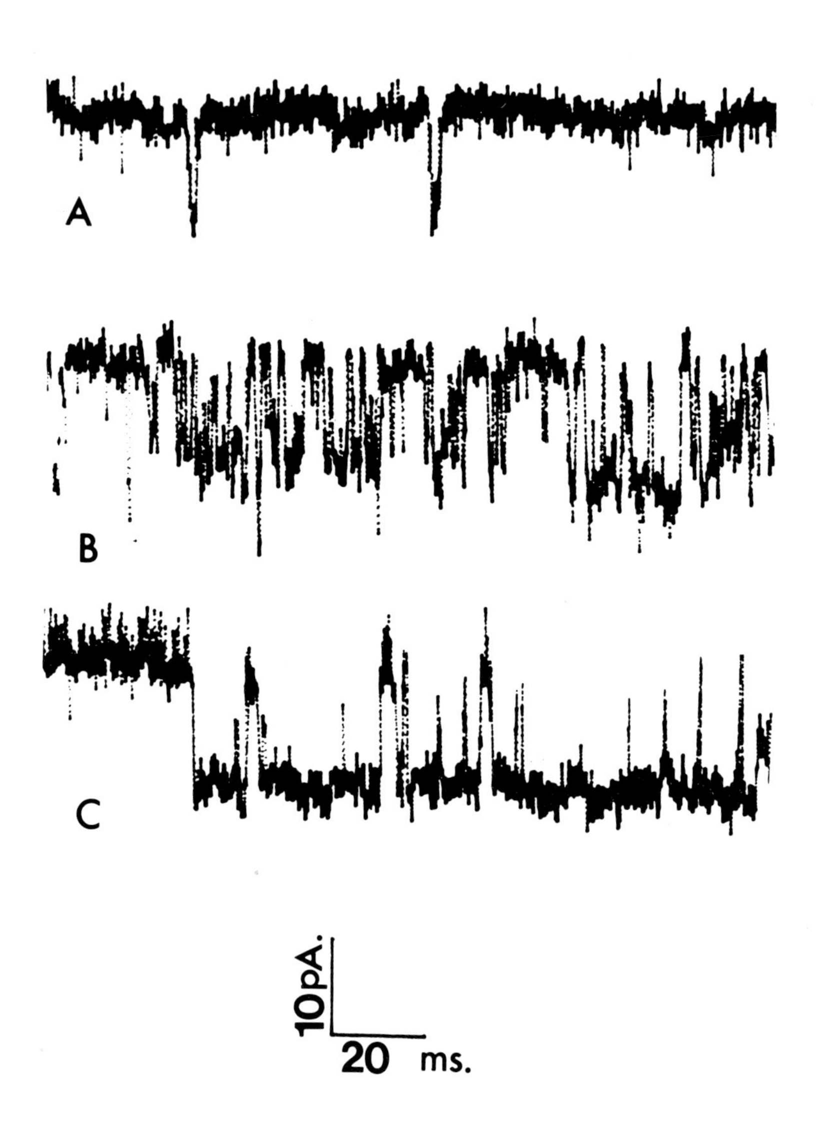

Figure 3.9: Single Channel Activity of N1E-115 Cells Induced by 10^{-7}M 5HT
(A) Single events with mean closed time greater than mean open time.
(B) Multiple events with similar mean open and closed times. (C) Persistent activity with mean open period much greater than the mean closed period. Membrane potential −100 mV, temperature 20°C. Records supplied by Prof. P.N.R. Usherwood.

acts relatively selectively on H_2 receptors. Also the responses were selectively blocked by the H_1 receptor antagonist mepyramine but not by metiamide, an H_2 receptor blocker. Tricyclic antidepressants also blocked the increase in cyclic GMP induced by histamine. This antagonism appeared to be competitive (Richelson, 1978b). Histamine responses on N1E-115 cells could desensitize, but this was not associated with the increased intracellular levels of cyclic GMP and was specific for histamine receptors, there being no cross-desensitization with carbachol responses (Taylor and Richelson, 1979).

Opiate Receptors

There has obviously been interest in the actions of morphine and related drugs on neural function for a very long time and interest in this area of research has been greatly stimulated by the more recent discovery of endogenous peptides with opiate properties. If neuroblastoma cells have functional receptors for morphine-like drugs and for enkephalins, elucidation of the molecular mechanisms of these compounds would be facilitated. Unfortunately, despite the large number of reports on the subject, there have not been any great advances in the understanding of the actions of opiate molecules on nerves based on tissue culture studies.

Specific receptors for morphine and other related drugs were postulated to be present on neuroblastoma × glioma hybrid clones NG108-5 and NG108-15 by Klee and Nirenberg (1974). This was based on studies of the binding of tritiated dihydromorphine, which was found to be saturable, stereospecific and displaceable by a number of morphine-like drugs. However, this study did not include any kind of functional assay to determine what physiological charge accompanied activation of such binding sites. Although these hybrid cells bound about 20 fmoles dihydromorphine/mg protein, which was equivalent to about 3×10^5 sites/cell, no specific binding was detected on either of the parent clones, the mouse neuroblastoma line N18TG2 and the rat glioma C6BU-1.

In a subsequent study, Sharma *et al.* (1975) discovered that morphine lowered the basal levels of cyclic AMP in NG108-15 cells and also reduced the ability of prostaglandin E_1 to raise cyclic AMP content. As these actions of morphine were prevented by the morphine antagonist naloxone, they would seem to be mediated by specific opiate receptors. Also there was reasonable agreement between concentrations of morphine-like drugs needed to decrease adenylate cyclase activity and those effective at displacing bound naloxone (Sharma *et al.*, 1975).

Although Traber *et al.* (1974) had shown earlier that morphine antagonized the actions of prostaglandin E_1 on neuroblastoma clone N4TG3, it had not been demonstrated that this was a specific effect involving opiate receptors. Morphine has also been reported to depolarize NG108-15 cells (Traber *et al.*, 1975b). Once again it is not certain that this action of morphine is on specific opiate receptors. The ionic basis for the effect was not determined and there do not appear to have been any subsequent studies on this potentially interesting finding.

Following the discovery of endogenous opiate peptides, the properties of the morphine receptors on cloned nerve cells were reinvestigated. NG108-15 cells bound enkephalins, endorphins and β-lipotropin fragments more selectively than they bound morphine and its analogues (Brandt *et al.*, 1977; Wahlström *et al.*, 1977). The endogenous opiates also decreased the effect of prostaglandin E_1 on cyclic AMP levels but it is not known whether they affect membrane permeability. A mouse neuroblastoma clone, N4TG1 was also found to bind enkephalins and β-endorphin, though not β-lipotropin (Chang *et al.*, 1978). There was evidence from competition-binding studies on N4TG1 cells that enkephalins and morphine analogues bound differently to the receptor site or that there were different forms of the opiate receptor. Subsequent work has led to the suggestion of multiple forms of the opiate receptor and N4TG1 neuroblastoma cells were found to have the subtype that has a higher affinity for enkephalins than for morphine or naloxone (Chang and Cuatrecasas, 1979). The enkephalin or δ-receptor has been characterized in NG108-15 cells and in NCB-20 cells (McLawhon *et al.*, 1981). In NCB-20 cells, ^{3}H-[D-Ala^2, D-Leu^5] enkephalin bound to a single high affinity site with a dissociation constant of 3×10^{-9} M. Maximum binding was 400 fmoles/mg protein in NCB-20 cultures and 90 fmoles/mg protein in NG108-15 cells. Investigation of binding of benzomorphan drugs indicated that there were separate sites in addition to the enkephalin receptors (McLawhon *et al.*, 1981); these may represent the K- and σ-opiate receptors that have been postulated by other workers.

Although binding studies reveal a multiplicity of forms of opiate receptors, some of which are present on cloned nerve cells, their biological significance remains obscure. Binding studies can also show specific sites that are not apparently functional. For example, several hybrid lines including TCX17 bind naloxone in a selective manner (Blosser *et al.*, 1976) but morphine has no effect on the prostaglandin-induced increase in the cyclic AMP content of these cells (Blosser *et al.*, 1978). Similarly, despite the initial reports to the contrary, specific binding of opiates (Law *et al.*, 1979) and β-endorphin (Hammonds and Li, 1981) has been detected on N18TG2 neuroblastoma cells, although opiates do not affect basal cyclic AMP levels or the action of prostaglandin E_1 (Law *et al.*, 1979).

There have been few mechanistic studies of the effects of opiates on cultured neuroblastoma cells, and most of these have concentrated on the means whereby levels of cyclic AMP are altered. Several possibilities have been suggested. The enkephalin-induced decrease in adenylate cyclase activity of NG108-15 cells required the presence of GTP and Na^+ ions (Blume *et al.*, 1979) and it was proposed that the coupling of opiate receptors to adenylate cyclase requires additional, unidentified membrane components. A different view was taken by Koski and Klee (1981) who found that opiates increased GTPase activity of NG108-15 cells and these authors therefore suggested that hydrolysis of GTP was responsible for the inhibition of adenylate cyclase. Finally, opiates have been demonstrated to increase cyclic GMP levels in N4TG1 neuroblastoma cultures (Gwynn and Costa, 1982), and the increase in cyclic GMP may have an inhibitory effect on adenylate cyclase. However, the importance of changes in cyclic nucleotide levels in nerve cells is more assumed than understood, and hence concentration of research on these effects may hinder advances in the knowledge of opiate action.

Three studies do reveal effects of opiates that are not related to cyclic nucleotides. Dawson *et al.* (1979) found that long-term incubation with morphine and enkephalin was associated with a decrease in the synthesis of membrane glycoproteins and gangliosides, and Hazum *et al.* (1980) found that clustering of enkephalin receptors on the surface of N4TG1 neuroblastoma cells was increased by incubation with morphine. In both cases the significance of the changes is not yet known. Also, enkephalins have been shown to alter the subcellular distribution of calmodulin in NG108-15 cells (Baram and Simantov, 1983).

Prostaglandin Receptors

The actions of prostaglandins are widespread throughout the body but their physiological role is usually uncertain. The influence of prostaglandins on the central nervous system is an example of this uncertainty. Despite this, or perhaps because of it, there have been many studies of the actions of prostaglandins on neuroblastoma cells. As the most common response is an increase in the intracellular concentration of cyclic AMP, often prostaglandins are used only as a stimulatory agent in experiments designed to examine the mechanisms by which other compounds may regulate cyclic AMP production. Such studies have already been described in the appropriate subsections.

Gilman and Nirenberg (1971) showed that prostaglandin E_1, but not prostaglandin $F_{2\alpha}$, increased the cyclic AMP content of several mouse neuroblastoma clones (N4TG1, N-10, N-18 and S-20). The concentration of prostaglandin E_1 for 50 per cent maximal effect was about 10^{-7}M. Inhibition of phosphodiesterase activity by theophylline increased the prostaglandin effects by two or three times. Subsequently, prostaglandin E_1 stimulation of cyclic AMP levels or of adenylate cyclase activity has been found in N1E-115 (Matsuzawa and Nirenberg, 1975), N4TG1 (Minna and Gilman, 1973), NS-20 neuroblastoma cells (Blume and Foster, 1976), NG108-15 (Traber *et al.*, 1975b), TCX11 (Blosser *et al.*, 1978; Myers *et al.*, 1978) and NCB-20 hybrid cells (MacDermot *et al.*, 1979; Blair *et al.*, 1980).

During differentiation of TCX11 cells, there was a change in the sensitivity to prostaglandin E_1: undifferentiated cells were more sensitive, with a maximal effect occurring at 10^{-7}M (Blosser *et al.*, 1978), whereas differentiated cells were stimulated by higher concentrations and a maximal response had not been reached at 10^{-6}M (Myers *et al.*, 1978). NCB-20 cells were found to be about 150 times more sensitive to prostacyclin (PGI_2) than to prostaglandin E_1 (Blair *et al.*, 1980). Labelled prostacyclin bound to a single class of sites, each cell having about 2.6×10^5 sites. From competition studies, the concentrations needed for 50 per cent inhibition of binding were 1.7×10^{-8}M for prostacyclin and 1.4×10^{-7}M for prostaglandin E_1 (Blair and MacDermot, 1981). Another prostaglandin, prostaglandin D_2 was more active than prostaglandin E_1 at increasing cyclic AMP levels in N1E-115 cells (Shimizu *et al.*, 1979).

Long-term exposure of NG108-15 cultures to low concentrations of prostaglandin E_1 resulted in a decrease in the activity of adenylate cyclase in these cultures and in a lower responsiveness to subsequent exposure to higher concentrations of prostaglandin E_1 (Kenimer and Nirenberg, 1981). This effect was slowly reversed by the removal of the prostaglandin, recovery being associated with protein synthesis.

There are some inconsistencies in the reports of electrophysiological actions of prostaglandins on neuroblastoma cells. Traber *et al.* (1975b) found that prostaglandins E_1 and $F_{1\alpha}$ hyperpolarized NG108-15 cells, whereas Christian *et al.* (1978a) reported that cells in the same clone were depolarized by prostaglandin $F_{2\alpha}$ but not by prostaglandin E_1, and MacDermot *et al.* (1979) also found that prostaglandin E_1 did not affect the membrane potential. Kondo *et al.* (1981) reported that NG108-15 cells were depolarized by prostaglandin D_2 and prostacyclin, and that prostaglandins E_1 and $F_{2\alpha}$ were 5-10 times less effective. The action of prostaglandin D_2 appeared to be dependent on external Ca^{2+} ions (Kondo *et al.*, 1981).

Adenosine Receptors

In N1E-115 neuroblastoma cultures, adenosine decreased cyclic GMP and increased cyclic AMP levels (Matsuzawa and Nirenberg, 1975), and 2-chloroadenosine stimulated the activity of adenylate cyclase in homogenates of NS-20 neuroblastoma cultures (Blume and Foster, 1976). The actions of two potent adenosine analogues (5′-*N*-ethylcarboxamide-adenosine and N^6-phenylisopropyladenosine) have also been studied on the adenylate cyclase activity of PC12 phaeochromocytoma cells (Guroff *et al.*, 1981). These compounds increased adenylate cyclase activity and their actions were competitively antagonised by methylxanthines. Guroff *et al.* (1981) suggested that PC12 cells had specific adenosine receptors.

4 SKELETAL MUSCLE

Since the earliest days of tissue culture, attempts have been made to maintain skeletal muscle in isolation from the body (Harrison, 1907, 1910; Lewis, 1915; Lewis and Lewis, 1917b). Early studies used organ culture or explants of muscle, but often such tissue suffered from 'dedifferentiation', i.e. loss of characteristic structures of the tissue and reappearance of the cytological features typical of an earlier state of differentiation. Despite this difficulty, Harrison (1907, 1910) observed muscle fibres, some of which were striated, growing out of an explant of the myotomes of frog embryos. The Lewises described the behaviour of explants of chick embryo leg muscle in culture. Some of the outgrowths from the explants contracted spontaneously in the absence of nerves, although cross-striations could not be seen in such myogenic cells (Lewis, 1915). No mitotic figures were observed in the nuclei of the long fibres, but dividing uninuclear cells were noted (Lewis and Lewis, 1917b). After the establishment of enzyme dissociation techniques (Moscona, 1952; Rinaldini, 1959), studies were made on embryonic skeletal muscle grown in monolayer cell cultures (see Murray, 1960, for a review of early work), and the sequence of events during differentiation of skeletal muscle in culture has since been studied in some detail. After enzymatic or mechanical dissociation the cultures consist of spherical uninuclear cells. These settle on to the surface of the culture dish and after a few hours different types of cells can be distinguished. Some cells appear extremely flattened and have an irregular multipolar shape; these cells are usually said to be fibroblasts (Figure 4.1a). Other cells have a marked bipolar shape with a prominent central nucleus and scanty cytoplasm (Figure 4.1a); these cells are thought to be myogenic in origin and are termed 'myoblasts'. The uninuclear cell population goes through a period of replication. After a time some cells stop dividing and fuse to form multinucleated cells called 'myotubes'. The multinuclear myotubes are often highly branched and the cell nuclei are usually in a central position (Figure 4.1b). These cells continue to grow and develop in culture, forming mature cross-striated muscle fibres which have nuclei aligned along the cell membrane (Figure 4.1c, d).

Successful cultures have been prepared from skeletal muscles of several species, including chick, mouse, rat and human. Different

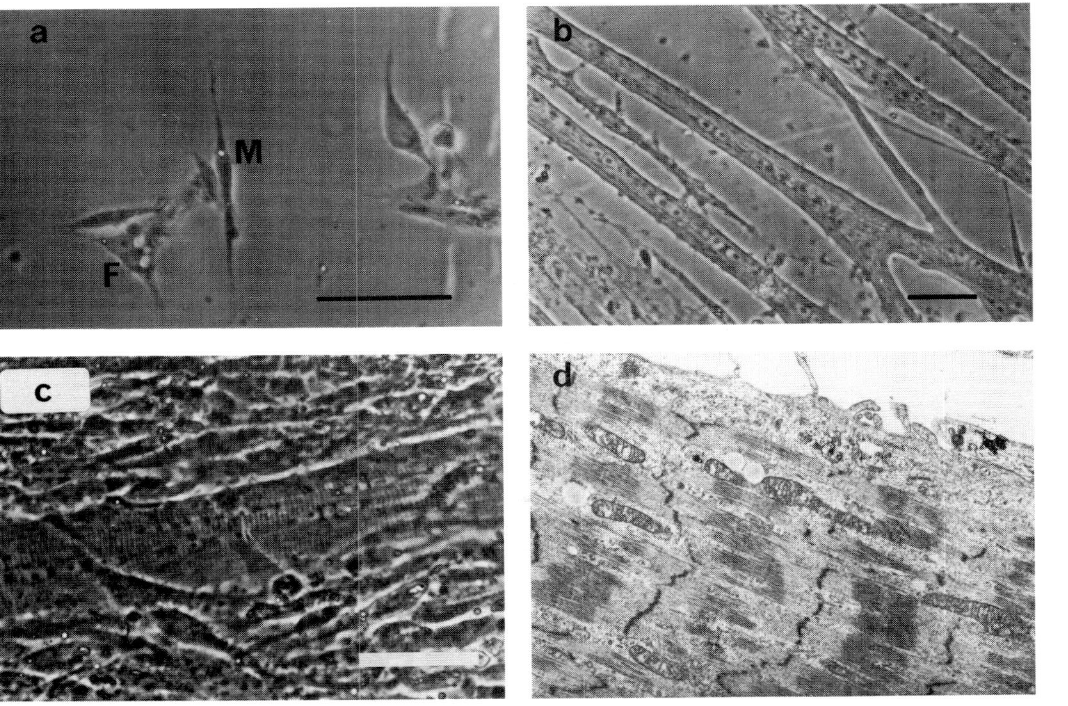

Figure 4.1: Skeletal Muscle Cells in Culture (a) 20 h culture of chick embryo leg muscle, showing myoblasts (M) and fibroblasts (F). (b) 4-day culture of chick embryo leg muscle. Note the branching of the myotubes and central position of the nuclei. (c) Large cross-striated myotube in a 6-day culture of chick embryo leg muscle. Calibrations in (a)-(c) = 50 μm. (d) Electron micrograph of a human muscle cell in culture. Magnification, 11,000. Supplied by Dr C. Maunder, Hammersmith Hospital, London.

culture methods were reviewed by Hauschka (1972); more recent work has concerned methods for growing human muscle in dissociated cell cultures (Yasin *et al.*, 1977; Blau and Webster, 1981). One of the attractions of applying culture methods to skeletal muscle is the ability to use human tissue to investigate diseases such as the muscular dystrophies. So far, progress has been disappointing (for a review, see Witkowski, 1977).

Skeletal muscle cells have also been grown from unusual starting tissues, including cardiac muscle (Kimes and Brandt, 1976b), thymus (Wekerle *et al.*, 1975), pineal (Freschi *et al.*, 1979) and pituitary glands (Brunner and Tschank, 1982). The significance of these findings is not clear, although skeletal muscle-like cells in the thymus may play a role in the pathogenesis of myasthenia gravis.

Electrical Properties and Drugs Acting on Ion Channels

Passive Membrane Properties

A number of workers have measured membrane potentials in cultured muscle cells at different stages from myoblast to muscle fibre. There is general agreement that, in primary muscle cultures, membrane potentials are low in myoblasts and small myotubes and that potentials increase with development in culture. Myoblasts have resting membrane potentials of around –10 mV in cell cultures from chick (Dryden *et al.*, 1974), rat (Fambrough and Rash, 1971; Ritchie and Fambrough, 1975b) and human muscle (Harvey *et al.*, 1979). With chick, rat and mouse (Powell and Fambrough, 1973) myotubes there is a rapid increase in potential to values of –40 to –60 mV. Still higher values (between –70 and –95 mV) have been reported for fibres in cultures of chick embryo muscle (Kano and Shimada, 1971; Hooisma *et al.*, 1975). In cultures of human fetal muscle, fibres can have resting potentials of –80 to –90 mV (Harvey *et al.*, 1980) but development is slower than in cultures of other species (Figure 4.2).

Although a developmental change in membrane potential apparently takes place, the underlying mechanisms are obscure. The electrolyte content of muscle changes with growth, and this is presumably related to the alterations of the membrane potential. There is, however, little information about the ionic permeability of the muscle membrane at different stages of growth. From tissue culture studies it has been suggested that the rise in membrane potential results from a change in the ratio of the permeabilities to Na^+ and K^+ (Dryden *et al.*, 1974; Ritchie and Fambrough, 1975b). In chick muscle cultures, the intracellular concentration of K^+ remained constant at about 140 mM in myoblasts and myotubes while Na^+ concentration decreased from 56 to 22 mM (Dryden *et al.*, 1974). Catterall (1975b) estimated that the internal Na^+ concentration of chick embryo myotubes was 13 mM, which was the same as in rat myotubes (Ritchie and Fambrough, 1975b). Although the latter authors did not find a change in Na^+ ion concentrations at different stages, they did report that the ratio of Na^+ to K^+ permeabilities changed from 0.4 for immature myotubes with resting potentials of –24 mV, to 0.07 for mature cells with resting potentials of –55 mV, indicating that the membrane became relatively less permeable to Na^+. A Na^+:K^+ permeability ratio of 0.14 was calculated for cultured human myotubes (Merickel *et al.*, 1981). The resting conductance of chick myotubes falls from about 1.2 $mScm^{-2}$ to around 0.2 $mScm^{-2}$ during the first week in culture, but the resting K^+ conductance does not appear to change (Thomson and Dryden, 1980). These authors suggested that, in order to account for the previously measured change in the ratio of Na^+:K^+ permeabilities, young myotubes must have a substantial resting Na^+ conductance which decreases during development.

Changes in the responses to alteration of the external K^+ concentration

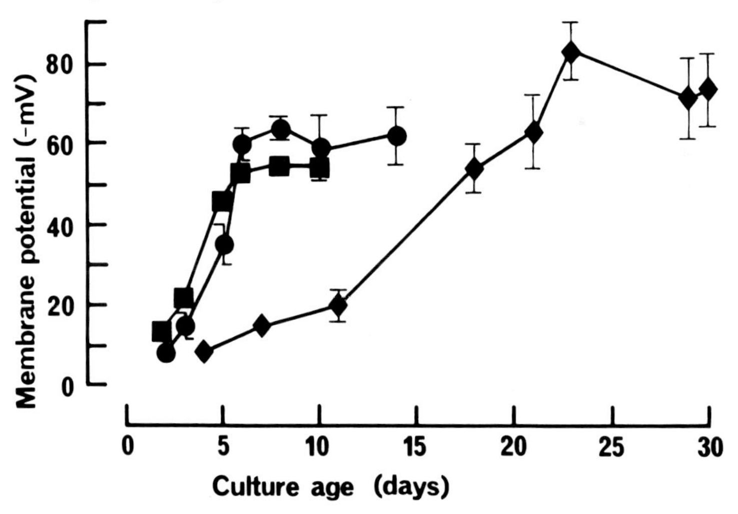

Figure 4.2: Development of Resting Membrane Potential in Skeletal Muscle Cultures Derived from Chick Embryo (●), Neonatal Rat (■) and Human Fetal (♦) Muscle

Each point represents the mean ± standard error of the mean of values from at least six different cultures.

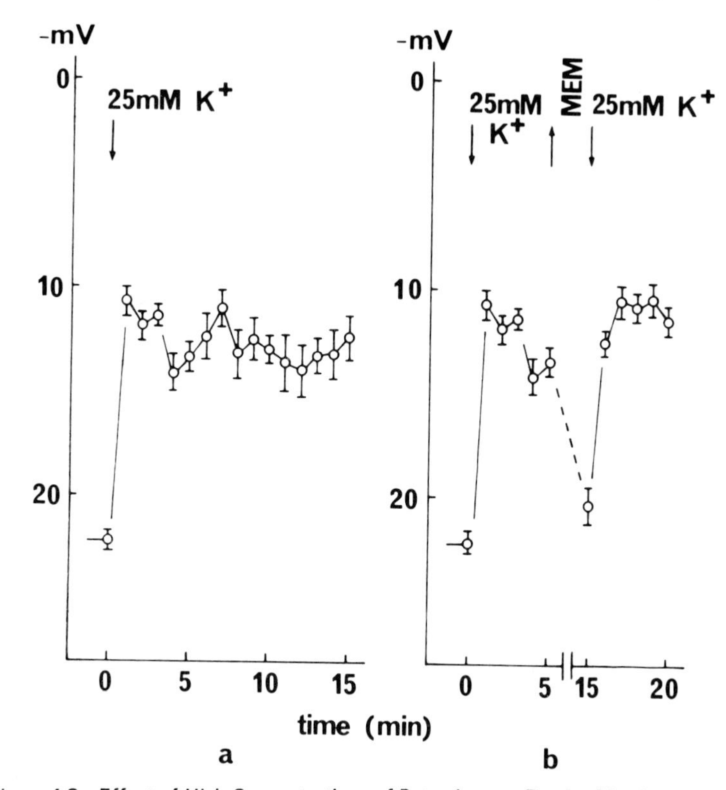

Figure 4.3: Effect of High Concentrations of Potassium on Resting Membrane Potentials of Chick Embryo Myotubes in Cell Cultures
Ordinate gives membrane potential values (in negative mV). (a) 25 mM K^+ applied for 15 min. (b) 25 mM K^+ applied for 5 min. and reapplied after 10 min. in MEM. Each point represents the mean ± standard error of the mean of between 12 and 40 individual measurements made on six different cultures. From Harvey and Dryden (1974a), with permission.

suggest that there is a developmental change in membrane K^+ permeability. Both chick and rat myoblasts and immature myotubes were found to be relatively insensitive to changes in external K^+ (Dryden *et al.*, 1974; Ritchie and Fambrough, 1975b) but older cells with higher resting membrane potentials have an increased K^+ sensitivity which is similar to that of adult fibres. Exposure to high K^+ concentrations produces a rapid and sustained fall in membrane potential that is readily reversed on return to normal solutions (Figure 4.3).

In cells of the rat myogenic cell line, L_6 (Yaffe, 1968), the relationship between resting potential and morphological maturity is quite different from that of cells in primary cultures. L_6 myoblasts had an average resting membrane potential of about −70 mV which was not significantly different from the average value for myotubes (Kidokoro, 1973, 1975a). Additionally, there was no developmental change in membrane ionic permeabilities: the membranes of both myoblasts and myotubes were permeable to K^+ with little permeability to Na^+ and Cl^- (Kidokoro, 1975b). The unusual properties of the membrane of L_6 myoblasts may be a consequence of adaptation of the cells to continuous maintenance in culture.

The values obtained for passive membrane constants for chick, mouse, rat and human myotubes are in general agreement, allowing for different analytical methods (Table 4.1). However, technical difficulties and the individual variations between myotubes have precluded a detailed study of membrane constants during development. In young myotubes it is difficult to insert two microelectrodes that would allow more precise measurement of membrane responses to injected current, and the highly branched nature of some more mature cells leads to complex non-linear responses because of variations in the electrotonic spread of current. Muscle cells with simpler geometry have been obtained by treatment of myotubes with colchicine to give 'myosacs' (Fukada *et al.*, 1976a, b) and by growing myoblasts on a surface (Sylgard resin) that prevents cell adhesion so that spherical 'myoballs' are formed (Fischbach and Lass, 1978a, b). More extensive use of such preparations may aid future studies of the development of the electrical properties of the muscle membranes.

The values obtained for the membrane constants of cultured cells are intermediate between those of innervated and denervated muscle or intermediate between values from twitch and tonic chick muscle fibres (Table 4.1). A direct demonstration of the lack of differentiation of cultured fibres into mature fibre types came from Purves and Vrbová (1974) who prepared cultures from the anterior and posterior latissimus dorsi muscles of 15-day chick embryos. *In vivo* the membrane properties of fibres in the singly innervated posterior latissimus dorsi become markedly different from those of the multiple innervated latissimus dorsi but no differences in membrane potential, input resistance or time constant were found between myotubes grown from the two different muscles.

In primary cultures, the relationship between applied current and voltage response is variable: some myotubes respond linearly over a wide range of applied currents, whereas others respond with a delayed increase in conductance (i.e. delayed rectification). Myoblasts of the rat

Table 4.1: Membrane Constants of Cultured and Non-cultured Muscle Cells

Preparation	Membrane potential (–mV)	Input resistance (MΩ)	Specific membrane resistance (Ωcm²)	Specific membrane capacitance (μF/cm²)	Membrane time constant (ms)	Length constant (μm)	Reference
Chick embryo myotubes	8 – 50	4 – 68	2,639 ± 369	3.9 ± 0.46	9.95 ± 1.72	295 – 1,085	Fischbach *et al.* (1971)
Chick embryo myotubes	49 – 64	1 – 2	180 – 720	–	2.5 – 9.7	–	Harris *et al.* (1973)
Chick embryo myotubes	76 ± 2.5	5.1 ± 1.4	3,300 ± 720	10.3 ± 2.2	22.3 ± 4.1	1,000 ± 190	Engelhardt *et al.* (1976)
Chick anterior latissimus dorsi	51 ± 1	0.78 ± 0.08	4,388 ± 785	8.2 ± 1.0	35.0 ± 6.0	1,780 ± 170	Fedde (1969)
Chick posterior latissimus dorsi	57 ± 1.7	0.25 ± 0.03	561 ± 50	7.0 ± 0.6	3.7 ± 0.11	680 ± 38	Fedde (1969)
Mouse embryo myotubes	38 ± 1.2	2.25 ± 0.16	694 ± 75	8.4 ± 1.2	5.4 ± 0.9	609 ± 50	Powell and Fambrough (1973)
Mouse embryo myotubes	60 ± 2.8	10.2 ± 2.4	1,855 ± 531	4.3 ± 1.4	5.8 ± 1.3	–	Christian *et al.* (1977)
Mouse G8 myotubes	49 ± 2.2	7.5 ± 1.7	2,555 ± 467	4.1 ± 0.9	8.7 ± 1.0	–	Christian *et al.* (1977)
Rat embryo myotubes	51 ± 1.5	~ 9	1,046 ± 158	4.0 ± 0.3	4.9 ± 0.5	–	Ritchie and Fambrough (1975b)
Rat L_6 myoblasts	37 – 62	53 – 140	8,100 ± 939	1.0 ± 0.09	7.4 ± 0.7	–	Kidokoro (1975a)
Rat L_6 myotubes	50 – 67	80 – 120	12,300 ± 1,878	4.7 ± 0.9	58.0 ± 17.3	–	Kidokoro (1975a)
Rat extensor digitorum longus	77 ± 0.2	0.41 ± 0.01	545 ± 11	2.8 ± 0.04	1.5 ± 0.003	530 ± 10	Albuquerque and McIsaac (1970)
Rat denervated (15d) extensor digitorum longus	54 ± 1.3	0.82 ± 0.1	1,250 ± 41	5.1 ± 0.4	6.4 ± 0.3	700 ± 30	Albuquerque and McIsaac (1970)
Human adult myotubes	48 ± 0.4	–	7,780 ± 676	3.8 ± 0.3	22.9 ± 1.8	938 ± 280	Merickel *et al.* (1981)
Human intercostal biopsies	86 ± 0.9	~0.5	4,074 ± 258	4.8 ± 0.5	18.9 ± 1.2	2,244 ± 108	Elmqvist *et al.* (1960)

Note:
Where possible, values are given as means ± standard error of the mean.

myogenic cell line L_6 do not show delayed rectification although myotubes do, suggesting that a change in membrane ion channels occurs following fusion (Kidokoro, 1975a). L_6 myotubes also have much higher input resistances than myotubes in primary cultures, presumably because of their smaller size.

Action Potentials and Drugs Acting on Ion Channels

Development of Action Potential Mechanisms. The ability to generate action potentials is acquired early in the development of muscle fibres *in vivo* but there are developmental changes in the characteristics of the action potential. For example, in 13-16-day chick embryos, the muscle action potential took the form of a long plateau which lasted 100-200 ms (Kano, 1975). This response was insensitive to tetrodotoxin and was probably a result of Ca^{2+} inflow. From 17 days to hatching, the action potential was biphasic with an initial rapid 'spike' response carried by Na^+ and then the long plateau. After hatching, the action potential consisted only of the short Na^+-dependent spike, which could be abolished by tetrodotoxin.

Similar differences in the action potential of individual cells have been seen in chick embryo muscle developing in cell culture, although a developmental sequence is not so obvious. Kano *et al.* (1972) found that some myotubes had a monophasic spike response and others had a 2-phase spike-plateau response (Figure 4.4). Tetrodotoxin abolished spike but not plateau responses, which were shown to be blocked by Mn^{2+} (Kano and Shimada, 1973). Tetrodotoxin-sensitive action potentials were also noted in mature myotubes in explant cultures of chick embryo muscle (Harris *et al.*, 1973). Although immature cells with low membrane potentials could generate action potentials after hyperpolarization (Fischbach *et al.*, 1971), the maximum rates of rise increased with the age of the culture between 4 and 14 days (Kano and Yamamoto, 1977). Developmental changes were also studied by Spector and Prives (1977) who found that fast spike potentials could be generated only in localized patches of membrane on young myotubes with large areas of membrane displaying passive responses. In older cultures which had more developed myotubes, fast rising action potentials could be generated over the entire cell. Concurrently, the maximum rate of rise increased from 10-20 V/s to 200 V/s. Also the action potentials in immature myotubes were insensitive to tetrodotoxin while those in mature cells were tetrodotoxin-sensitive. Similar changes happen in L_6 and mouse myotubes (Land *et al.*, 1973). Biochemical evidence for an increase in the number of Na^+ channels with growth in

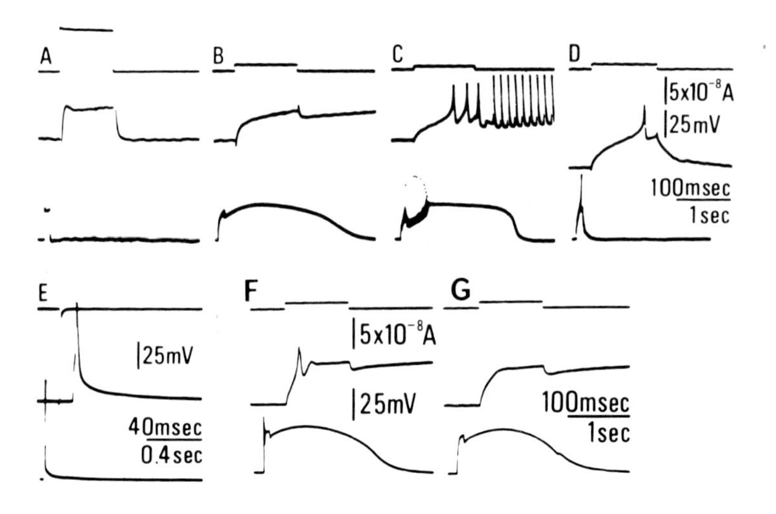

Figure 4.4: Intracellular Potential Changes in Cultured Chick Embryo Muscle Fibres

Top trace, reference level (0 mV) for the middle trace and current passed through the cell membrane. Middle trace, fast sweep speed record of membrane potential. Bottom trace, slow sweep speed record of membrane potential. (A) 17-day culture, no electrogenic responses. (B) 27-day culture, plateau response alone. (C) 21-day culture, spike and plateau responses. (D) 19-day culture, spike response alone. (E) Response of innervated muscle fibre. Stimulation to nerve element elicits spike response alone. (F and G) Same cell showing the effect of tetrodotoxin (10^{-8} g/ml): (F) control; (G) after addition of tetrodotoxin. The spike response is abolished but the plateau response remains. From Kano *et al.* (1972), with permission.

culture has been provided from studies of $^{22}Na^+$ uptake and tetrodotoxin binding (Catterall, 1980; Frelin *et al.*, 1981).

Additional to the rapid Na^+-dependent spike and the Ca^{2+}-dependent plateau response, a third type of regenerating depolarizing response in cultured chick myotubes has been described by Fukada *et al.* (1976a, b). This is a prolonged depolarization occurring after the initial spike and dependent on the intracellular concentration of Cl^-. The significance of this chloride conductance is obscure and it is not known if it occurs during normal development *in vivo*.

Na^+ Channels. Even as myoblasts, muscle cells appear to have functional Na^+ channels. L_6 myoblasts can generate small action potentials that vary in amplitude with the extracellular Na^+ concentration

(Kidokoro, 1973, 1975a, b), and a low level of $^{22}Na^+$ uptake can be detected in cultures of chick embryo myoblasts (Frelin *et al.*, 1981).

The properties of Na^+ channels in myotubes in primary cultures have been studied using voltage (Fukada *et al.*, 1976a) and patch clamp techniques (Sigworth and Neher, 1980). Depolarization results in an increase in inward current which is blocked by tetrodotoxin or by substitution of Na^+ by tetramethylammonium. The average single channel conductance was estimated to be 18 pS at about 20°C.

Other attempts to characterize the Na^+ channel in membranes of cultured muscle cells have made extensive use of agents that activate (veratridine, batrachotoxin, grayanotoxin) or block (tetrodotoxin, saxitoxin) the ionophore. Uptake of $^{22}Na^+$ by cells of the rat myogenic cell line L_8 and by myotubes in primary cultures of chick embryo muscle was stimulated by veratridine (Catterall and Nirenberg, 1973). The increase in Na^+ uptake was assumed to result from flow through action potential channels and it was implied that uptake was blocked by tetrodotoxin. In a later study (Catterall, 1976) it was found that both myoblasts and myotubes of the rat cell line L_5 responded to veratridine and batrachotoxin by an increased uptake of $^{22}Na^+$. The response was larger in myotubes than in myoblasts, agreeing with findings on the amplitude of action potentials (Kidokoro, 1973). The stimulation of Na^+ uptake was relatively resistant to tetrodotoxin and saxitoxin. Using L_6 myotubes, Sastre and Podleski (1976) demonstrated that prolonged administration of veratridine caused a Na^+-dependent depolarization which could be inhibited by high concentrations of tetrodotoxin and saxitoxin. Short exposures to veratridine prolonged the decay phase of the action potentials which were shown to be Na^+-dependent, but relatively resistant to tetrodotoxin or saxitoxin.

More quantitative studies (Catterall, 1980; Lawrence and Catterall, 1981a, b; Lombet *et al.*, 1982) examined the interactions of batrachotoxin and veratridine with scorpion toxin and sea anemone toxin ATx II in cultures of chick embryo and fetal rat muscle and of the L_5 cell line. These compounds had actions similar to those previously seen in neuroblastoma cells (see Chapter 3) although there were quantitative species differences (see Table 4.2). Lombet *et al.* (1982) found that rat myotubes in primary cultures appeared to have high and low affinity sites for tetrodotoxin. The low affinity site was revealed by inhibition of $^{22}Na^+$ uptake induced by veratridine and ATx II, whereas the high affinity site was only apparent in competition binding studies with a labelled tetrodotoxin derivative.

Table 4.2: Sensitivity of Cultured Muscle Cells from Different Sources to Agents Acting on Na^+ Channels

Preparation	Tetrodotoxin	Veratridine		Batrachotoxin		Scorpion toxin*	ATxII*	Reference
	(EC_{50},nM)	(EC_{50},μM)	(Max. effect, nmole Na^+ uptake/ min/mg)	(EC_{50},μM)	(Max. effect, nmole Na^+ uptake/ min/mg)	(EC_{50},nM)	(EC_{50},μM)	
Chick embryo myotubes	4–6	10	22	1.5	143	30	70	Catterall (1980) Lombet *et al.* (1982)
Rat embryo myotubes	500–2,000	–	–	0.3	n.d.	20	5	Lawrence and Catterall (1981b)
	1.6†	–	–	–	–	–	–	Lombet *et al.* (1982)
Rat L_5 myotubes	600–1,000	140	44	1.7	380	60	more than one class of site	Lawrence and Catterall (1981a) Catterall (1976)
Rat L_6 myotubes	1,000	300	67	–	–	–	–	Stallcup (1977)
Mouse G8 myotubes	200	100	8	–	–	–	–	Stallcup (1977)

Notes:
* Determined by ability of scorpion toxin and ATxII to increase the effect of veratridine or batrachotoxin.
† Determined by inhibition of [^{3}H] ethylenediamine tetrodotoxin binding.

Other Ion Channels. Relatively little work has been carried out on the other ion channels (for Ca^{2+} and K^+) of cultured skeletal muscle. Although Ca^{2+}-dependent action potentials can be recorded in chick embryo muscle cells in culture, Fukada *et al.* (1976a) found that the amplitude of the Ca^{2+} current that could be recorded under voltage clamp conditions was always too small to permit thorough analysis.

Two forms of K^+ conductance have been studied in cultures of skeletal muscle. The K^+ channels responsible for rectification were found to be blocked by tetraethylammonium, 4-aminopyridine and Ba^{2+} (Barrett *et al.*, 1981) and single channel recordings have been made of the unitary K^+ currents of the anomalous rectifier (Ohmori *et al.*, 1981). In addition, there is another K^+ current that is more slowly activated and is responsible for after-hyperpolarizations following action potentials (Barrett *et al.*, 1981). This K^+ conductance is not blocked by tetraethylammonium. It is dependent on Ca^{2+} and it can be blocked by Mn^{2+} and Ba^{2+}, or by intracellular injection of the Ca^{2+} chelating agent EGTA. Single channel recordings from patches of membrane isolated from rat myotubes reveal that this K^+ channel is activated by depolarization and by intracellular Ca^{2+} (Barrett *et al.*, 1982). The single channel conductance of about 200 pS is 10-20 times larger than that of the anomalous rectification K^+ channel.

Acetylcholine Receptors

In comparison to the morphological and physiological studies, the pharmacological properties of skeletal muscle in tissue culture have, until recently, been almost ignored. A description of the effects of several drugs on the mechanical activity of explants of chick embryo skeletal muscle (Sacerdote de Lustig, 1942, 1943) was probably the first reported pharmacological study. Physostigmine and low concentrations of acetylcholine activated the spontaneous contractility of the explants, whereas atropine and high concentrations of acetylcholine paralysed the muscle. Adrenaline paralysed innervated skeletal muscle explants (Sacerdote de Lustig, 1942). Several other substances (curare and erythrina alkaloids, cobra and rattlesnake venoms, strychnine and veratrine) were found to inhibit the activity of innervated skeletal muscle explants, but to activate and then block denervated muscle. The curare and erythrina alkaloids and cobra venom blocked only the paralysing action of acetylcholine, while strychnine and veratrine could activate cultures paralysed by physostigmine, acetylcholine, curare or erythrina alkaloids (Sacerdote de Lustig, 1943).

Murray (1960) devised a system to quantify the action of drugs on the spontaneous activity of explanted skeletal muscle. Her studies indicated that, in fetal rat muscle explants, procaine, veratrine, tubocurarine and ryanodine were inhibitory, although tubocurarine in low concentrations could be stimulatory. ATP and acetylcholine had little effect, but caffeine stimulated activity.

The techniques used by Sacerdote de Lustig and Murray are of little value for more precise pharmacological investigations and some of the effects noted defy explanation by standard mechanisms. Cultured skeletal muscle as a pharmacological preparation only became a possibility with the development of microelectrode techniques capable of measuring the intracellular membrane potential of skeletal muscle cells grown in monolayers.

Microapplication of acetylcholine has demonstrated that cultured skeletal muscle fibres are sensitive over their entire surface (Dryden, 1970; Fischbach, 1970; Fambrough and Rash, 1971; Harris *et al.*, 1971; Kano *et al.*, 1971). Fetal skeletal muscles prior to innervation and denervated muscles are sensitive to acetylcholine over the whole muscle surface. Since muscle in culture is generally of fetal origin and is grown in the absence of nerves, it is not surprising that acetylcholine sensitivity is distributed over the fibre membrane.

The ability to study the consequences of receptor activation has been greatly enhanced with the advent of patch clamp techniques for recording directly from single receptors (Neher and Sakmann, 1976a; Neher *et al.*, 1978). In turn, the refinement of such techniques (Hamill *et al.*, 1981) has also been helped by the characteristics of cultured muscle cells. These cells, with their relatively low density of acetylcholine receptors spread throughout their surface and their lack of overlying connective tissue, are almost custom-designed for patch clamp recording.

The ability to detect sensitivity to acetylcholine by iontophoretic application is complemented by use of radioactively-labelled snake toxins to estimate the numbers of binding sites in cultures and to illustrate, by autoradiography or horseradish peroxidase labelling, the distribution of the receptors. The availability of such specific labels for acetylcholine receptors has also allowed the metabolic turnover of the receptor protein to be characterized.

Development of Acetylcholine Receptors

Many workers have used tissue culture techniques to enable them to study the development of acetylcholine sensitivity in skeletal muscle and the mechanisms controlling the production and distribution of acetylcholine receptors. It has been thought that fusion of myoblasts is the biochemical signal that triggers the production of many muscle-specific proteins, including cholinoceptors. In support of this, no responses to acetylcholine could be seen in undifferentiated uninucleated cells in cultures obtained from chicken (Dryden, 1970), rat (Fambrough and Rash, 1971), mouse (Powell and Fambrough, 1973) and human muscle (Harvey *et al.*, 1979) and sensitivity appears after formation of multinucleated cells. Binding of α-neurotoxins (Patrick *et al.*, 1972; Prives and Paterson, 1974) and autoradiography (Powell and Fambrough, 1973; Sytkowski *et al.*, 1973; Blau and Webster, 1981) also revealed a very low level of acetylcholine receptors on myoblasts and a rapid increase in receptor numbers following fusion (Figure 4.5). Nevertheless, fusion is not absolutely essential for the production of acetylcholine receptors because a small proportion of myoblasts do differentiate to become sensitive to acetycholine (Fambrough and Rash, 1971; Dryden *et al.*, 1974) and this proportion can be increased by blocking fusion with low Ca^{2+} medium or 5-fluorodeoxyuridine (Fambrough and Rash, 1971). Moreover, there is evidence that myoblasts may have higher levels of acetylcholine receptors than previously believed. Teng and Fiszman (1976) reported that α-bungarotoxin would bind to myoblasts obtained by mechanical dissociation of thigh muscle from chick embryos, although most of the specific binding appeared to be intracellular. Smilowitz and Fischbach (1978), however, did find that intact myoblasts derived from chick embryo muscle bound α-bungarotoxin after the muscle had been dissociated mechanically. There was no binding if trypsin was used to obtain the cell suspension. It is not known if the myoblasts that bound α-bungarotoxin were also sensitive to acetylcholine.

Some myoblasts of the rat myogenic cell line L_6 responded to acetylcholine with a slow hyperpolarization (Harris *et al.*, 1971; Patrick *et al.*, 1972; Steinbach, 1975). This response was apparently not mediated by a nicotinic cholinoceptor since it was not inhibited by the presence of an α-neurotoxin (Patrick *et al.*, 1972) or tubocurarine (Steinbach, 1975). A similar response could be elicited by the application of the muscarinic agonist, acetyl-β-methylcholine (Steinbach, 1975), but there is no information on the ability of muscarinic receptor antagonists such as atropine to block the hyperpolarization. A prolonged hyperpolarization in response to acetylcholine was found in primary cultures of rat muscle cells when grown in the presence of tubocurarine (Fambrough and Rash, 1971), but the mechanism

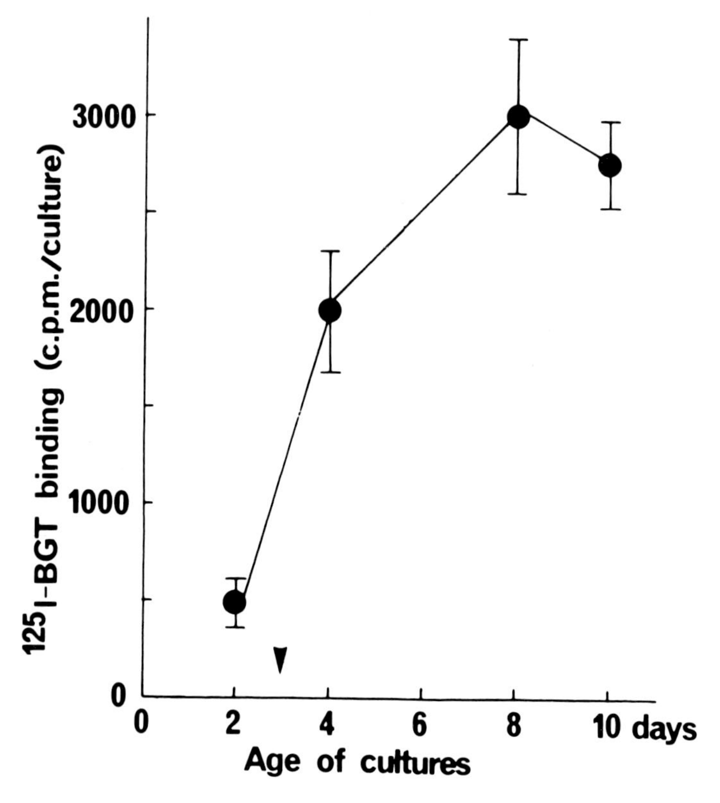

Figure 4.5: Binding of Radioactive Labelled α-bungarotoxin to Chick Embryo Skeletal Muscle Cultures

Counts per minute per culture are shown on the ordinate. Each point is the mean ± standard error of the means of between 5 and 9 cultures. Arrowhead indicates the approximate time of onset of rapid cell fusion.

involved and the relationship of this response to that seen in L_6 myoblasts is unknown. Although some myoblasts in primary cultures of chick embryo muscle hyperpolarized in response to local application of high concentrations of acetylcholine, the direction of the potential change was assumed to be because the resting membrane potential was more positive than the reversal potential; in myoblasts with higher resting potentials, acetylcholine produced a depolarization (Dryden *et al.*, 1974). This explanation does not hold for L_6 myoblasts because the resting potentials of L_6 myoblasts were between –60 and –70 mV (Steinbach, 1975) which is greater than even the most mature cells studied by Dryden *et al.* (1974). Moreover, tubocurarine blocked both the hyperpolarization and depolarization responses of the chick myoblasts, suggesting that a nicotinic receptor was involved (Dryden *et al.*, 1974). The response of the L_6 myoblasts may arise from an interaction with a membrane component not usually found in muscle cells but which results from membrane alterations during the prolonged subculturing of the cell line. It would be interesting to know if other myogenic cell lines produce similar responses, and whether the response persists after formation of L_6 myotubes.

Leaving aside the question of the precise number of cholinoceptors on individual myoblasts, it is apparent that following fusion there is a dramatic increase in toxin binding sites (Figure 4.5) (Patrick *et al.*, 1972; Sytkowski *et al.*, 1973; Prives and Paterson, 1974; Devreotes and Fambrough, 1975; Prives *et al.*, 1976; Teng and Fiszman, 1976; Spector and Prives, 1977; Shainberg and Brik, 1978) and in acetylcholine sensitivity (Fambrough and Rash, 1971; Powell and Fambrough, 1973; Dryden *et al.*, 1974; Harvey *et al.*, 1979; Blau and Webster, 1981) (Figure 4.6). There is also an increase in receptor numbers in cultures of the non-fusing muscle cell line BC_3H1 after the cells stop dividing (Patrick *et al.*, 1977).

On young myotubes, acetylcholine sensitivity and toxin binding sites

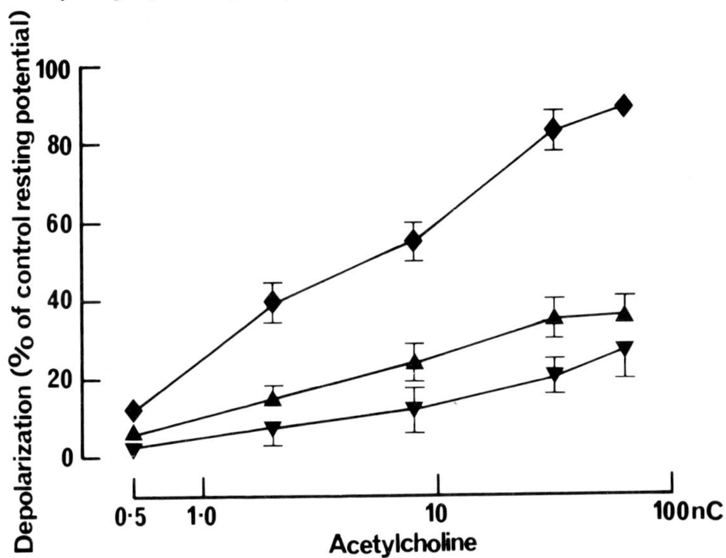

Figure 4.6: Responses to Acetylcholine in Myotubes in Explant Cultures from a 12-week Human Fetus at 24 Days (▲), 31 Days (▼) and 47 Days (◆) in Culture Standard errors bars are shown unless smaller than the symbols (n = 8—13). From Harvey *et al.* (1979), with permission.

are uniformly distributed, but with further development local areas of high sensitivity and high toxin binding can be found (Vogel *et al.*, 1972; Fischbach and Cohen, 1973) (Figure 4.7). The clusters of binding sites, or 'hot spots', become more numerous as the myotubes develop and have about ten times as many receptors per unit membrane as other regions of the myotube (Sytkowski *et al.*, 1973). Regions of high acetylcholine sensitivity have been shown to correspond closely to areas of high toxin binding (Hartzell and Fambrough, 1973; Land *et al.*, 1977; Yee *et al.*, 1978) (Figure 4.7).

Receptor clusters detected by toxin binding or by sensitivity to acetylcholine have been found in primary cultures of chick (Vogel *et al.*, 1972; Fischbach and Cohen, 1973), mouse (Christian *et al.*, 1978b), rat (Axelrod *et al.*, 1976; Land *et al.*, 1977) and *Xenopus* muscle (Gruener and Kidokoro, 1982; Kidokoro and Gruener, 1982), but not in human muscle cultures (Askanas *et al.*, 1977; Harvey *et al.*, 1979; Blau and Webster, 1981) nor in rat myogenic cell lines (Harris *et al.*, 1971; Vogel *et al.*, 1972; Land *et al.*, 1977; Podleski *et al.*, 1978; but see Kidokoro and Patrick, 1978). 'Hot spots' have been described in myotubes in a mouse cell line (Noble *et al.*, 1978) and in innervated rat L_6 myotubes (Harris *et al.*, 1971). Clusters of intramembrane particles thought to correspond to acetylcholine receptors have also been identified on non-innervated muscle fibres cultured from *Xenopus* embryos (Peng and Nakajima, 1978) and on cultured chick myotubes (Cohen and Pumplin, 1979). The average density of receptor sites differs between species (Table 4.3).

It must be noted that myotubes developing *in vivo* do not accumulate several separate patches of receptors as happens in culture (Bevan and Steinbach, 1977; Burden, 1977a) but despite this, the appearance of 'hot spots' on cultured fibres provides a convenient system for the study of mechanisms involved in the stabilization of groups of receptors. Use of fluorescent derivatives of α-neurotoxins allows receptors to be visualized on living cells and enables their motion to be recorded. Receptors which are diffusely distributed are freely mobile in the membrane, whereas the receptors in clusters are restricted in their movement (Axelrod *et al.*, 1976). About 90 per cent of receptors were found in this study to be uniformly distributed on the myotube membrane, with the remaining 10 per cent being found in clusters. Initially, it was thought that there was no exchange of receptors into or out of clusters (Axelrod *et al.*, 1976), but more recent evidence suggests that diffusely distributed receptors may be able to join clusters (Stya and Axelrod, 1983). The structural integrity of the membrane 'hot spots'

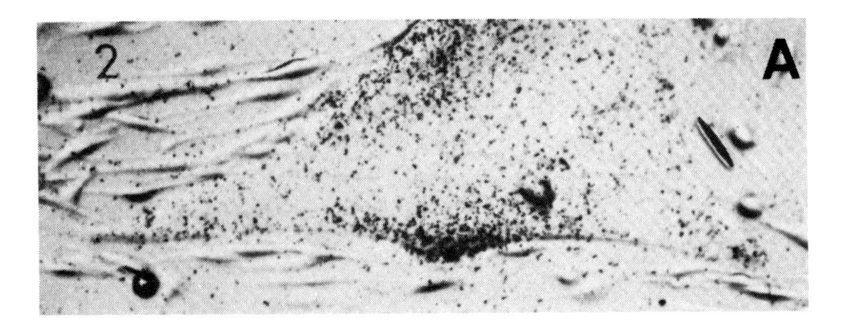

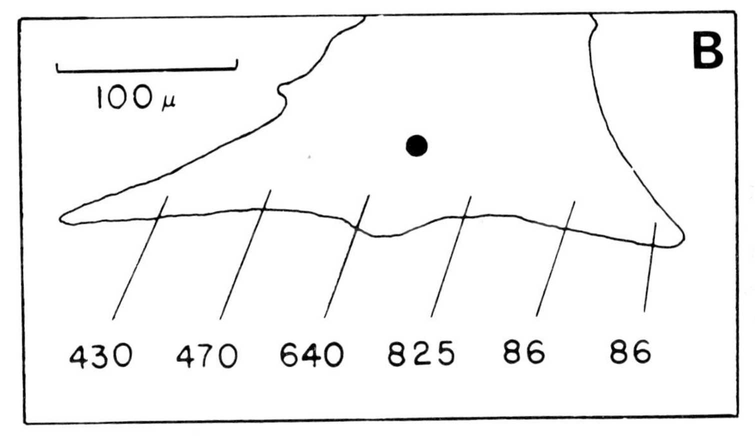

Figure 4.7: Spatial Distribution of Acetylcholine (ACh) Sensitivity and α-bungarotoxin (α-BGT) Binding Sites on Cultured Rat Muscle

(A) After the ACh sensitivities were measured, the culture was incubated in [^{125}I] α-BGT, rinsed, and autoradiography was performed. The exposure time was 19 h. (A) is a phase-contrast photomicrograph of the same cell shown in outline in (B) and illustrates the distribution of the α-BGT binding sites.

(B) The ACh sensitivity at points along a large rat myotube in culture was measured electrophysiologically. ACh sensitivities are expressed millivolts per nanocoulomb. The filled circle in the centre of the cell outline indicates the position of the recording electrode. Since the distance (the space constant) over which the ACh response decays to 1/e of its initial amplitude due to electrotonic decay is more than the length of this cell, the effect of electrotonic decay on the ACh potential has been neglected in these measurements. Since the ACh pulse was actually between 10^{-11} and 10^{-10} coulombs, the source radius was about 5 μm thus limiting the spatial discrimination between points. From Hartzell and Fambrough (1973), with permission.

Table 4.3: Density of Acetylcholine Receptors on Cultured and Non-cultured Muscle Cells

Preparation		Receptor density (sites/μm^2)	Reference
Chick embryo myotubes	(cluster)	9,000	Sytkowski *et al.* (1973)
	(diffuse)	900	
Rat embryo myotubes	(cluster)	8,000	Axelrod *et al.* (1976)
	(diffuse)	2,000	
Xenopus embryo myotubes			
	(cluster)	890	Kidokoro and Gruener (1982)
	(diffuse)	104	
Rat L_6 myotubes		20–100	Patrick *et al.* (1972); Podleski *et al.* (1978)
Mouse sternomastoid endplates		30,500	Fertuk and Salpeter (1976)
Rat diaphragm endplates		13,000	Fambrough and Hartzell (1972)
Rat diaphragm (denervated)		635	Fambrough (1974)

was investigated more thoroughly in other studies by Axelrod *et al.* (1978a, b). The effects of enzymes, drugs and physical alterations were tested on the lateral movements of receptors and cluster formation, and on movements of bound concanavalin A and lipid probes (Axelrod *et al.*, 1978a). Acetylcholine receptor patches were thought to be stabilized by an immobile cell structure made of non-receptor molecules which bound cholinoceptors selectively. However, the nature of this structure has not been defined, although it has been suggested that collagen may be involved (Kalcheim *et al.*, 1982). Lipid molecules could move laterally in areas of membrane containing both high and low densities of receptor. Acetylcholine receptor mobility outside or stability inside patches was not affected by agents which acted on extrinsic cell surface proteins or on cytoplasmic microfilaments and microtubules (Axelrod *et al.*, 1978a). Indeed, colchicine treatment to produce 'myosacs' did not prevent the development of hot spots (Fukada *et al.*, 1976a, b). Receptor stabilization cannot be controlled entirely by membrane fluidity as increasing the fatty acid content of the membrane had little effect on receptor mobility (Axelrod *et al.*, 1978b). Clusters on *Xenopus* myoballs were dispersed by treatment with trypsin but not with cytochalasin B, hyaluronidase or metabolic inhibitors (Orida and Poo, 1981).

With increasing age in culture, there is a gradual disappearance of receptor clusters (Axelrod *et al.*, 1976; Prives *et al.*, 1976) which has been ascribed to an effect of muscle contractility. Fewer clusters were found on cells with well-developed cross-striations (Prives *et al.*, 1976) and receptor numbers and acetylcholine sensitivities were higher and 'hot spot' area greater in cultures in which spontaneous electrical activity had been blocked by tetrodotoxin or local anaesthetics (Cohen and Fischbach, 1973; Shainberg *et al.*, 1976; Cohen and Pumplin, 1979). Although Axelrod *et al.* (1976) reported that 'twitching and non-twitching' myotubes both lost receptor clusters, the extent of the contractile activity and the morphological maturation of the cells was not detailed. It would be interesting to know whether a local anaesthetic would have prolonged the existence of the receptor clusters in this culture system.

Receptor Turnover

The availability of specific and almost irreversible labels for nicotinic acetylcholine receptors plus the experimental convenience of the tissue culture system has allowed the synthesis and degradation of acetylcholine receptors in skeletal muscle to be analysed (see Fambrough, 1979).

Cholinoceptors in cultured skeletal muscle are synthesized by an energy-dependent process that can be inhibited by blockade of protein synthesis, e.g. with cycloheximide, or by inhibition of RNA synthesis with actinomycin D (Hartzell and Fambrough, 1973). After treatment with such inhibitors there is a delay of about 2-3 h before the incorporation of new receptors into the surface membrane is blocked, indicating that there is an internal pool of newly-synthesized receptors. Some of these receptors are in the Golgi apparatus (Fambrough and Devreotes, 1978). New receptors are inserted into the membrane at a rate of about 90 receptors/μm^2/h. Degradation of receptors is a random, energy-dependent, proteolytic process (Devreotes and Fambrough, 1975). The half life of receptors on chick and rat myotubes was estimated at 22 h (Figure 4.8), but this figure has been revised to 17 h by Gardner and Fambrough (1979), who demonstrated that the life of the receptor-α-bungarotoxin complex studied previously was significantly longer than that of receptors labelled with ^{2}H, ^{13}C, ^{15}N amino acids. Degradation of receptors involves movement of the receptors to the inside of the cell (Devreotes and Fambrough, 1975; Axelrod *et al.*, 1976). By use of protease inhibitors and chloroquine (to inhibit lysosomal function) it has been shown that degradation of

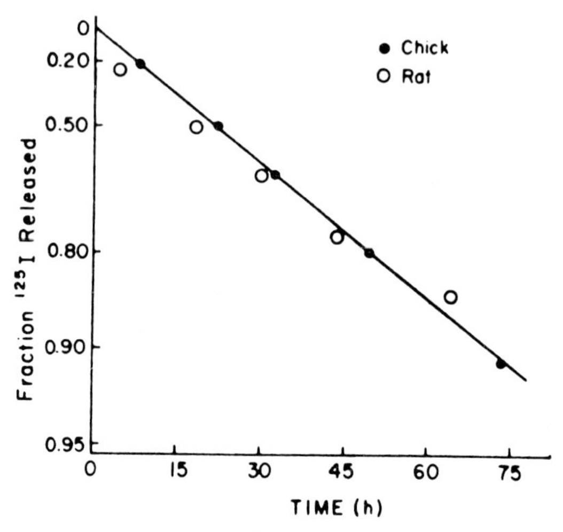

Figure 4.8: Acetylcholine Receptor Metabolism in Cultured Muscle Cells
Degradation of acetylcholine receptors determined by release of ^{125}I into the culture medium after labelling with [^{125}I] α-bungarotoxin. The half time is 22 h. From Devreotes and Fambrough (1975), by copyright permission of the Rockefeller University Press.

acetylcholine receptors does not follow the same pattern as that of average cell protein (Libby *et al.*, 1980; Libby and Goldberg, 1981). Acetylcholinesterase also has a different half life (about 50 h), although it appears to be inserted into the membrane by a process similar to that for cholinoceptors (Rotundo and Fambrough, 1980a, b).

Receptor turnover has also been estimated for muscle *in vivo*. Receptors at neuromuscular junctions are metabolized at a very slow rate with a half life of about seven days, whereas the half life of receptors at extrajunctional areas of denervated muscle is about 10-20 hours (Berg and Hall, 1975; Chang and Huang, 1975). During development there is a transition from rapid to slow turnover (Burden, 1977a, b; Steinbach *et al.*, 1979); such a change has not been reported in cultured muscle.

The availability of techniques to measure synthesis and degradation of cholinoceptors should allow study of the effects of drugs and other treatments on receptor metabolism. As has been discussed above, electrical activity of muscle cells in culture is associated with a reduction in acetylcholine sensitivity and in cholinoceptor numbers. The effects of electrical stimulation on receptor turnover may be mediated by changes of the intracellular concentrations of calcium and cyclic nucleotides (see Changeux and Danchin, 1976; Lömo, 1976). Treatments thought to increase the concentration of cyclic GMP selectively decreased the rate of receptor synthesis in cultures of chick embryo muscle, whereas increase in cyclic AMP concentrations led to an increase in receptor synthesis accompanying an increase in total protein synthesis (Betz and Changeux, 1979). However, other workers have been unable to confirm the effects of cyclic GMP (Blosser and Appel, 1980). Incubation of cultures of embryonic chick muscle or of a mouse cell line G_8 with carbachol caused a decrease in acetylcholine receptors (Noble *et al.*, 1978; Gardner and Fambrough, 1979). This decrease was brought about by a reduction in the rate of receptor synthesis rather than a change in degradation. Blockade of myotube depolarization increases receptor synthesis, apparently by a calcium-dependent but cyclic nucleotide-independent mechanism (McManaman *et al.*, 1982; see also Birnbaum *et al.*, 1980). Receptor degradation is, however, increased by a factor of about 2 by exposure of cultured myotubes to acetylcholine receptor antibody from myasthenic serum (Appel *et al.*, 1977; Heinemann *et al.*, 1977; Kao and Drachman, 1977; Prives *et al.*, 1979).

Functional Properties of Acetylcholine Receptors

When acetylcholine is applied to cultured muscle fibres, the membrane potential is depolarized in a concentration dependent manner (Figure 4.9). The size of the response increases with hyperpolarization (Figure 4.10) and the reversal potential has been estimated to be close to

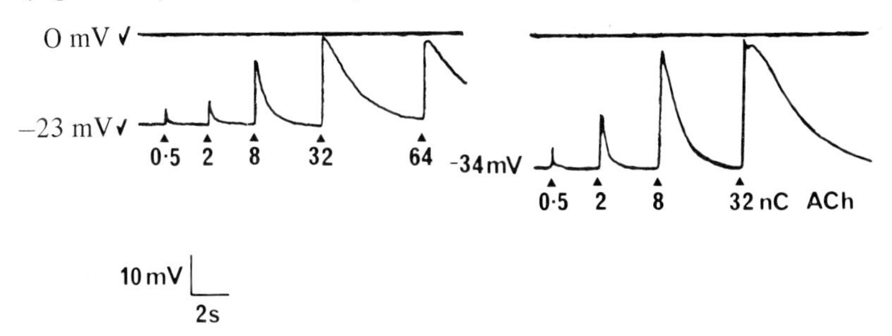

Figure 4.9: Responses of Two Myotubes Derived from Human Fetal Muscle to Iontophoretic Application of Acetylcholine
Increasing doses of acetylcholine (expressed as nanocoulombs, nC, charge passed) produced increasing depolarizations. Note that both cells depolarized close to 0 mV although the resting membrane potentials differed by 11 mV. From Harvey *et al.* (1979), with permission.

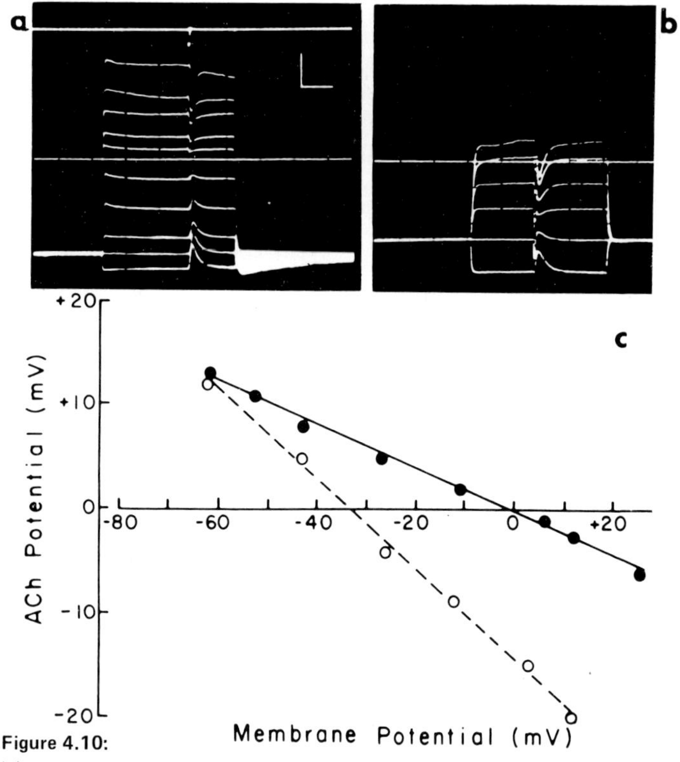

Figure 4.10:
(a) Oscilloscope traces for measurement of the acetylcholine (ACh) reversal potential of a cultured rat myotube in standard medium. The ACh potential was recorded through an intracellular microelectrode as the membrane potential was displaced with a second intracellular current-passing electrode. The duration of the polarizing current was 400 ms. The duration of the iontophoretic pulse was 5 ms. The vertical calibration represents a 20 mV displacement for the recording electrode (hyperpolarizations in the downward direction) and a current of 5×10^{-8} A for the iontophoretic pipette (top trace, outward current in the downward direction). The horizontal calibration represents 100 ms. The middle trace is zero potential obtained at the end of the experiment by removing the recording electrode from the cell. The temperature was 37°C. (b) Oscilloscope traces for the measurement of an ACh reversal potential in medium containing 20 mM Na^+ and 5.3 mM K^+. The current from the iontophoretic pipette was not recorded in this trace. All other details are the same as in (a). However, these measurements were not recorded from the same myotube shown in (a). (c) A plot of the ACh response *vs*. membrane potential for the traces shown in (a) (●) and (b) (○). The reversal potentials obtained from this plot were 0 mV in the control medium (136 mM Na^+, 5.3 mM K^+) and –33 mV in the medium containing low Na^+ (20 mM Na^+, 5.3 mM K^+). In the latter case isotonicity was maintained by replacing NaCl with sucrose. From Ritchie and Fambrough (1975a), by copyright permission of the Rockefeller University Press.

0 mV (Figure 4.10) in chick, rat and human myotubes (Harris *et al.*, 1973; Ritchie and Fambrough, 1975b; Sachs and Lecar, 1977; Harvey *et al.*, 1979). Activation of receptors causes a reduction in membrane resistance which is associated with a large increase in the rate of influx of sodium ions into the myotube (Catterall, 1975b). The receptor ion channel is also permeable to potassium, but not to chloride ions (Ritchie and Fambrough, 1975a; Steinbach, 1975; Huang *et al*, 1978). Since the agonist-induced increase in sodium flux is not inhibited by tetrodotoxin, the channel must have properties different from those of the action potential channel (Catterall, 1975b). From a study of the permeability of a large series of compounds, Huang *et al.* (1978) concluded that the receptor ion channel was a water-filled pore containing hydrogen-accepting groups and a negatively-charged site. The ionic selectivity did not change during early development of rat and chick myotubes since the reversal potential for acetylcholine responses was constant at 0 to –2 mV (Ritchie and Fambrough, 1975b).

Fundamental properties of acetylcholine-activated ion channels on cultured skeletal muscle have been studied with noise analysis (Figure 4.11), voltage jump (Figure 4.12) and single channel (Figure 4.13) recording techniques. In voltage-clamped cultured cells the mean single channel conductance, γ, of acetylcholine-activated channels has been estimated to be 25-70 pS (Table 4.4), which is similar to values determined on adult muscle. More recently, patch clamp recordings revealed that acetylcholine-activated channels on rat myoballs were not a homogeneous population, all with the same conductance (Hamill and Sakmann, 1981). There appeared to be two discrete populations with conductances of 25 and 35 pS, and each channel could also exist in a lower conducting state (10-13 pS) (Figure 4.13). It is possible that the two types of channel correspond to 'junctional' and 'extrajunctional' types of channel that have been described in, for example, adult human muscle in organ culture (Cull-Candy *et al.*, 1979). Sugiyama *et al.* (1982) have shown that rat and mouse (but not chick) myotubes in culture have two different forms of acetylcholine receptor that can be distinguished by isoelectric focusing. The isoelectric points for these two types of receptor are similar to those found previously for junctional and extrajunctional receptors in rat diaphragm muscle.

Carbachol and suberyldicholine also activate channels with similar conductances (although different open times; see below) to those operated by acetylcholine (Table 4.4). Additionally, tubocurarine has been found to cause single channel openings in human (Jackson *et al.*,

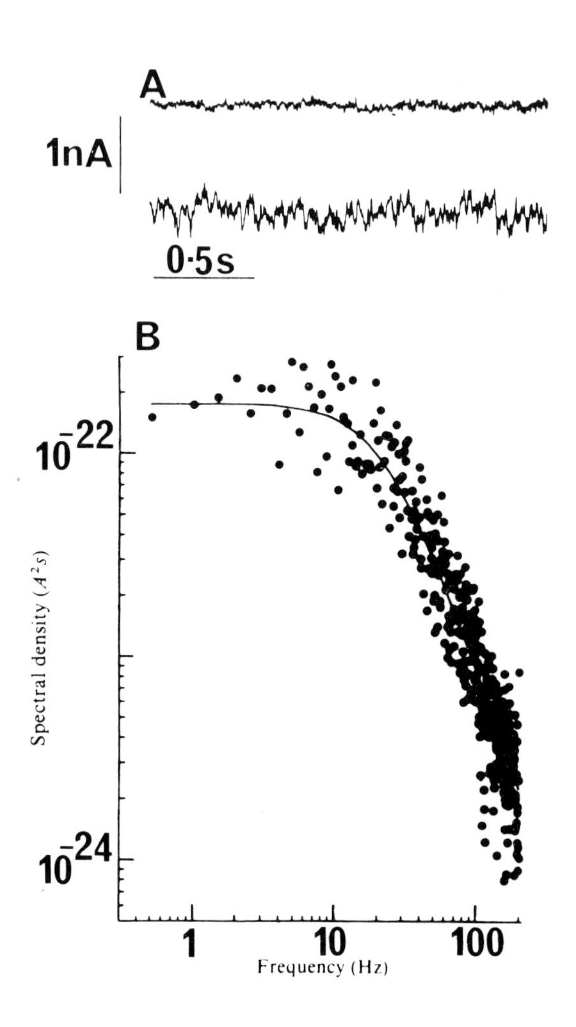

Figure 4.11: Acetylcholine-induced Current Noise in Cultured Human Myotubes (A) Digitized records of current noise before (a) and during (b) an acetylcholine-induced current of 2.7 nA amplitude. Cell was voltage clamped at –100 mV; temperature, 23°C. (B) Power spectrum of acetylcholine-induced current noise obtained by fast Fourier transforms performed on records such as shown in (A). Holding potential, –80 mV; temperature, 23°C; mean acetylcholine-induced current, 3.5 nA; τ, 6.92 ms; γ, 47.9 pS. From Bevan *et al.* (1978), with permission. Copyright 1978, Macmillan Journals Ltd.

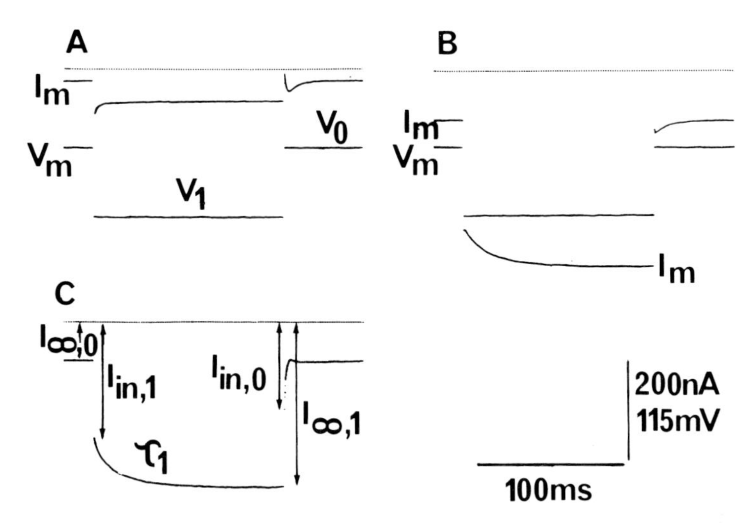

Figure 4.12: Voltage Jump Experiment to Show the Effect of Acetylcholine (ACh) on a Cultured Rat Myoball
The membrane potential, V_m, was voltage clamped to a holding potential, V_0, of −40 mV. In each record the dotted line represents zero current. Each record shows the effect of a 130 ms voltage jump to V_1, in this case, −120 mV. (A) The control response. Note the small anodal break inward current, during which V_m was constant. (B) The same voltage jump recorded in the presence of iontophoretically applied ACh. Note that the negative holding current increased in ACh and that the current relaxed to an equilibrium value after jumping to −120 mV. The magnitude and time course of V_m was not affected by ACh, indicating that the gain of the clamping amplifiers was sufficiently high. (C) The effect of digitally subtracting I_m recorded in (A) from that recorded in (B). $I_{\infty,0}$ is the ACh-induced current at V_0. $I_{in,1}$ is the instantaneous current induced at the beginning of the voltage jump. τ_1 is the relaxation time constant at V_1, in this case, 21.1 ms. The steady state current at V_1 is $I_{\infty,1}$ and the instantaneous current at the end of the jump is $I_{in,0}$. From Horn and Brodwick (1980), by copyright permission of the Rockefeller University Press.

1982a) and rat myotubes (Trautmann, 1982). In rat cells, three populations of events with differing conductances were found, and the agonist action of tubocurarine was blocked by α-bungarotoxin. Although tubocurarine has not been reported to depolarize cultured muscle cells, it has been found to depolarize intercostal muscles from fetal rats (Ziskind and Dennis, 1978). The significance of the agonist action of tubocurarine and its relation to normal development is not known at present.

Voltage jump techniques (see Figure 4.12) have been applied to internally-perfused myoballs in order to study the flow of ions through acetylcholine-activated channels (Horn and Brodwick, 1980; Horn *et al.*, 1980). Evidence was presented in support of a model of the channel which contained two asymetrically-distributed energy barriers for permeating ions (Horn and Brodwick, 1980). Further evidence for the asymmetrical nature of the acetylcholine receptor ion channel was provided by the finding that two quaternary derivatives of the local anaesthetic lignocaine could block channels if applied externally but not internally (Horn *et al.*, 1980). Rather similar results were described recently for QX314, *N*-ethyl- and *N*-octyl-guanidine (Farley and Narahashi, 1983).

Noise analysis studies on voltage clamped cultured muscle (Figure 4.11) can also yield values for the mean channel open time, τ (Table 4.4). In keeping with results on adult preparations, τ increases with hyperpolarization, with an *e*-fold change for about a 100 mV change in holding potential. With acetylcholine, typical values for τ (recorded at about –70 mV and 24°C) are 4-12 ms. These are consistently longer than values obtained from adult neuromuscular junctions, which are generally about 1-2 ms. Receptors in denervated mammalian and amphibian muscle have longer open times, and during development of rat muscle there is a transition from long to short open times (Sakmann and Brenner, 1978; Fischbach and Schuetze, 1980). This developmental change does not appear to occur in cultures of mammalian or avian muscle even after innervation (Fischbach and Schuetze, 1980; Schuetze, 1980), but it should be noted that the open times of acetylcholine receptors in chick muscle do not alter markedly during development (Schuetze, 1980; Harvey and van Helden, 1981). However, in myotubes cultured from *Xenopus* embryos there is a developmental change in mean channel open time (Brehm *et al.*, 1982). In young cultures, τ (determined from extracellular noise experiments) was about 3 ms at room temperature; after a few days in culture, the noise spectra comprised two Lorentzian functions indicating the presence of two populations with different mean open times. With age, the proportion of 'fast' channels (τ about 0.7 ms) increased. However, innervation of these myotubes had the unexpected effect of suppressing this developmental change in channel open times.

Although the value of τ is thought to be dependent on the agonist used, no one has conducted a systematic study of the effects of a wide range of agonists on cultured muscle. This could be useful in view of the findings of multiple conductance states with different lifetimes (Hamill and Sakmann, 1981) and very rapid closings during single channel opening (Colquhoun and Sakmann, 1981); these phenomena may relate to the different efficacies of different drugs. Different open times were recorded in cultured human muscle when acetylcholine, carbachol and suberyldicholine were used (Jackson *et al.*, 1982a). Analysis of the patch clamp recordings revealed that the cholinoceptor ion channel may have open states with two different durations (see Table 4.4); the longer form differed markedly in duration depending on the agonist used.

The effect of temperature on the behaviour of acetylcholine receptor ion channels has been determined in attempts to establish the influence of membrane fluidity on channel properties. There is general agreement that τ increases with decreasing temperature. For example, in human myotubes, τ was 2.4 ± 0.2 ms at 37°C and 12.2 ± 1.3 ms at 23°C (Bevan *et al.*, 1978). However, there is controversy about the effect of temperature changes on γ. Fischbach and Lass (1978b) reported that in chick myoballs γ decreased with decreasing temperatured and that there was a sharp break at about 20°C, which was interpreted as reflecting a phase transition in the membrane. In similar experiments, no such effect was found (Sachs and Lecar, 1977; Nelson and Sachs, 1979).

Pharmacological Properties of Acetylcholine Receptors

There is no evidence that development of skeletal muscle in the artificial environment of cell culture leads to any aberrant receptor types. Adrenaline, glutamate and GABA have no effect on the membrane potential of rat muscle in culture (Obata, 1974) and adrenaline, noradrenaline, dopamine, histamine, 5-hydroxytryptamine and GABA do not alter the resting potential of myotubes derived from chick embryo muscle (Harvey and Dryden, 1974b). Similar negative results were obtained with myotubes derived from rat pineal glands (Freschi *et al.*, 1979). In contrast, skeletal muscle grown in culture responds to any nicotinic agonist, including acetylcholine, carbachol, nicotine, decamethonium, suxamethonium, tetramethylammonium and dimethylphenylpiperazinium (Harvey and Dryden, 1974b) (Figure 4.14). Most evidence indicates that cultured muscle does not have muscarinic cholinoceptors since the muscarinic agonists pilocarpine, acetyl-β-methylcholine (methacholine) (Harvey and Dryden, 1974b) and bethanecol (Dryden *et al.*, 1974) do not affect myotubes in primary cultures of chick embryo muscle, and acetyl-β-methylcholine does not affect myotubes of the rat myogenic cell line L_6 (Steinbach, 1975). However, another muscarinic agonist, oxotremorine, and

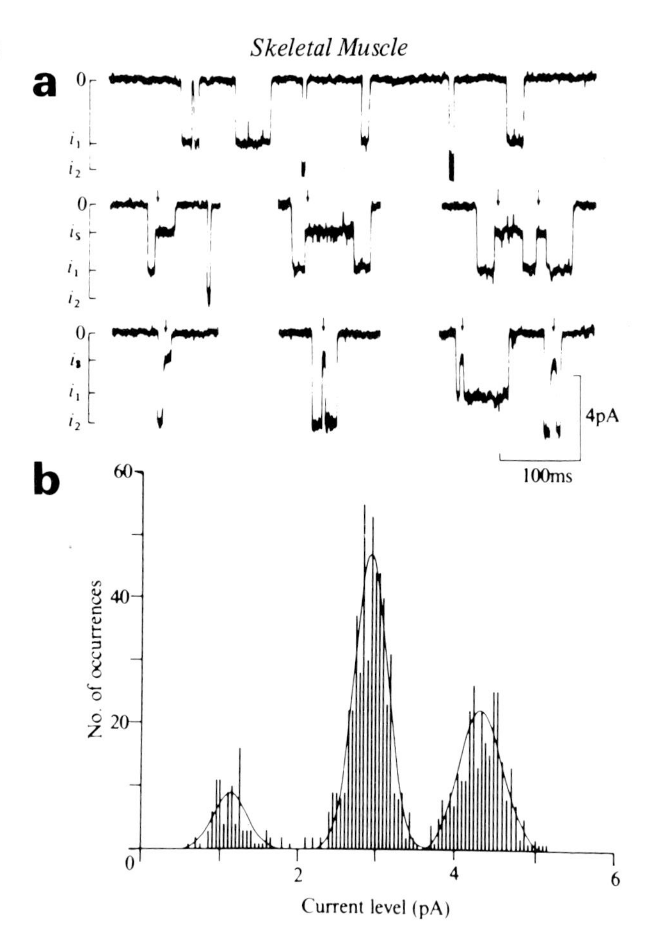

Figure 4.13: Single Channel Recordings from a Patch of Membrane from a Cultured Myoball

(a) Oscilloscope traces showing acetylcholine receptor channel currents recorded at low frequency from an isolated patch of sarcolemma. The top trace is a continuous record and indicates the typical current behaviour. Two distinct classes of elementary currents (i_1, i_2) are evident. A small proportion of currents from both classes display more complex waveforms and examples are shown in the middle and lower traces. Some currents jump down to a sublevel (i_s) of about 1 pA and then either return to the resting level or jump back to the original main level. A small proportion of events show repetitive fluctuations between the two levels (last example in middle trace). In none of the 809 events recorded in this experiment did the sublevel precede the initial jump to the main level. In two cases the sublevel appeared in apparent isolation. Membrane potential −100 mV; temperature, 8°C. Day 14 myoball, four days after colchicine treatment; 1,000 Hz filtering. (b) Histogram showing the relative occurrence of the three different current levels for the patch described in (a). The curves are gaussian fits to the amplitudes which had mean values of 1.1, 2.9 and 4.3 pA. A total of 809 discrete current steps were recorded: 509 were in the class that had an average amplitude acetyl-β-methylcholine have been found to depolarize rat myotubes (Dhillon and Harvey, 1982). These responses are probably mediated by nicotinic cholinoceptors as they are blocked by low concentrations of nicotinic antagonists. Nevertheless, there would appear to be some different characteristic of the receptors on the rat myotubes as oxotremorine and methacholine were without effect on chick muscle cells grown under identical culture conditions (Figure 4.15).

The activation of cholinoceptors on cultured muscle cells is transient despite the continued presence of the agonist (Harvey and Dryden, 1974a; Catterall, 1975b; Ritchie and Fambrough, 1975b; Steinbach, 1975; Sine and Taylor, 1979). This phenomenon occurs on prolonged exposure to agonist of receptors at adult motor endplates and is termed 'desensitization' (Katz and Thesleff, 1957). In cultures of chick embryo muscle, desensitization is more pronounced in older cultures which had higher resting membrane potentials (Harvey and Dryden, 1974a) and Catterall (1975b) suggested that desensitization increased with morphological development. Since desensitization is known to vary with membrane potential (Magazanik and Vyskočil, 1970), it seems likely that this increase in rate is a consequence of the greater membrane potentials of the more fully developed cells. Although desensitization could be induced in myotubes of the rat cell line L_6 (Steinbach, 1975), it was reported to be absent in primary cultures of rat muscle (Ritchie and Fambrough, 1975b). More recent studies on rat muscle in primary cell culture (Dhillon and Harvey, unpublished) reveal that desensitization to acetylcholine does take place but its time course is much longer than with chick muscle (Figure 4.16). Interestingly, desensitization can be induced more readily in rat myotubes by exposure to the muscarinic agonists oxotremorine and acetyl-β-methylcholine (Figure 4.16).

The specificity of the receptors on cultured muscle has also been assessed by measuring the effectiveness of various antagonists. As has been found in adult skeletal preparations (Beranek and Vyskočil, 1967, 1968), high concentrations of muscarinic antagonists are required to block the responses of muscle in culture to acetylcholine (Harris *et al.*, 1973; Fischbach and Cohen, 1973; Harvey and Dryden, 1974b;

---

of 2.9 pA and 298 events were of the 4.3 pA class. Of these, 73 and 17 events, respectively, displayed the sublevel of average amplitude 1.1 pA. Three events with an average amplitude of 2.9 pA displayed a sublevel of about 2 pA, but because of the low frequency of occurence of this level it was not studied. From Hamill and Sakmann (1981), by permission. Copyright 1981, Macmillan Journals Ltd.

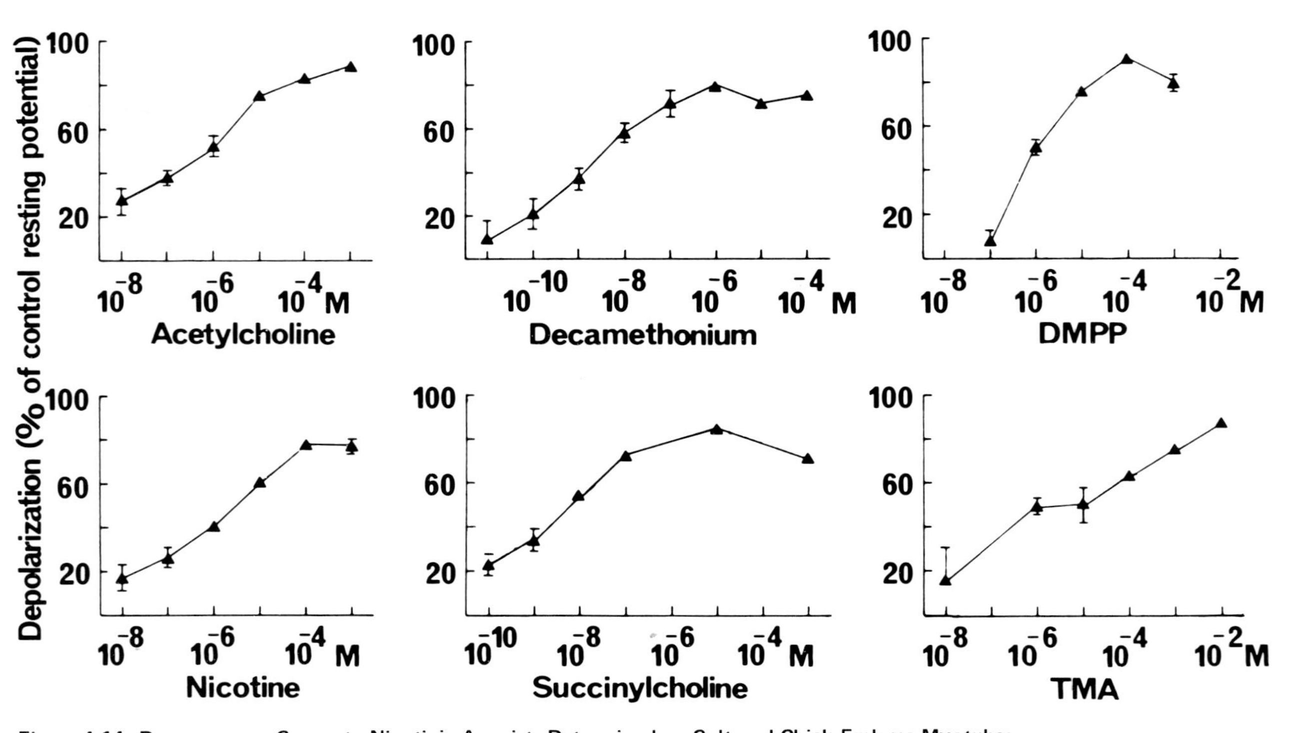

Figure 4.14: Dose-response Curves to Nicotinic Agonists Determined on Cultured Chick Embryo Myotubes
TMA is tetramethylammonium. From Harvey and Dryden (1974b), with permission.

Table 4.4: Acetylcholine Receptor Channel Properties in Cultured and Non-cultured Muscle Cells

Preparation	Recording Technique	Agonist	Temperature (°C)	Holding potential (−mV)	Single channel conductance (pS)	Single channel open time (ms)	Reference
Chick embryo myotubes	current noise	acetylcholine	35	50	70	1.1	Sachs and Lecar (1973)
Chick embryo myoballs	current noise	acetylcholine	~25	70	25 – 40	~7	Fischbach and Lass (1978a)
Chick embryo myotubes	extracellular noise	acetylcholine	24	65 – 70	–	3.5 ± 0.5	Schuetze *et al.* (1978)
Chick embryo myotubes	extracellular noise	carbachol	21	67	–	~2.2	Schuetze (1980)
Chick embryo myotubes	single channel	suberidyldicholine	22 – 24	?	60	13	Nelson and Sachs (1979)
Chick posterior latissimus dorsi	current noise	acetylcholine	23	40	37 ± 3	4.2 ± 0.6	Harvey and van Helden (1981)
Chick anterior latissimus dorsi	current noise	acetylcholine	23	40	35 ± 1	5.0 ± 0.4	Harvey and van Helden (1981)
Rat embryo myotubes	extracellular noise	acetylcholine	21	60 – 70	–	4.4	Fischbach and Schuetze (1980)
Rat myotubes	single channel	acetylcholine	23 – 25	65	35	–	Horn and Patlak (1980)
Rat embryo myoballs	single channel	acetylcholine	5 – 8	50 – 120	10	24	Hamill and Sakmann (1981)
					25	36	
					35	12	
Rat embryo myotubes	single channel	carbachol	22	90	49	3.2	Jackson and Lecar (1979)
Rat diaphragm	current noise	acetylcholine	20	80	25	1.2	Colquhoun *et al.* (1977)
Human myotubes	current noise	acetylcholine	23	70	40 ± 1.3	12.2 ± 1.3	Bevan *et al.* (1978)
Human myotubes	single channel	acetylcholine	21 – 23	90 – 100	45 ± 1	0.7 ± 0.1*	Jackson *et al.* (1982)
						4.5 ± 0.8*	
		carbachol	21 – 23	90 – 100	43 ± 3	~0.4*	
						1.5 ± 0.1*	
		suberidyldicholine	21 – 23	90 – 100	44 ± 2	0.8 ± 0.1*	
						9.3 ± 0.7*	
Human fibres in organ culture	current noise	acetylcholine	21	80	20 – 25 (junctional)	1.6 ± 0.1	Cull-Candy *et al.* (1979)
					11 – 16 (extra-junctional)	3.5 ± 0.3	
Xenopus embryo myotubes	extracellular noise	acetylcholine	20 – 22	~80	–	3.2 ± 0.7	Brehm *et al.* (1982)
Frog cutaneous pectoris	current noise	acetylcholine	8	80	23 ± 2	3.1 ± 0.2	Neher and Sakmann (1976b)
Frog denervated cutaneous pectoris	current noise	acetylcholine	20 – 25	65 – 80	15 ± 2	11 ± 1.6	Neher and Sakmann (1976b)

Notes:

* In most cells, fast and slow components could be distinguished.
Where possible, values are given as means ± standard error of the mean.

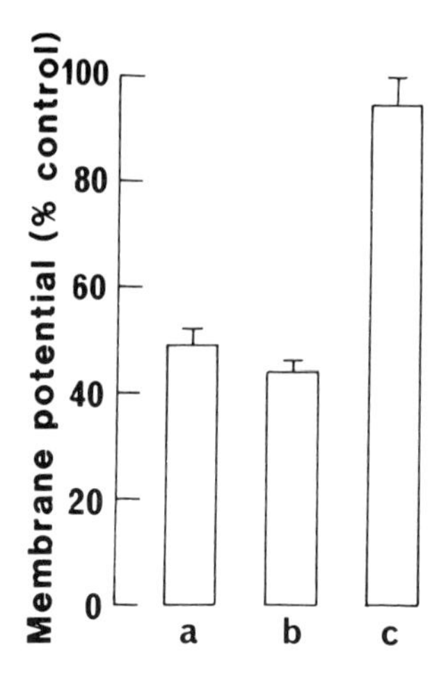

Figure 4.15: Effect of Acetyl-β-methylcholine (Methacholine, 500 μM) on the Membrane Potential of Myotubes Cultured from Neonatal Rat Skeletal Muscle (a) or Thymus Tissue ('Thymotubes') (b), and from Chick Embryo Muscle (c) Methacholine depolarizes the rat but not the chick cultures. From an unpublished experiment of D.S. Dhillon.

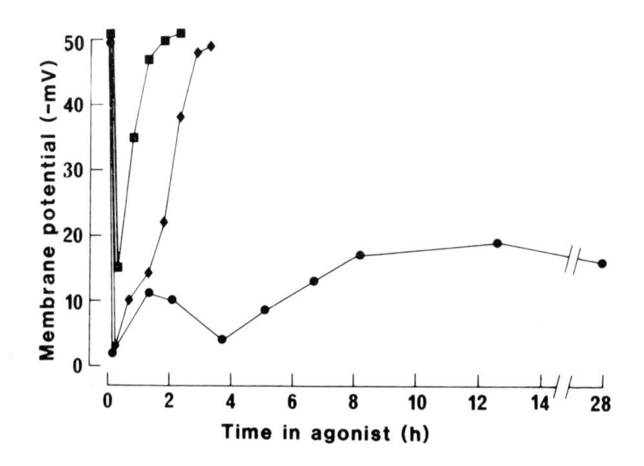

Figure 4.16: Desensitization of Cultured Rat Myotubes by (■) Acetyl-β-methylcholine (Methacholine), (◆) Oxotremorine and (●) Acetylcholine (ACh) Each point represents the mean membrane potential of at least twelve cells. Desensitization is revealed by the spontaneous recovery of the membrane potential after the initial depolarization. With methacholine and oxotremorine, a second addition of drug after recovery of membrane potential was without effect on the membrane potential. From an unpublished experiment of D.S. Dhillon.

Table 4.5: Affinity Constants for Antagonists on Cultured and Non-cultured Muscle Cells

Antagonist	Preparation	Test method	Affinity Constant	Reference
Tubocurarine	chick embryo myotubes	block of carbachol depolarization	10^7	Harvey and Dryden (1974a)
	chick embryo myotubes	inhibition of toxin binding	$\sim 10^6$	Vogel *et al*. (1972)
	rat myotubes	block of ACh depolarization	2×10^6	Dhillon and Harvey (1982)
	innervated rat myotubes	block of endplate potentials	2.5×10^6	Kidokoro and Heinemann (1974)
	rat L_6 myotubes	inhibition of toxin binding	5×10^6	Patrick *et al*. (1972)
	BC_3H1 cells	inhibition of toxin binding	3×10^6	Sine and Taylor (1979)
	chick biventer cervicis	block of ACh contractures	4×10^6	Harvey, unpublished
Gallamine	chick embryo myotubes	block of carbachol depolarization	3×10^6	Harvey and Dryden (1974b)
	BC_3H1 cells	inhibition of toxin binding	1.5×10^5	Sine and Taylor (1979)
	chick biventer cervicis	block of ACh contractures	1.2×10^6	Harvey, unpublished
Chandonium	chick embryo myotubes	block of ACh depolarization	$10^7 - 10^8$	Harvey *et al*. (1975)
	chick biventer cervicis	block of carbachol contractures	10^8	Teerapong *et al*. (1979)
Atropine	rat myotubes	block of ACh depolarization	3.1×10^4	Dhillon and Harvey (1982)
	chick biventer cervicis	block of ACh contractures	1.6×10^4	Harvey, unpublished

Catterall, 1975b). Atropine is at least 1,000 times less effective at the cholinoceptors of cultured muscle than on muscarinic receptors (Table 4.5). In contrast, low concentrations of nicotinic antagonists block responses to acetylcholine (Dryden, 1970; Fischbach, 1972; Hartzell and Fambrough, 1973; Obata, 1974; Harvey and Dryden, 1974a, b; Steinbach, 1975). These drugs act as classical competitive antagonists (Figure 4.17). Snake venom α-neurotoxins such as erabutoxin b and α-bungarotoxin also reduce acetylcholine responses and the binding of such toxins can be inhibited by tubocurarine (Vogel *et al.*, 1972; Patrick *et al.*, 1972; Sine and Taylor, 1979). The affinity of several nicotinic antagonists for cholinoceptors on muscle in culture is similar to that found in more conventional preparations of skeletal muscle (Table 4.5).

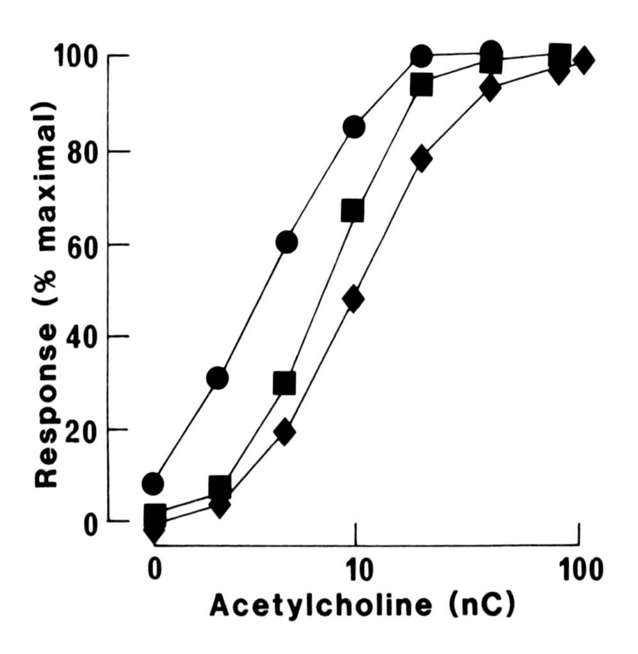

Figure 4.17: Antagonism of Acetylcholine-induced Depolarization of Cultured Rat Myotubes by Tubocurarine
○, control; ■, in the presence of 0.5 μM; and ◆, 1.0 μM tubocurarine. The pA_2 value for tubocurarine from this experiment is 6.2.

5 CARDIAC MUSCLE

The first account of cardiac muscle in tissue culture was published by Burrows (1910) who had been working in Harrison's laboratory. Using plasma clots as a substrate, he grew small explants of tissues taken from 60 h chick embryos, and observed that 'outgrowths from the heart contracted rhythmically along with the portion of the heart from which they arise.' Thus, one of the first uses of tissue culture was to provide evidence for the myogenic origin of cardiac contractions. In these explant cultures it was not possible to determine whether the rhythmicity was a property of intact groups of cells or inherent to single cells. However, a few years later Lake (1915/16) reported that single cells which had migrated from explants of mammalian heart muscle (Figure 5.1) could contract spontaneously at rates different from that of the explant itself.

Most of the very early work on cardiac muscle in culture was concerned with the origin of the rhythmic contractions, but explant cultures were also used for pharmacological investigations. Markowitz (1931) demonstrated that acetylcholine and adrenaline had their characteristic effects on the rate of beating of explants from embryonic chick heart. Since these cultures were not innervated, Markowitz concluded that the results provided evidence for 'an intermediary receptive substance' which was the target for the cardioactive drugs. In this study, the effects of drugs on heart rate were monitored visually, but it was not long before an attempt was made to use more sophisticated techniques to record the activity of the cultured heart muscle. The first use of microelectrodes was reported by Hogg *et al.* (1934), who described spontaneous action potentials recorded from explants of fetal rat ventricle with glass microelectrodes with 2-5 μm tips. They were unable to record intracellularly because the electrodes produced too much membrane damage. It was not until about 20 years later that Mettler *et al.* (1952), Fänge *et al.* (1956) and Crill *et al.* (1959) demonstrated that action potentials could be recorded from cultured heart cells with intracellular electrodes (see Figure 5.5).

Explant cultures are easy and convenient to prepare but they do not offer much advantage over standard *in vitro* preparations of heart muscle. For this reason a great deal of effort has been given to developing methods for growing cardiac muscle as cell cultures.

Figure 5.1: Camera Lucida Drawing of an Explant Culture of Rabbit Heart Muscle
The main explant (A) and the outgrowing cells (C) and (F) beat spontaneously. From Lake (1915/16), with permission.

Development of Methods

Although Rous and Jones had described in 1916 the use of trypsin to obtain suspensions of viable single cells (Rous and Jones, 1916), it was not until the 1950s that enzyme dissociation began to be widely used for establishing primary cultures (Moscona, 1952). The conditions used by Fänge *et al.* (1956) appear to be excessively severe and since then the development of satisfactory dissociation procedures has been a continual compromise between obtaining a high cell yield and avoiding damage to the cells.

The most commonly used dissociation techniques employ trypsin in calcium- and magnesium-free salt solution, thereby creating two problems: trypsin is known to damage cells, and exposure to calcium-free solutions can cause morphological changes in cardiac muscle when calcium is re-introduced: this is the so-called 'calcium paradox', which involves disruption of the basement membrane and excessive entry of calcium to the heart cells (Frank *et al*, 1977; Goshima *et al.*, 1980). Thus it would seem sensible to establish a different method of dissociation. As Masson-Pévet *et al.* (1976) demonstrated that collagenase-treated cells from neonatal rat hearts were less damaged and recovered more quickly in culture than trypsin-treated cells, collagenase would appear to be more suitable for use with cardiac muscle. Also, Kitzes and Bern (1979) and Norwood *et al.* (1980) obtained successful cultures with Viokase or pancreatin but there do not appear to be any reports to date with Dispase (a neutral protease preparation). In addition, it would be useful to try to perform the dissociation in the presence of calcium and hence avoid damage caused by the 'calcium paradox'.

Many species have been used, but the most common are the rat and chicken. Harary and Farley (1963) described a method for dissociating heart muscle from neonatal rats which depended on multiple incubations with trypsin, and DeHaan (1967) published a similar procedure for use with embryonic chick hearts. Most subsequent methods are variations on the same theme which has been successful with cardiac tissue from fetal mice (Goshima, 1976), neonatal hamsters (Coetzee *et al.*, 1977), adult rats (Schwarzfeld and Jacobson, 1981) and human fetuses (Halbert *et al.*, 1973).

Although it is not difficult to obtain cultures in which the individual cells beat spontaneously (Figure 5.2), there have been a number of problems associated with cardiac cell culture. These are: the cultures overgrow with fibroblast-like cells; the cardiac cells tend to lose spontaneous activity with time; the muscle cells lose or do not develop the typical morphology of cardiac muscle; the cells lose or do not develop mature electrophysiological properties; and the cultured cells do not always show characteristic responses to drugs. These points merit some consideration since they are obviously important in any attempt to use cultured cardiac muscle as a model for physiological and pharmacological studies. Many studies have been concerned with improving

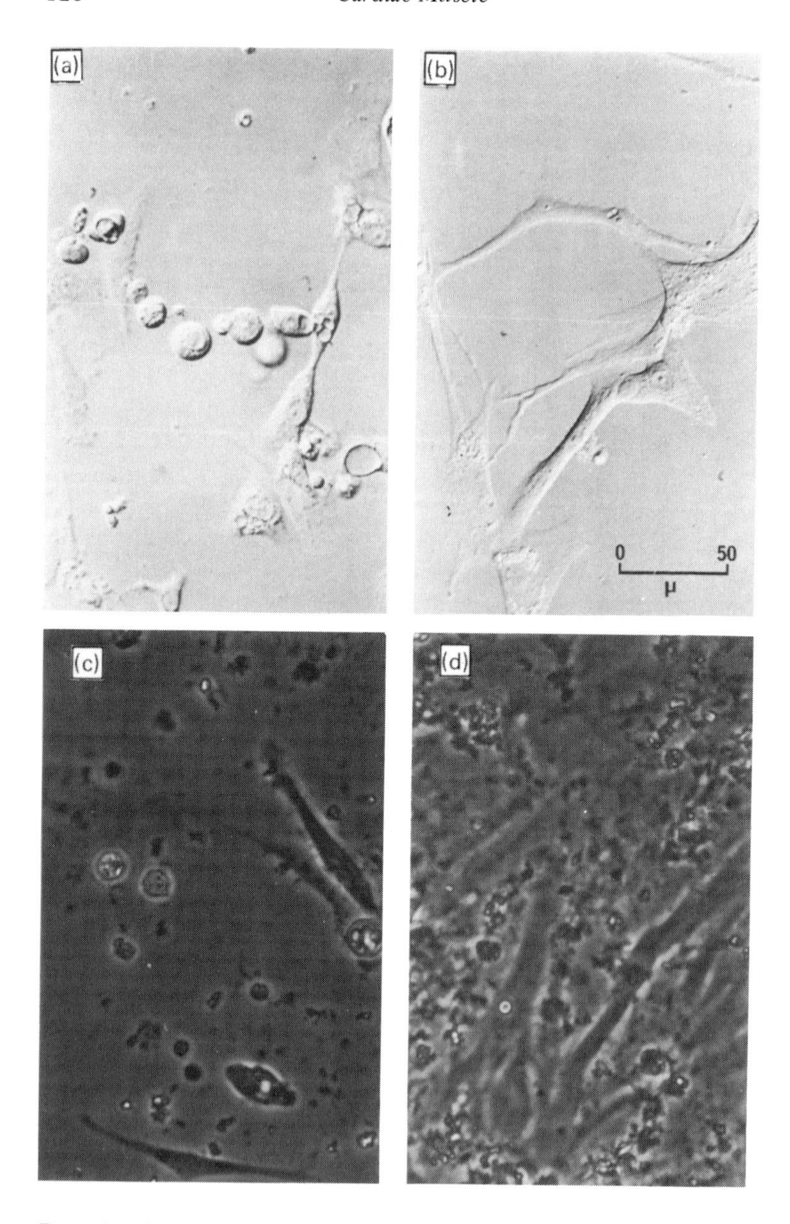

Figure 5.2: Embryonic Chick Heart Cells in Dissociated Cell Cultures
(a) Nomarski differential interference contrast of 24 h culture and (b) 48 h culture. (c) and (d) Phase contrast micrographs of beating cells in isolation and in clumps (4-day cultures). Scale bar = 50 μm. (a) and (b) from Koidl *et al.* (1980), with permission. Copyright Academic Press Inc. (London) Ltd.

culture methods but none has been comprehensive or systematic. However, it is possible to draw some conclusions and make some general recommendations about which techniques should offer the best chance of success.

In theory, absolutely pure myogenic cultures could be obtained by growing cardiac myoblasts in cloning conditions and maintaining the daughter cells as a continuous cell line. In practice, this technique has not been successful. Although Carrel in 1912 derived a cell line from embryonic chick heart and showed that cells could be maintained apparently indefinitely in culture, the cells were fibroblast-like (see Witkowski, 1979). A more recent line, the so-called Girardi heart cells derived from human heart muscle, is also composed of cells which lack the morphological and physiological properties of cardiac muscle (see, for example, Mølstad *et al.*, 1978). A more interesting 'failure' was described by Kimes and Brandt (1976b). Serial passaging of cells from rat hearts produced one cell line that had properties of skeletal rather than cardiac muscle.

Another approach to decreasing fibroblast contamination is to separate the different cell types before establishing the cultures. Some sort of differential centrifugation would no doubt be possible, but a far simpler technique is to rely on the fact that fibroblasts settle out of suspension and attach to a culture plate faster than myoblasts. Such differential plating was described by Pollinger (1970) for chick embryo heart cells and by Blondel *et al.* (1971) for neonatal rat cells. The proportion of myoblasts in the cultures can be increased to 80-90 per cent.

A third means of limiting the numbers of fibroblasts is to establish the cultures and then expose them to a drug that blocks cell division. Differentiated cardiac muscle cells do not usually divide while fibroblasts replicate continuously so that fibroblasts will be more sensitive to the antimitotic drug. This method is often used with cultures of skeletal muscle but does not appear to be popular with cardiac muscle. However, Coetzee *et al.* (1977) showed that treatment with 5-bromo-2′-deoxyuridine not only prevented fibroblast overgrowth but also stimulated muscle differentiation. Although the mechanism of this effect is unclear, it is a useful bonus and suggests that the technique is worth wider consideration.

The proportion of cells which beat spontaneously in culture can vary widely, depending on the type of tissue and the dissociation conditions, but most authors comment that the number of beating cells declines with age in culture. Not all mature cardiac muscle cells would normally be spontaneously active, so some of the 'degenerating' cultures could actually have been maturing functionally. However, it does seem likely that, in many cases, loss of activity points to some defect in the culture conditions. The cessation of beating could be due either to a membrane defect such that pacemaker potentials could not stimulate action potentials, or to a metabolic defect whereby the cells ran out of sufficient energy stores. Unfortunately, there does not appear to have been a comprehensive study of the membrane properties of cultured heart cells before and after they lost spontaneous activity. Overgrowth by fibroblasts may be expected to reduce electrical excitability by forming low resistance pathways that act as 'current sinks', drawing current from the excitable cardiac cells into the inexcitable fibroblasts.

Experiments with added hormones indicate that deficiencies in culture media may be involved in loss of activity. For example, McCarl *et al.* (1965) reported that hydrocortisone and deoxycorticosterone could increase the number of beating cells in young cultures and re-establish spontaneous activity in older, quiescent cultures. The mechanism of this effect, however, remains obscure. A similar type of action was found by Wildenthal (1970) with insulin, either alone or in combination with hydrocortisone. When whole hearts from fetal mice were maintained as organ cultures, spontaneous beating was lost after about 14 days when cultured in medium supplemented with calf serum. When insulin was added to the medium, spontaneous beating persisted for 21 days; and when the medium contained serum, insulin and hydrocortisone, spontaneous beating continued for more than 30 days. Insulin stimulates uptake of amino acids by cultured heart cells (Elsas *et al.*, 1975) and can also produce membrane hyperpolarization (Lantz *et al.*, 1980), but it is not known whether either effect contributes to the maintenance of rhythmic activity.

Some breakdown of the internal organization of the cells is to be expected because of the trauma caused by the tissue dissociation. However, if the cells recover completely in culture it would be expected that morphological differentiation would take place, but such development *in vitro* does not always appear to happen. Once again, it may be possible to modify the culture medium to increase differentiation. Morphological development of cultured chick embryo heart cells can apparently be stimulated by establishing the cultures in Leibovitz L-15 medium plus 10 per cent fetal calf serum and then changing to a modified medium (no histidine and additional methionine) containing only 0.1 per cent serum (Jones *et al.*, 1978).

Many early studies of the electrical properties of cardiac cells in culture found that the cells did not develop the characteristics expected of mature cardiac muscle and in some cases the properties of the cells reverted to those of less developed muscle. (These findings are reviewed in more detail below, in *Action Potentials*.) Some of these effects have been ascribed to contact of muscle cells with fibroblasts (Pollinger, 1973; Lompre *et al*., 1979; Jourdan and Sperelakis, 1980) and some to the effects of trypsinization. Additionally, electrophysiological maturation in culture may be stimulated by aggregation of cardiac muscle cells. Both cylindrical strands (Purdy *et al*., 1972) and spherical reaggregates (McLean and Sperelakis, 1976; Nathan and DeHaan, 1978) are reported to have more fully differentiated electrical properties (see review by Lieberman *et al*., 1981). However, other workers have reported similar differentiation in normal monolayer type cultures (Athias *et al*., 1979; Lompre *et al*., 1979; Robinson and Legato, 1980) so that it is not certain whether aggregation of cells into distinct clusters is essential for their maturation.

Electrical Properties and Drugs Acting on Ion Channels

The heart *in vivo* relies on electrical currents to initiate the contractions of each individual cell and to synchronize the mechanical activity of the tissue as a whole. Since an understanding of the electrical phenomena in cardiac muscle is necessary in order to examine the mechanisms of action of many drugs such as antidysrhythmics, much effort has concentrated on establishing the basic properties of membrane ion channels, including how they change during development.

Action Potentials

The activity of the heart is associated with action potentials which result from the flow of ions through channels in the heart cell membranes, the channels being reasonably selective with respect to the ions which can permeate. Action potentials are an averaged signal composed of contributions from different individual currents and affected by the passive electrical properties of the cell. It is therefore difficult to study the separate roles of particular types of ionic channel unless voltage clamp techniques are used. Unfortunately, cardiac tissue, with its complex geometry and close cell-to-cell contact, provides great technical difficulties for voltage clamp studies. As will be seen below, tissue culture provides simpler preparations that are very useful in the analysis of the electrical properties of cardiac cells.

The cardiac action potential is often discussed in terms of five phases

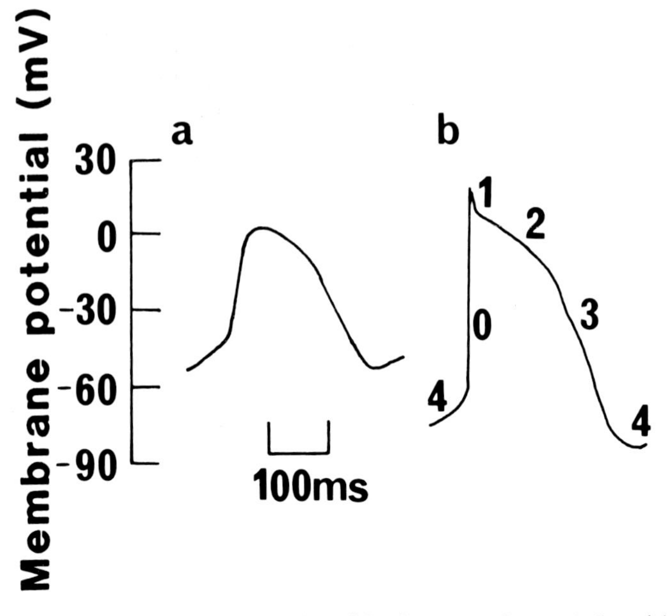

Figure 5.3: Schematic Representation of Cardiac Action Potentials from (a) Sinoatrial Node and (b) Purkinje Cells (The numbers refer to the different phases of the action potential

(Figure 5.3) and each phase of the action potential is associated with different combinations of ionic currents. The shape and properties of the action potentials vary in different regions of the heart, e.g. the sinoatrial node, atrial muscle, atrioventricular node, Purkinje fibres and ventricular muscle (Figure 5.3). However, the main currents are carried by Na^+, K^+ and Ca^{2+}. The rapid depolarization during phase 0 is a result of an inward current carried by Na^+ ions. This current is similar to that responsible for the spike of nerve and skeletal muscle action potentials. It is blocked by tetrodotoxin and saxitoxin (Figure 5.4) and abolished by removal of Na^+ from the external solution. These so-called fast Na^+ channels are voltage-sensitive, being activated by small depolarizations in the region –90 mV to –70 mV, and are also voltage-dependent in that they are inactivated at potentials less negative than about –50 mV. The plateau stage of the action potential (phase 2) is a balance between a slow inward current and an opposing outward current. The inward current is probably carried by both Na^+ and Ca^{2+}. The Ca^{2+} current turns on more slowly than the fast inward Na^+ current of phase 0, is not blocked by tetrodotoxin, but can be blocked by ions such as Co^{2+} and Mn^{2+}, and by drugs such as verapamil and D600 (Figure 5.4). The slow inward current is largely responsible for the depolarization of pacemaker cells. The outward current is carried by K^+ and it can be suppressed by tetraethylammonium (Figure 5.4). Additionally, it has been suggested that phase 1 (the rapid repolarization) is associated with a current carried by Cl^- ions but there is some doubt about this (discussed by DeHaan, 1980).

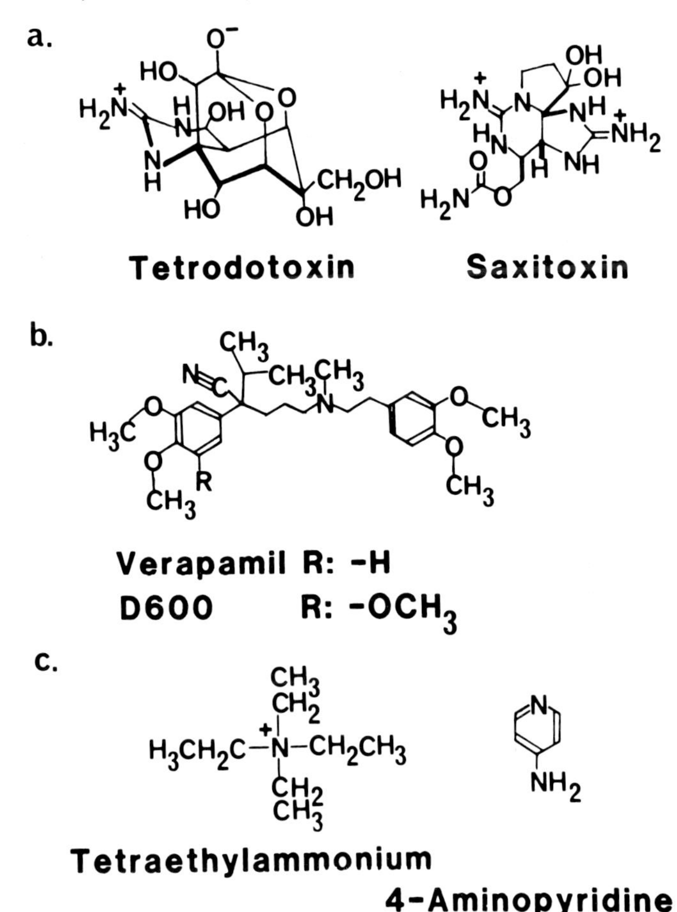

Figure 5.4: Structures of Compounds Use to Block Specific Ion Channels in Nerve and Muscle Membranes
(a) Action potential Na^+ channel blockers: tetrodotoxin and saxitoxin. (b) Slow Ca^{2+} channel antagonists: verapamil and D600. (c) K^+ channel blockers: tetraethylammonium and 4-aminopyridine.

In vivo, the resting membrane potential becomes more negative with age. For example, in ventricular muscle from chick embryos, membrane potential increased from about –35 mV at two days of incubation to about –70 mV after 12 days' incubation (Sperelakis and Shigenobu, 1972), the increase being associated with an increase in the resting permeability to K^+ ions. The characteristics of the action potential also change with age (for reviews, see Spitzer, 1979; DeHaan, 1980). Action potentials recorded from heart tissue from early embryos (2-3 day chick embryos or 9-10 day rat fetuses) are similar to pacemaker potentials of mature hearts, in that they are generated from low resting potentials (around –40 to –50 mV) and they have slow rates of rise (maximum rate of depolarization, $\dot{V}_{max}$ is less than 20 V/s). With development, the maximum rate of depolarization increases to 100-200 V/s and a distinct plateau phase becomes apparent.

It does seem likely that very young hearts have few functional fast Na^+ channels, that the action potential is carried initially by ions moving through the slow Na^+/Ca^{2+} channel, and that the relative dependence on the fast Na^+ channels increases with age. A more complete understanding of the development of ion channels in different types of cardiac cells awaits application of voltage clamp techniques, probably allied to use of tissue cultures.

For many years it was thought that cardiac cells in culture could not develop the electrical properties of mature cells and that cells taken from more developed stages reverted to a less differentiated form under culture conditions. More recent findings demonstrate that fully differentiated cardiac cells can be cultured.

The first studies in which microelectrodes were used to record intracellularly from cultured cardiac muscle demonstrated the presence of spontaneous, rhythmically-occurring action potentials (Mettler *et al.*, 1952; Fänge *et al.*, 1956; Crill *et al.*, 1959). The action potentials had relatively slow rates of rise and the cells had low resting potentials (about –40 mV to –50 mV; Figure 5.5). Higher potentials were subsequently obtained in cultures from 6-8-day chick embryo ventricle by Lehmkuhl and Sperelakis (1963), who noticed that cells in small clumps had significantly higher potentials than isolated cells (around –70 mV compared to –60 mV). In fact, the resting potentials recorded from these 'clump' cells were not different from those obtained from intact tissue of the same age. However, action potentials of the cultured cells still had a lower maximum rate of rise than those of non-cultured cells.

Similar differences between isolated and grouped cells were found by Lieberman (1967) and by Pappano and Sperelakis (1969) who

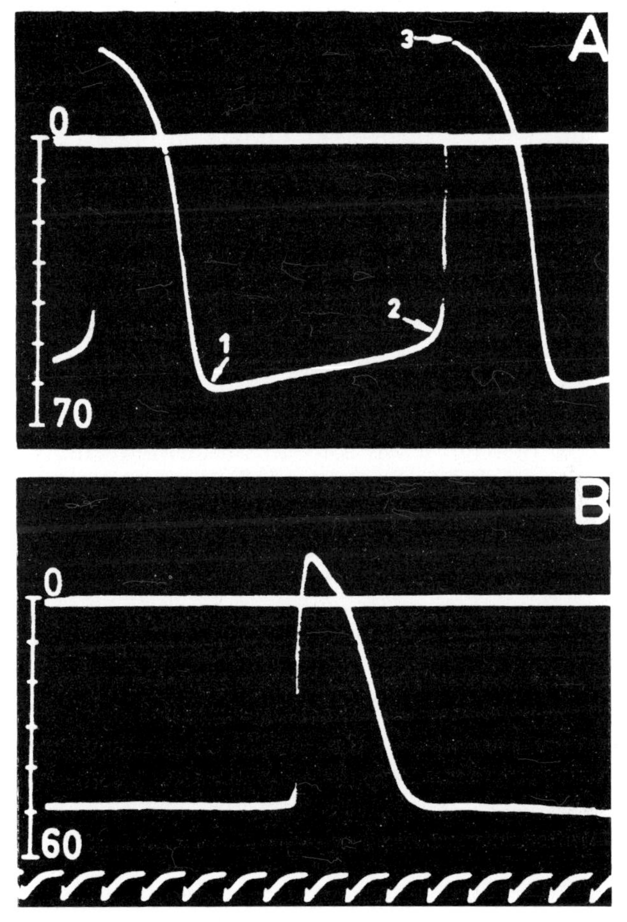

Figure 5.5: Intracellularly-recorded Action Potentials from Spontaneously Beating Heart Cell Clusters in Dissociated Cell Cultures from (A) 5- or (B) 8-day-old Chick Embryos

24 h cultures. Vertical calibration in mV, time marker 0.1 s. From Fänge *et al.* (1956), with permission.

suggested that the increase in membrane potential in grouped cells was caused by an increase in the resting permeability of the membrane to K^+. Although tetrodotoxin did not inhibit the action potentials in these cultures, veratridine (which decreases the inactivation of fast Na^+ channels) increased the duration of action potentials (Sperelakis and Pappano, 1969). This effect was blocked by tetrodotoxin, suggesting that fast Na^+ channels were present, although not normally contributing to the spontaneous action potentials. Veratridine was not completely specific as it also caused depolarization which was not prevented by tetrodotoxin, and Sperelakis and Pappano (1969) postulated that it could be due to an action of veratridine on potassium permeability.

Since the physical arrangement of cells in culture appears to influence the state of development, McDonald *et al.* (1972) compared the activity of intact hearts from 2-12-day chicken embryos with that of single cells and aggregates cultured from hearts at corresponding ages. The spontaneous beating of hearts from 2-4-day embryos was not inhibited by tetrodotoxin, but sensitivity to tetrodotoxin developed between days 4 and 7. Very few of the spontaneously active single cells in culture were affected by tetrodotoxin, and there was no change in sensitivity with age. In contrast, the aggregates of single cells in culture were similar to intact hearts in their responsiveness to tetrodotoxin. However, it was not clear whether aggregates prepared from young hearts that were insensitive to tetrodotoxin could develop sensitivity after prolonged culturing.

These results would suggest that cells which can aggregate in culture can form more normal interactions which lead to them obtaining mature characteristics. However, when hearts from young, tetrodotoxin-insensitive chick embryos were maintained in organ culture they did not develop sensitivity to tetrodotoxin, even after two weeks in culture (Sperelakis and Shigenobu, 1974) nor when cultured in the chorioallantoic membrane of host eggs (Shigenobu and Sperelakis, 1974). Cell culture of ventricular muscle from older chick embryos apparently resulted in a loss of differentiated properties which did not return after a few days in culture (McLean and Sperelakis, 1974). This loss was not attributed to trypsinization.

Sensitivity to tetrodotoxin could be retained or induced if the culture medium contained insulin (Le Douarin *et al.*, 1974) or messenger RNA isolated from adult heart (McLean *et al.*, 1976, 1977). The mechanism of action of insulin is not known, since Le Douarin *et al.* (1974) only measured the ability of tetrodotoxin to stop the spontaneous beating of isolated cells. However, insulin can hyperpolarize cardiac cells in culture (Lantz *et al.*, 1980) and this may have switched on tetrodotoxin-sensitive, fast Na^+ channels. The mechanism of messenger RNA is also unknown, but it seems to be effective in both organ and reaggregate cultures. In the presence of messenger RNA from heart muscle, there is an increase in membrane potential, the action potentials have a faster $\dot{V}_{max}$ (around 100 V/s) and they become sensitive to tetrodotoxin (McLean *et al.*, 1977). These effects were not produced by messenger RNA isolated from liver but they could be prevented by co-incubation with cycloheximide to inhibit protein synthesis. Earlier work had provided evidence that protein synthesis was required to restore tetrodotoxin sensitivity in cells damaged by trypsinization (Sachs *et al.*, 1973).

Other work indicates that functional differentiation can be obtained without such supplementation of the growth medium. Most, but not

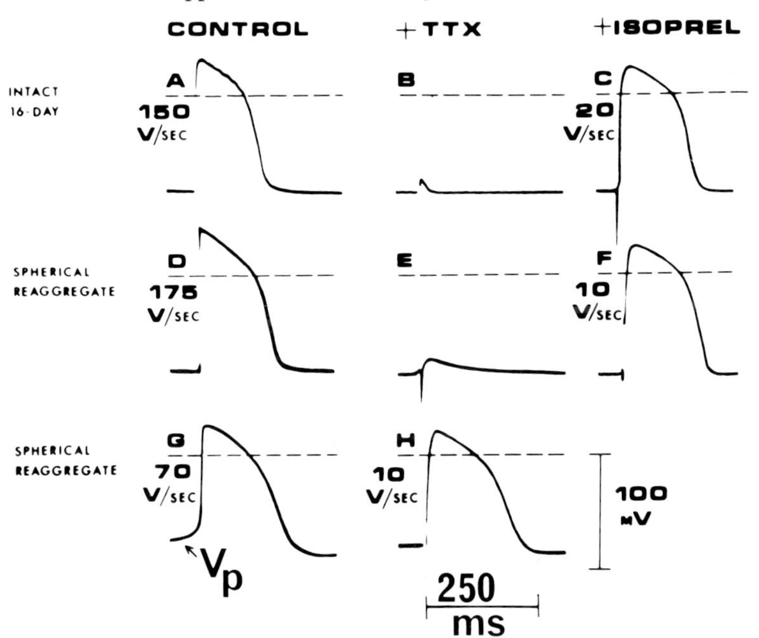

Figure 5.6: Comparison of Tetrodotoxin (TTX) Sensitivity of Cells in Intact 16-day-old Chick Embryo Ventricle with Cultured Spherical Reaggregates of Cells Isolated from 16-day Embryonic Hearts

(A-C) Recordings from one cell in an intact heart. (A) Control action potentials had fast rates of rise ($\dot{V}_{max}$; 150 V/s). (B) After TTX (1 μg/ml), all excitability was abolished despite a tenfold increase in stimulus intensity. (C) Isoprenaline (isoprel; 10^{-6}M) produced a slowly rising action potential in the TTX-blocked heart. (D-F) Recordings from one non-pacemaker cell in a spherical reaggregate. (D) Control action potential with $\dot{V}_{max}$ of 175 V/s. (E) Action potential was abolished by TTX (0.05 μg/ml). (F) Isoprenaline induced a slow response. (G-H) Recordings from a pacemaker cell in a reaggregate. (G) Control action potential with an intermediate $\dot{V}_{max}$ of 70 V/s and pacemaker potential (V_p) (H) TTX (2 μg/ml) suppressed automaticity, but stimulation elicited an action potential with $\dot{V}_{max}$ reduced to 10 V/s. Time and voltage calibrations in panel (H) apply throughout. Field stimulation applied in all panels except (G). From McLean and Sperelakis (1976), with permission.

all, of these studies have used aggregate cultures of one form or another. High membrane potentials and fast-rising, tetrodotoxin-sensitive action potentials were recorded from reaggregate cultures derived from ventricular muscles of 11-16-day chick embryos (McLean and Sperelakis, 1976; Bernard and Couraud, 1979; Nathan and DeHaan, 1979; Ebihara *et al.*, 1980) and from fetal (Nathan, 1981) or neonatal (Jourdan and Sperelakis, 1980) rat ventricle muscle (Figure 5.6). Similar properties could be found in monolayer cultures prepared without special reaggregation steps so long as fibroblast contamination was reduced (Athias *et al.*, 1979; Kitzes and Bern, 1979; Lompre *et al.*, 1979; Robinson and Legato, 1980). Although these authors all used rat cells, it is also possible to obtain similar differentiation with monolayer cultures from 8-day chick embryo hearts (Jones *et al.*, 1978).

The most recent work on electrical properties of heart cells in tissue culture has involved voltage clamp methods. Cell aggregates were used in these studies, although Undrovinas *et al.* (1980) have described a method for clamping single, isolated cardiac muscle cells, and the giant cells formed by cell fusion, which were described by Goshima *et al.* (1979), would certainly appear to be suitable for voltage clamp techniques. It is also possible to use patch clamp methods to record from single ionic channels (e.g. Colquhoun *et al.*, 1981; Reuter *et al.*, 1982).

The first indication that multicellular groups of cardiac cells in culture could be suitable for voltage clamping was provided by Crill *et al.* (1959) who demonstrated that the response to injections of current into small cell clusters was independent of the distance which separated the recording electrode from the current source. Later, DeHaan and Fozzard (1975) showed that there was very good electrical coupling with little decrement with distance in specially prepared spherical reaggregates of single cells from chick embryo hearts. Nathan and DeHaan (1978, 1979) confirmed that such spherical reaggregates could be voltage clamped, although with certain technical limitations (Clay *et al.*, 1979; see review by Lieberman *et al.*, 1981).

It has been confirmed that functional differentiation of fast Na^+ channels can take place in culture (Nathan and DeHaan, 1978). Cells were taken from 3-day chick embryo hearts and grown as 100-250 μm diameter spherical clusters. After two days in culture, action potentials were not inhibited by tetrodotoxin and had a low V_{max}. However, after a further three days in culture, the action potentials had a fast rising phase (V_{max} 100-150 V/s) and were sensitive to tetrodotoxin (Figure 5.7). The Ca^{2+} channel antagonist D600 had little effect on the rising phase of the action potentials but it did reduce the height and duration of the plateau (Figure 5.7). Voltage clamp studies revealed that cells in the first two days of culture had a slow inward current that was not blocked by tetrodotoxin, but the cells in older cultures had two inward currents - one fast, the other slow. The development of the fast current depended on protein synthesis. In a more detailed analysis of the inward currents of similar aggregates prepared from 7-day chick embryo ventricle (Nathan and DeHaan, 1979) it was confirmed that the fast inward current was suppressed by tetrodotoxin and shown that the slower inward current could be decreased by D600 (Figure 5.8). However, use of a third intracellular electrode to monitor the adequacy

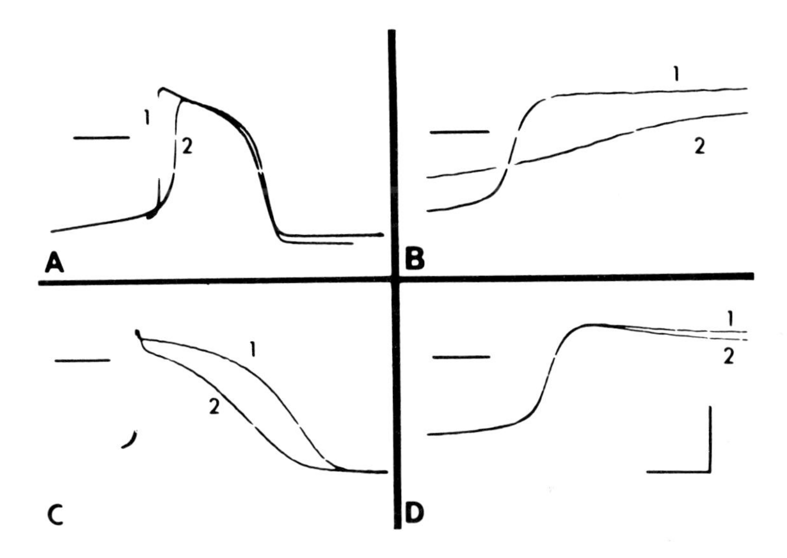

Figure 5.7: Effect of Tetrodotoxin and D600 on Action Potentials from a Small Spherical Aggregate of Chick Embryo Heart Cells in Culture

Each panel shows a control record (1) and an action potential after drug treatment (2). (A) and (B) Effect of tetrodotoxin (10 μg/ml). (C) and (D) Effect of D600 (0.5 μg/ml). $\dot{V}_{max}$ of control action potentials were 153 V/s (B, trace 1) and 137 V/s (D, trace 1). Tetrodotoxin reduced $\dot{V}_{max}$ to 12 V/s (B, trace 2), but did not affect the plateau (A); whereas D600 reduced the amplitude and duration of the plateau (C, trace 2) without affecting $\dot{V}_{max}$ (D, trace 2). Vertical scale, 50 mV for all records; horizontal scale, 100 ms in (A), 1 ms in (B), 40 ms in (C) and 1 ms in (D). Horizontal lines at the start of each record represent 0 mV. From Nathan and DeHaan (1979), by copyright permission of the Rockefeller University Press.

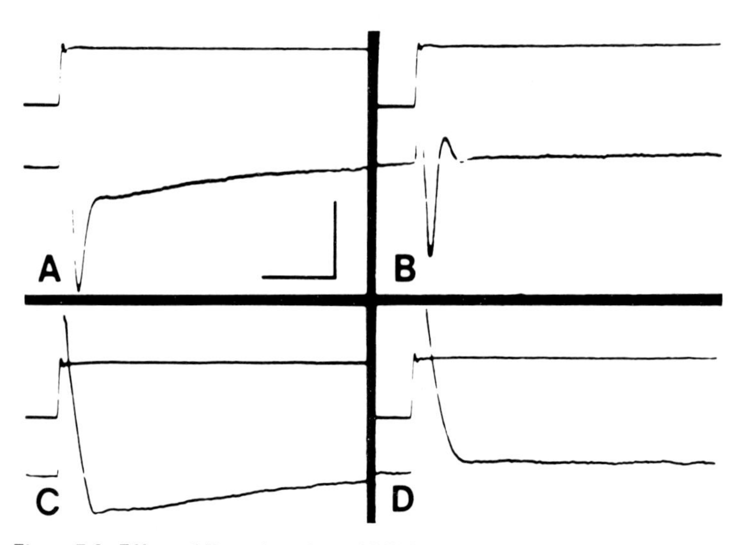

Figure 5.8: Effect of Tetrodotoxin and D600 on Membrane Currents of Aggregates of Chick Embryo Heart Cells in Culture
The voltage clamp step was from the holding potential of −60 mV to 0 mV. Upper traces in each panel represent membrane potential and lower traces indicate the membrane current with inward currents shown by downward deflections. (A) Control. (B) D600 (0.5 μg/ml). (C) Tetrodotoxin (1.25 μg/ml). (D) D600 (0.5 μg/ml) and TTx (1.25 μg/ml). Note that D600 blocks the late inward current, whereas tetrodotoxin blocks the fast initial inward current. Horizontal scale, 10 ms; vertical scale, 80 mV and 0.5 μA. From Nathan and DeHaan (1979), by copyright permission of the Rockefeller University Press.

of the voltage clamp revealed that there was a significant loss of voltage control during the peak of the fast inward current. Ebihara *et al.* (1980) used slightly smaller reaggregates (55-90 μm diameter) and managed to control more effectively the voltage during peak current flow. They found that the initial fast inward current followed classical Hodgkin-Huxley kinetics, but was about four times faster than the equivalent current in squid giant axon.

These results point the way to a new dimension in our understanding of the electrophysiological properties of heart cells. It should be possible to obtain a precise description of the behaviour of the many ion channels that function in the cardiac cell membrane and to analyse more accurately the mode of action of drugs on the heart. In addition, culture systems would appear to be ideally suited for experiments designed to elucidate the mechanisms controlling the development of the ion channels. Since cells can be cultured from different regions of the heart (Norwood *et al.*, 1980), factors which make cells in different areas develop their characteristic properties could be analysed. Indeed, the studies of Shrier and Clay (1980; Clay and Shrier, 1981) have begun to probe this interesting area. They have shown that the pacemaker currents recorded from reaggregates of chick ventricle cells decrease with age of the embryonic source. This may be correlated with loss of spontaneous activity of ventricular cells with age.

Drug Actions on Ion Channels

It might be expected that the advantages of an isolated system with easy accessibility and good visualization would have appealed to many pharmacologists with an interest in the heart. However, this does not appear to be so as there have been relatively few studies of the actions of drugs on cultured heart muscle.

The Pharmacology of the Fast Inward Current. As explained earlier, the fast Na^+ current of cultured heart muscle is sensitive to tetrodotoxin (e.g. Galper and Catterall, 1978; Nathan and DeHaan, 1979), but few authors have studied quantitatively the relationship between concentration and effect. Half maximal inhibition of veratridine-stimulated Na^+ uptake was obtained with 1.5×10^{-9}M tetrodotoxin in chick embryo heart cultures (Galper and Catterall, 1978) which agrees very closely with the apparent dissociation constant of 6.6×10^{-9}M calculated by Fosset *et al.* (1977) in a similar culture system. These values suggest that tetrodotoxin has similar affinities for its binding sites on cultured chick embryonic heart cells, on intact chick heart (IC_{50} about 10^{-8}M; Iijima and Pappano, 1979), and on nerve and skeletal muscle (about 10^{-8}M; Catterall, 1980). However, Nathan (1981) found that cultured fetal rat heart was about five times less sensitive to tetrodotoxin than intact muscle; rat muscle is generally less sensitive compared with chick muscle (Catterall and Coppersmith, 1981).

Many other agents can affect the function of fast Na^+ channels (see review by Catterall, 1981a; and Chapter 2) and some of these have been used to analyse the properties of this channel in cultured heart muscle. Studies with veratridine have been referred to earlier (Sperelakis and Pappano, 1969; Fosset *et al.*, 1977; Galper and Catterall, 1978), and agents with actions similar to those of veratridine include toxin ATx II from the sea anemone, *Anemonia sulcata* (De Barry *et al.*, 1977) and toxin II from the scorpion *Androctonus australia* (Bernard and

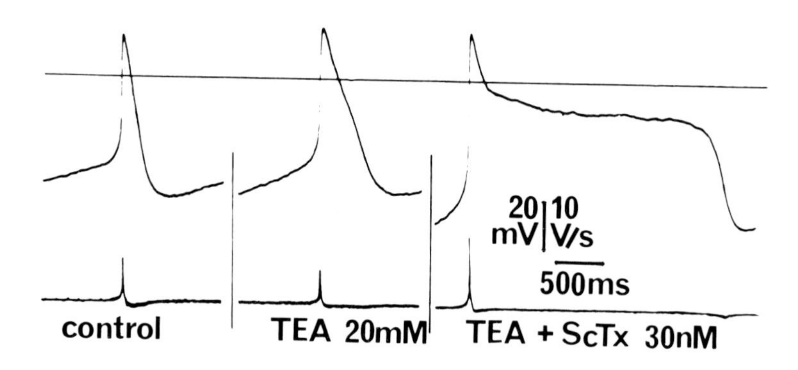

Figure 5.9: Effects of Tetraethylammonium (TEA) and Scorpion Toxin (ScTX) on Action Potentials of Reaggregate Cultures of Chick Embryo Heart Muscle
Upper records show the membrane potential with 0 mV being indicated by the horizontal line. Lower records show the rate of rise of the action potentials. Prolongation of the action potentials is seven times greater with scorpion toxin in combination with TEA than with TEA alone. From Bernard and Couraud (1979), with permission.

Couraud, 1979). Both these toxins increase the duration of action potentials (Figure 5.9) by slowing the inactivation of fast Na^+ channels, an effect that is blocked by tetrodotoxin. By providing a system in which it is possible to correlate ion flux studies (e.g. Catterall and Coppersmith, 1981) with electrophysiology, cultured cardiac muscle forms a useful preparation for the analysis of the functioning of the fast Na^+ channel.

An interesting use of cultured cardiac muscle was described by Hasin *et al.* (1980). They cultured neonatal rat heart cells in a medium that led to a reduction in the cholesterol content of the cell membranes. This did not cause any obvious morphological changes but the cholesterol depleted cells had higher V_{max} values and were less sensitive than control cells to tetrodotoxin. Increasing the cholesterol content of chick embryo heart cells in monolayer cultures has also been shown to result in faster rising action potentials (Renaud *et al.*, 1982). In these cultures, however, the cells became highly sensitive to tetrodotoxin.

The Pharmacology of the Slow Inward Current. The slow inward current carried by Ca^{2+} and Na^+ ions can also be affected rather selectively by drugs. As mentioned previously, the slow current is reduced by the slow channel blocking agents, verapamil and D600 (Figures 5.7 and 5.8). Blockade by verapamil is associated with about 40 per cent decrease in total Ca^{2+} uptake (Barry and Smith, 1982); Na^{+}/Ca^{2+} exchange is probably responsible for the fraction that is insensitive to verapamil. The slow inward current can also be blocked by a series of metal ions, including Mn^{2+}, Co^{2+} and La^{3+} (De Barry *et al.*, 1977). These are not completely specific in their action, and they can produce depolarization by affecting potassium permeability (Kitzes and Bern, 1979). La^{3+} has also been shown to block $^{45}Ca^{2+}$ uptake in cultures of chick embryo ventricular cells (Barry and Smith, 1982).

The functioning of the slow inward current can be enhanced both in intact (Shigenobu and Sperelakis, 1972) and in cultured heart preparations (McLean and Sperelakis, 1976; Josephson and Sperelakis, 1977). This effect can be brought about by drugs stimulating β-adrenoceptors, or by agents such as the methyl xanthines which increase the intracellular concentration of cyclic AMP. Such treatment can induce spontaneous slow action potentials in heart cells that were blocked by tetrodotoxin (Figure 5.6). Further evidence for the role of cyclic AMP was provided by studies on the effect of 5′-guanylimidodiphosphate (Gpp(NH)p) (Josephson and Sperelakis, 1978). Gpp(NH)p, which is thought to stimulate adenylate cyclase directly, also induced slow action potentials in reaggregate cultures of cells from chick embryo ventricles. These action potentials were blocked by Mn^{2+} and verapamil but were not affected by propranolol, indicating that β-adrenoceptors were not involved. However, the precise mechanism by which cyclic AMP controls the number or activity of slow inward channels remains to be elucidated, although it has been postulated that a protein constituent of the slow channels must be phosphorylated in order to become available for activation by a change in voltage (Azuma *et al.*, 1981). The fundamental properties of cardiac Ca^{2+} channels can be studied with patch clamp techniques (Reuter *et al.*, 1982) so that it should be possible to perform more testing experiments on cultured preparations in order to elucidate the influences that modulate their behaviour.

Drugs and K^{+} Channels. Comparatively little work has been done on the actions of drugs on outward potassium currents in cultured heart cells. Considering the recent resurgence of interest in aminopyridine-like compounds, this is perhaps surprising. However, tetraethylammonium, a standard K^{+} channel blocker, does prolong action potentials in cultured cardiac muscle (Josephson and Sperelakis, 1978) (Figure 5.9). Maruyama *et al.* (1980) demonstrated that tetraethylammonium was more effective when applied intracellularly, suggesting that the drug normally has to penetrate the cell membrane in order to reach its binding site on the K^+ channel. Once again, the ease with which drugs can be applied to the inside of a cardiac muscle cell highlights the potential usefulness of the culture system in pharmacological studies. More recently, patch clamp techniques have been used to record single channel K^+ currents from dissociated heart cells (DeFelice and Clapham, 1981). These currents were 1-2 pA in amplitude and about 300 ms duration at –60 mV and 32°C; they were abolished by 4-aminopyridine.

Cardiac Glycosides. Although these drugs are not usually thought to act directly on membrane ion channels, they do cause a redistribution of ions across cardiac cell membranes and this may indirectly affect the function of ion channels.

Investigation of the effects of cardiac glycosides was among the first attempts to use cultured heart muscle for pharmacology (for review, see Wollenberger, 1964). This was because effects on rate and rhythm can be observed easily in individual cells. Although the characteristic effects of the glycosides are reproduced in culture, these studies have not contributed much to a fuller understanding of the mechanism of action of the cardiac glycosides. Okarma *et al*. (1972) used a complex of digoxin and albumin to demonstrate that the cardiac glycosides acted on the outside of cells. Galper and Catterall (1978) and McCall (1979) demonstrated that ouabain increased $^{22}Na^+$ uptake by cultured chick and rat heart cells. The concentration required for half the maximal effect was around 10^{-6}M in both studies. Using labelled ouabain, McCall (1979) calculated that there were 1.6×10^6 ouabain binding sites/cell which corresponds to an estimated density of sodium pumps of $720/\mu m^2$.

In an attempt to clarify the mechanism whereby ouabain can generate arrhythmic activity, Goshima and Wakabayashi (1981a) made imaginative use of culture techniques. Knowing that there are marked species differences in sensitivity to cardiac glycosides, they prepared co-cultures of heart cells from mice (a relatively resistant species) and quail (a susceptible species). When cell pairs formed, they contracted synchronously, suggesting that they had established good contact. Such pairs were intermediate in their sensitivity to ouabain. Mouse and quail pairs that were fused by treatment with Sendai virus gave similar results. The authors suggest that ouabain blocked the Na^+ - K^+ ATPase of the susceptible cell, leading to a localized inflow of Na^+ which could be redistributed through gap junctions to the resistant cell which would then pump out the excess Na^+. Arrhythmias appeared when the intracellular concentration of Na^+ had been increased by more than 50 per cent by ouabain (Goshima and Wakabayashi, 1981a), but it is not clear why this should be so. Goshima and Wakabayashi (1981b) found that myofibrils remained in a contracted state during ouabain-induced arrhythmias, suggesting that the intracellular concentraion of Ca^{2+} was high. Comparison of Na^+ and Ca^{2+} uptake by the cultured cells revealed that the rate of Ca^{2+} uptake was stimulated by ouabain and by increased intracellular Na^+ concentrations. The authors concluded that ouabain inhibited the activity of the Na^+ - K^+ pump, causing an increase in intracellular Na^+ which stimulated the activity of a Na^+ - Ca^{2+} exchange system. This system transports Na^+ out of the cell while taking up Ca^{2+}. The build-up of Ca^{2+} inside the cell could eventually become high enough to generate arrhythmic contractions. Release of Ca^{2+} by caffeine is associated with asynchronous mechanical activity and also causes a transient inward current in voltage clamped ventricular reaggregates (Clusin, 1983). High intracellular Ca^{2+} concentrations have previously been shown to activate non-specific cationic channels for inward current, revealed by single channel recording from cultured rat heart cells (Colquhoun *et al.*, 1981).

Further evidence for a Na^+ - Ca^{2+} exchange system in cultured cardiac cells is provided in the studies of Couraud *et al.* (1976), Fosset *et al.* (1977) and De Barry *et al.* (1977). These workers found that scorpion toxin, veratridine and sea anemone toxin stimulated Ca^{2+} uptake as well as Na^+ uptake. However, blocking the Na^+ channels with tetrodotoxin also eliminated the increased Ca^{2+} uptake, suggesting that Ca^{2+} uptake was coupled to enhanced Na^+ uptake. More recently, Barry and Smith (1982) demonstrated that the amount of verapamil-insensitive Ca^{2+} uptake was linked to the intracellular Na^+ concentration.

Antidysrhythmic Drugs. Culturing heart cells provides an opportunity to prepare isolated cells which have characteristics of the different regions of the heart. These may then be used as models for testing drugs that modify electrical activity. However, little use has been made of such a system for examining the actions of antidysrhythmic compounds. Perhaps this has been because reliable culture methods have been slow to be developed.

The obvious way to produce cultures of different cell types is to prepare the cultures from distinct anatomical regions of the heart. With the less than ideal culture techniques previously available, this method was not completely successful. Fänge *et al.* (1959) did find that cells in cultures prepared from atria beat more rapidly than cells obtained from ventricles; the atrial cells were also more responsive to acetylcholine. Mercer and Dower (1966) found that almost as many ventricular cells as atrial cells could beat spontaneously in culture. This apparent generation of new 'pacemakers' in culture had also been noted by DeHaan (1967) and by Goshima (1974) who found that more cells in his mouse ventricle cultures beat spontaneously than in the corresponding atrial muscle cultures.

The question of whether characteristic regional differences could be maintained in culture was examined by Norwood *et al.* (1980). With cultures prepared from different parts of neonatal rat hearts, three distinct morphologies were found. The proportion of beating cells having a particular morphology was calculated: the majority of beating cells in cultures from atrial muscle and the atrioventricular node were of a different morphology from the beating cells in the ventricular cultures. The authors thought that the beating cells in their ventricular cultures might correspond to Purkinje fibres. In addition, the average rate of spontaneous contractions was significantly higher in atrial cultures than in ventricular cultures. Unfortunately, there has been no electrophysiological characterization of these different cell types.

Cells with action potentials similar to those of pacemaker cells or of ventricular cells were observed in well-differentiated cultures from neonatal rat hearts (Kitzes and Bern, 1979), and cells with different action potential properties have been produced by small variations of the culture conditions (Jones *et al.*, 1978). Most recently, Shrier and Clay (1982) have demonstrated that cells in reaggregate cultures prepared from chick embryo atria or ventricles differed in the shapes of the pacemaker potentials and that this could be related to different ionic currents recorded under voltage clamp.

Such cultures would appear to be ideal for study of the actions of antidysrhythmic drugs. However, the work that has been reported is rather disappointing, being descriptive and qualitative and not mechanistic. Studies in which the effects of quinidine, procainamide and phenytoin on the rate of beating or on arrhythmic activity were observed have shown that these antidysrhythmic drugs were generally depressant without consistently preventing arrhythmias (Cavanaugh and Cavanaugh, 1957; Mercer and Dower, 1966; Kleinfeld *et al.*, 1969; Goshima, 1977). Quinidine was shown to reduce $^{22}Na^+$ uptake and to slow the rate of beating in cultures of neonatal rat hearts (McCall, 1976). Since verapamil, but not tetrodotoxin, decreased Na^+ uptake and the frequency of contractions, and since quinidine had no effect following verapamil pretreatment, McCall (1976) concluded that quinidine could act on slow Na^+ - Ca^{2+} channels. Cultures have also been used to test for interactions between quinidine and digoxin (Horowitz *et al.*, 1982). The lack of any such interactions at the cellular level suggests that the clinical interference by quinidine on plasma levels of digoxin may be mediated by an effect on metabolism.

Acetylcholine Receptors

Some of the first studies of cardiac muscle in culture demonstrated that acetylcholine could decrease the rate of spontaneous activity (Markowitz, 1931; Fänge *et al.*, 1956) (Figure 5.10). This effect is mediated by muscarinic receptors because muscarinic, but not nicotinic, antagonists block the action of acetylcholine (Ertel *et al.*, 1971). Although this demonstrates qualitatively that cholinoceptors are present in cultured cardiac muscle, it does not necessarily mean that these receptors have properties identical to native muscarinic receptors of intact heart. We must consider whether the cultured receptor has the same affinity for drugs before we can determine whether studies on receptors in culture can reveal anything valid about the workings of this receptor system.

Affinity for Agonists and Antagonists

Although Ertel *et al.* (1971) found it difficult to get graded responses to acetycholine, other workers have produced dose-response curves for the chronotropic actions of muscarinic agonists (Lane *et al.*, 1977; Hermsmeyer and Robinson, 1977; Galper and Smith, 1978) (Figure 5.11). Estimates of potency varied, depending on whether the cells studied were isolated or in clusters (Lane *et al.*, 1977) and on the method of drug application (Hermsmeyer and Robinson, 1977). The potency of muscarinic drugs can also be estimated from competition experiments with labelled antagonists such as quinuclidinyl benzilate (QNB) (Figure 5.12). Values from work by several authors are summarized in Table 5.1. It is remarkable that there was a 10,000-fold difference in results obtained between single and grouped cells (Lane *et al.*, 1977). Although the higher value for carbachol is of the same order as that for acetylcholine reported by Hermsmeyer and Robinson (1977), it is hard to explain the mechanism for this apparent increase in sensitivity; perhaps observing cells in groups leads to detection of the most sensitive cell of the group. However, apart from these extremely

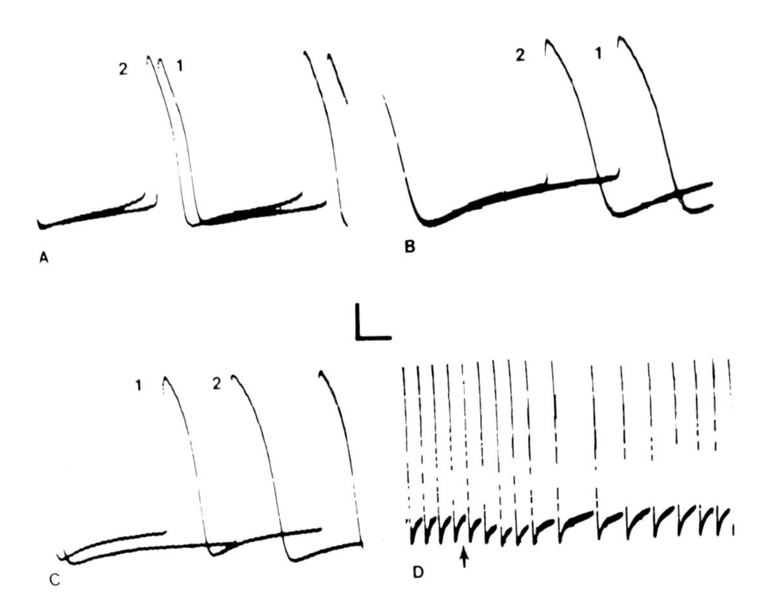

Figure 5.10: Effect of Noradrenaline and Acetylcholine on Spontaneous Action Potentials in Cultures of Chick Embryo Ventricular Muscle

(A) Superimposed traces of control (trace 1) and noradrenaline-affected (trace 2) action potentials. Application of 0.3 nM noradrenaline has caused an increase in the slope of the pacemaker depolarizing potential so that the action potential is triggered sooner than before. (B) As in (A), application of noradrenaline has increased the frequency of firing, although without noticeable increase in pacemaker slope (presumably because noradrenaline was acting on a neighbouring cell that is electrically coupled to the impaled cell). (C) 5.5 nM acetylcholine has decreased the frequency of firing by slowing the rate of depolarization of the pacemaker potential and by causing a slight hyperpolarization. (D) Slowing of the frequency of firing is caused by application of acetylcholine at the point marked by the arrow. Vertical calibration, 20 mV; horizontal calibration, 100 ms for (A-C) and 1 s for (D). From Hermsmeyer and Robinson (1977), with permission.

high values, the values obtained from culture for the effective concentrations of acetylcholine, carbachol, oxotremorine, atropine, hyoscine and QNB agree remarkably closely with corresponding values from non-cultured heart (Table 5.1). Thus, the culture process has not drastically altered the properties of the muscarinic cholinoceptors.

Density, Distribution and Turnover of Receptors

The distribution of cholinoceptors on cultured muscle is diffuse, as revealed by an autoradiographic study of labelled QNB binding

Table 5.1: Potency of Drugs on Muscarinic Cholinoceptors in Cultured and Non-cultured Cardiac Muscle

Drug	Cultured Muscle			Non-cultured Muscle		
	Preparation	Potency (M)*	Reference	Preparation	Potency (M)*	Reference
Acetylcholine	Fetal mouse heart in organ culture[a]	10^{-5}	Wildenthal (1974)	Chick embryo atria (18d)[a]	2×10^{-7}	Pappano and Skrowonek (1974)
	Chick embryo heart cell cultures[a]	5×10^{-7}	Ertel *et al.* (1971)	Guinea pig atria[c]	5×10^{-8}	Harvey (unpublished)
	Chick embryo ventricular cell cultures[a]	4×10^{-10}	Hermsmeyer and Robinson (1977)	Amphibian atria[b]	10^{-5}	Hartzell (1980)
Carbachol	Fetal mouse heart cell cultures[a]	2×10^{-6} (on single cells)	Lane *et al.* (1977)	Chick embryo heart[a]	3×10^{-8}	Galper *et al.* (1977)
		4×10^{-10} (on clustered cells)		Chick embryo heart[b]	10^{-5}	Galper *et al.* (1977)
	Chick embryo heart cell cultures[a]	10^{-5}	Galper and Smith (1978)	Amphibian atria[b]	10^{-4}	Hartzell (1980)
	Chick embryo heart cell cultures[b]	7×10^{-6}	Galper and Smith (1978)			
Oxotremorine	Chick embryo heart cell cultures[b]	10^{-7}	Galper and Smith (1978)	Amphibian atria[b]	1.5×10^{-6}	Hartzell (1980)
				Chick embryo heart[b]	10^{-7}	Galper *et al.* (1977)
Atropine	Chick embryo heart cell cultures[b]	3×10^{-9}	Galper and Smith (1978)	Chick embryo heart[b]	$1\text{-}4 \times 10^{-9}$	Galper *et al.* (1977)
				Amphibian atria[b]	3×10^{-9}	Hartzell (1980)
QNB	Chick embryo heart cell cultures[b]	10^{-9}	Galper and Smith (1978)	Amphibian atria[b]	5×10^{-10}	Hartzell (1980)
				Chick embryo heart[b]	10^{-9}	Galper *et al.* (1977)

Notes:

* Potency is the concentration of the drug required to produce either half its maximal effect or 50 per cent inhibition of labelled QNB binding.

a Effect on heart rate.

b Inhibition of QNB binding.

c Effect on force of contraction.

(Lane *et al.*, 1977). The number of sites on the cell surface increased from 400-800 sites/μm^2 after two days in culture and to 850-1,150 sites/μm^2 after seven days. Galper and Smith (1978) failed to detect an increase in total binding of QNB in developing cultures: it remained about 80 fmole/mg protein from day 4 to day 8 in culture, which was about half the amount determined in hearts developing *in vivo* (Galper *et al.*, 1977). The number of receptors per heart cell may have increased at the same time as the number of fibroblasts so that the total protein would contain a smaller contribution from myogenic cells.

The metabolism of nicotinic receptors of skeletal muscle has been extensively studied, often aided by use of muscle cells in culture (see Chapter 4). The synthesis and turnover of muscarinic receptors can also be analysed using cultured cardiac muscle, but this has only been attempted recently. Obata and Oide (1980) showed that the time-dependent increase in sensitivity to acetylcholine of chick embryo atria in explant or organ culture was dependent on RNA and protein synthesis, but not on DNA synthesis. Although this suggests that new receptors were being synthesized and incorporated into the cell membranes, QNB binding did not increase with age. Renaud *et al.* (1982) studied the turnover of QNB binding sites in chick embryo heart cell cultures after inhibition of protein synthesis. In immature cultures the half life of the binding sites was about 14 h, but in more differentiated cultures (grown in the presence of lipoprotein-deficient serum) there was no decrease in the numbers of sites in 20 h.

Mechanism of Action of Acetylcholine

Activation of muscarinic receptors leads to a decrease in heart rate and, at least in some species, to a decrease in the force of the heart beat. The chronotropic action is a consequence of a slowing of the rate of diastolic depolarization (phase 4) brought about by an increase in potassium permeability and/or a decrease in the slow inward current (Fiigure 5.10). The muscarinic receptors could be coupled directly to ion channels or act indirectly via a 'second messenger' such as cyclic AMP or cyclic GMP. Because of the technical difficulties of studying intact heart preparations, it is expected that cultured heart muscle will be used more frequently as a convenient model system for the analysis of the molecular mechanisms involved in muscarinic receptor function.

Such studies as have been published to date have been concerned principally with desensitization and the short-term regulation of the number of functional receptor sites (Galper and Smith, 1978, 1980; Galper *et al.* 1982). As found previously with acetylcholine (Ertel *et al.*,

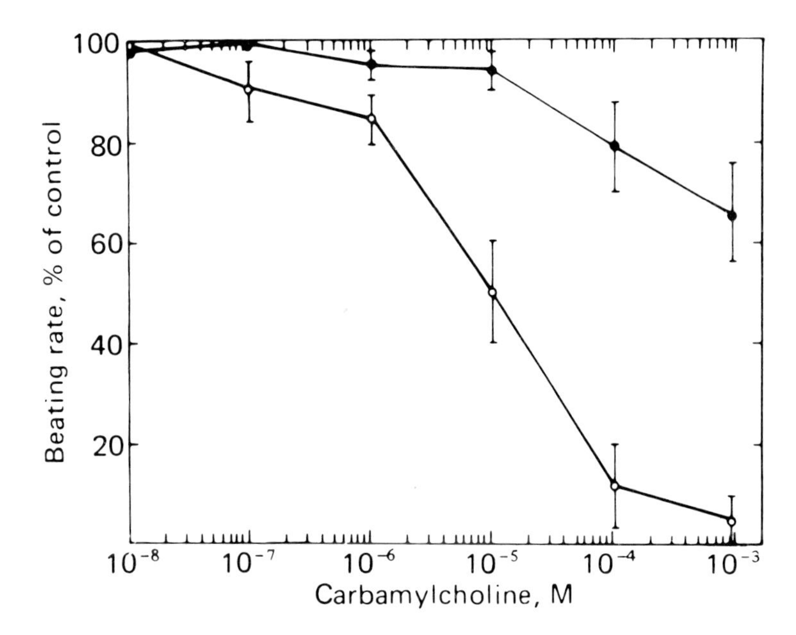

Figure 5.11: Effect of Carbachol (carbamylcholine) on the Beating Rate of Heart Cells in Culture Before and After Desensitization

In each experiment, control beating rates were determined in six confluent cultures. Each culture was then exposed to one of six concentrations of carbachol and beating rate was determined at 15 s intervals. The minimal beating rate at each concentration was plotted as a percentage of control (i.e. if a given concentration of carbachol did not alter beating rate, minimal beating rate was 100 per cent of control). Each dish was then incubated for 3 h in fresh medium containing 0.1 mM carbachol and washed three times with growth medium; the last wash was for 5 min. at 37°C in 5 per cent CO_2/95 per cent air. A post-treatment control rate of beating and the response of beating to a given concentration of carbachol were then determined. Data shown are mean (± S. D.) of 15 such experiments. ○, control cultures; ●, cultures after a 3 h incubation with 0.1 mM carbachol. From Galper and Smith (1978), with permission.

1971), the chronotropic response to carbachol is transient, suggesting that receptor desensitization occurred (Galper and Smith, 1978). Exposure of cultured heart cells from chicken embryos to 10^{-4}M carbachol (a concentration sufficient to elicit a near maximal response) for 1-3 h led to the frequency of beating being inhibited for a few minutes and then recovering to control values. A second addition of carbachol, even at ten times higher concentration, produced a greatly reduced response (Figure 5.11). In parallel experiments, QNB binding was found to be reduced to about 30 per cent of control levels by carbachol treatment. The relationship of the decrease in QNB binding sites and the loss in physiological responsiveness is not clear, but some aspects of the loss of QNB sites were studied in more detail (Galper and Smith, 1980; Galper *et al.*, 1982). It was confirmed that the reduction in binding was because there were fewer sites and not because of a decrease in affinity for QNB or a change in the kinetics of binding. Furthermore, the decrease in binding was biphasic, with about 26 per cent of the control level lost after 1 min. exposure to carbachol and a further 43 per cent lost in the next 2.5 h. It is possible, therefore, that the initial loss is equivalent to desensitization while the slower phase reflects a different process, perhaps internalization. After 1 min. conditioning exposure to carbachol, the ability of carbachol to displace bound QNB was reduced, 9.5×10^{-5}M instead of 1.5×10^{-5}M being required to displace 50 per cent of the QNB.

A similar sixfold decrease in the apparent affinity of carbachol for the receptor was brought about by exposure to Gpp(NH)p, GTP and GDP, but not to GMP or cyclic GMP. This effect of the guanine

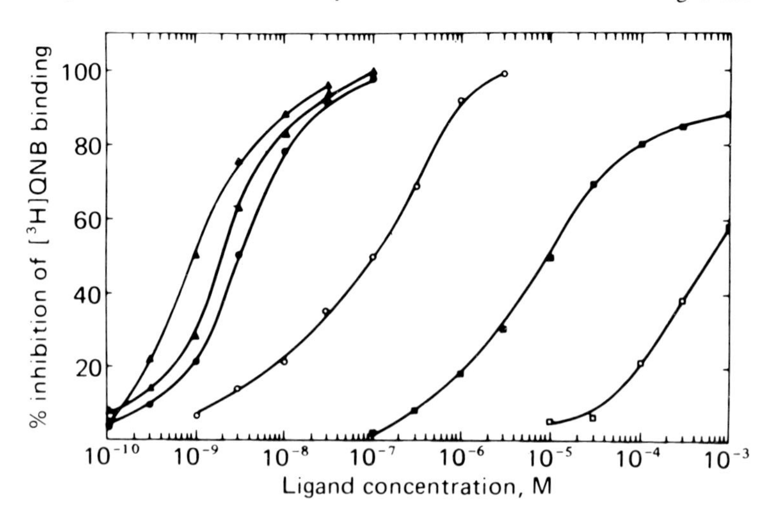

Figure 5.12: Inhibition of [^{3}H] QNB Binding by Muscarinic Agonists and Antagonists

Aliquots of a homogenate of heart cell cultures from hearts nine days *in ovo* were incubated for 1 h at room temperature in medium containing 1 nM [^{3}H] QNB and the concentrations of agonists and antagonists indicated. Each point represents the mean of three replicate determinations of bound [^{3}H] QNB. ▲, Unlabelled QNB; △, hyoscine; ●, atropine; ○, oxotremorine; ■, carbachol; □, tubocurarine. From Galper and Smith (1978), with permission.

nucleotides on carbachol affinity was not accompanied by a decrease in QNB binding. Administration of the conditioning exposure to carbachol together with Gpp(NH)p confirmed that the guanine nucleotide could maintain the receptors in a state in which they could bind normally with QNB but carbachol had a reduced affinity. As these binding studies have not yet been accompanied by a parallel electrophysiological investigation, the precise nature of the changes in receptor function can only be the subject of speculation.

Some work was performed on the rate of recovery of QNB binding sites after 3 h exposure to carbachol (Galper and Smith, 1980). There was a 3 h lag period followed by a slow recovery of the number of sites. Recovery could be prevented by cycloheximide, suggesting that protein synthesis was involved. The rate of reappearance of QNB binding sites may be related to the appearance of increased sensitivity to acetylcholine in developing atria (Obata and Oide, 1980), and it is interesting to note the superficial similarity of these processes to the rate of incorporation of newly synthesized nicotinic receptors in cultured skeletal muscle.

Noradrenaline Receptors

Catecholamines increase the rate and force of intact hearts but the molecular mechanisms underlying these effects are still incompletely understood. Catecholamines act on β-adrenoceptors to cause an increased rate of diastolic depolarization of pacemaker cells, an increase in inward Ca^{2+} currents and an increase in the intracellular concentration of cyclic AMP. The interrelationships of these events remain to be determined. With the advent of the voltage and patch clamp techniques for simple cultured systems of cardiac muscle the ionic basis of the actions of catecholamines could be readily explored. The ability to inject drugs and nucleotides directly to the inside of heart cell membranes could be useful in such studies.

Specificity and Affinity

Markowitz (1931) found that a few μg of adrenaline per ml produced a significant acceleration of the rate of beating of her explant cultures from chick embryo hearts. Without the availability of selective antagonists, it was difficult for her to analyse this effect further. However, other workers confirmed that cultured cardiac muscle contained adrenoceptors and that these were largely of the β-type. Wollenberger

(1964) found the threshold concentration of adrenaline to be as low as 2×10^{-9}M. Antagonists at β-adrenoceptors inhibited the response to adrenaline but phenoxybenzamine, an α-adrenoceptor antagonist, was ineffective. Graded, concentration-dependent effects on beating rate were obtained with noradrenaline and isoprenaline but no effect was found with the α-receptor agonist phenylephrine (Ertel *et al*., 1971). Isoprenaline responses were blocked by the β-antagonist propranolol but not by the α-antagonist phentolamine. Similar results with adrenaline, noradrenaline and isoprenaline were obtained by Fayet *et al*. (1974), who also demonstrated that the antagonism by propranolol was competitive (Figure 5.13).

Thus the results obtained (all from chick embryo cultures) are consistent with the cultured cardiac cells having a functional β-adrenoceptor system. However, using mouse heart cells, Lane *et al*. (1977) found that the chronotropic response to adrenaline or noradrenaline could not be completely inhibited by propranolol but only by a combination of phentolamine and propranolol. Approximately 60 per cent of the maximal response to adrenaline was blocked by a concentration of phentolamine that did not affect responses to isoprenaline. The presence of α-adrenoceptors was further indicated by responses to phenylephrine that were blocked by the α-antagonist phentolamine but not by the β-antagonist propranolol. Why these cultures should have two types of adrenoceptors is unclear, although there is some debate about the presence of α-adrenoceptors in intact heart. The culture system may be useful for further studies on the possible interchangeability of adrenoceptor subtypes.

The affinity of drugs for the adrenoceptors of cultured muscle has been assessed from dose-response curves for chronotropic activity and from competition experiments with labelled agonists and antagonists. Some of these results are summarized in Table 5.2. There is reasonable agreement between values obtained by different techniques and also with values derived for β-adrenoceptors in non-cultured cardiac tissue. Once again, it should be noticed that the method of agonist application may affect its apparent potency. A 1,000-fold difference in effectiveness of noradrenaline was found by Hermsmeyer and Robinson (1977). Although Ertel *et al*. (1971) did not appear to find desensitization a problem, desensitization of adrenoceptors can occur, and it seems sensible to make certain that the agonist reaches the receptors in its final concentration as rapidly as possible.

Binding studies have also led to estimates of the density of β-adrenoceptors on cultured cardiac muscle. Lefkowitz *et al*. (1974) using

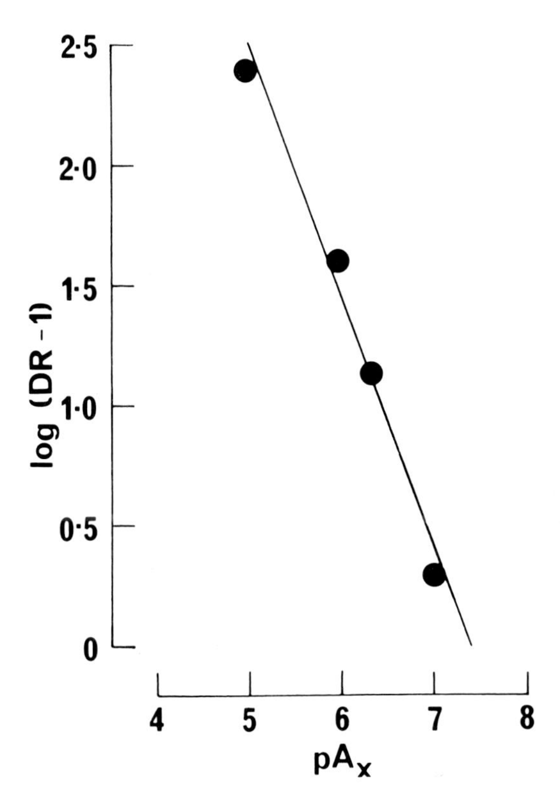

Figure 5.13: Arunlakshana and Schild Plot of the Inhibition by Propranolol of the Positive Chronotropic Effect of Noradrenaline on Chick Embryo Heart Cells in Culture
The logarithm of the dose ratio −1 is plotted against the negative logarithm of the molar concentration of propranolol. The pA_2 value from the plot is 7.4 and the line obtained by regression analysis has a slope of −1.05, which is consistent with competitive antagonism. Data from Fayet *et al.* (1974).

^{3}H-noradrenaline estimated that there were about 3×10^6 sites/cell, although these binding sites showed little stereoselectivity and propranolol was not very effective at reducing binding. From autoradiographic analysis of labelled alprenolol binding, Lane *et al.* (1977) calculated that there were 30-50 sites/μm^2, which is almost ten times fewer than found by Lefkowitz *et al.* (1974). Lau *et al.* (1980) compared binding of ^{125}I-hydroxybenzylpindolol to myogenic and fibroblast cells in their cultures. Myoblasts had 7,600 ± 2,100 sites/cell, which is closer to the values of Lane *et al.* (1977). The binding measured by Lau *et al.* (1980)

Table 5.2: Potency of Drugs on β-Adrenoceptors in Cultured and Non-cultured Cardiac Muscle

	Cultured Muscle			Non-cultured Muscle		
Drug	Preparation	Potency (M)*	Reference	Preparation	Potency (M)*	Reference
Noradrenaline	Chick embryo heart cell culture[a]	2×10^{-7}	Fayet *et al.* (1974)	Guinea pig atria[a]	$3\text{-}5 \times 10^{-8}$	Harvey, unpublished
	Chick embryo ventricular cell cultures[a]	8×10^{-10} to 3×10^{-6} †	Hermsmeyer and Robinson (1977)			
	Chick embryo heart cell cultures[b]	7×10^{-7}	Lefkowitz *et al.* (1974)			
	Fetal mouse heart in organ culture[a]	5×10^{-7}	Roeske and Wildenthal (1981)			
	Fetal mouse heart cell cultures[a]	3×10^{-7}	Lane *et al.* (1977)			
Adrenaline	Chick embryo heart cell cultures[a]	2×10^{-7}	Fayet *et al.* (1974)	Guinea pig atria[a]	10^{-8}	Harvey, unpublished
	Chick embryo heart cell cultures[b]	7×10^{-7}	Lefkowitz *et al.* (1974)			
	Fetal mouse heart cell cultures[a]	$10^{-9} - 10^{-8}$,	Lane *et al.* (1971)			
	Rat heart cell cultures[c]	5×10^{-7}	Tsai and Chen (1978)			
	Neonatal rat ventricular cell cultures[ab]	10^{-6}	Wollenberger and Irmler (1978)			
Isoprenaline	Chick embryo heart cell cultures[a]	2×10^{-8}	Ertel *et al.* (1971)	Guinea pig atria[a]	10^{-8}	Harvey, unpublished
	Chick embryo heart cell cultures[a]	2×10^{-8}	Fayet *et al.* (1974)			
	Chick embryo heart cell cultures[b]	7×10^{-7}	Lefkowitz *et al.* (1974)			
	Fetal mouse heart in organ culture[a]	10^{-7}	Wildenthal (1974)			
	Fetal mouse heart cell cultures[a]	10^{-9}	Lane *et al.* (1977)			
	Neonatal rat ventricular cell cultures[b]	2×10^{-7}	Lau *et al.* (1980)			
Propranolol	Chick embryo heart cell cultures[a]	3×10^{-8}	Fayet *et al.* (1974)	Guinea pig atria[a]	2.5×10^{-9}	Kaumann *et al.* (1980)
	Chick embryo heart cell cultures[b]	10^{-4}	Lefkowitz *et al.* (1974)			
Practolol	Neonatal rat ventricular cell cultures[b]	1.5×10^{-7}	Lau *et al.* (1980)	Guinea pig atria[a]	2.5×10^{-7}	Kaumann *et al.* (1980)
Alprenolol	Rat heart cell cultures[b]	10^{-9}	Tsai and Chen (1978)	Guinea pig atria[a]	2.5×10^{-9}	Kaumann *et al.* (1980)
Hydroxybenzylpindolol	Neonatal rat ventricular cell cultures[b]	9×10^{-11}	Lau *et al.* (1980)			

Notes.

* Potency in the concentration of the drug required to produce half its maximal effect or 50 per cent inhibition of labelled drug binding.

a Effect on heart rate.

b Inhibition of binding.

c Effect on cyclic AMP levels.

† Potency varied with different methods of drug application.

did show stereoselectivity, and from their competition experiments they concluded that the myogenic cells contained β_1-adrenoceptors and the fibroblasts had β_2-adrenoceptors.

Mechanism of Action of Noradrenaline

Little work has been done on the changes in ionic permeability that catecholamines cause in cultured cardiac muscle. Certainly, catecholamines can increase the frequency of firing of action potentials (Figure 5.10). Additionally, in chick embryo cultures in which fast Na^+ channels have been blocked by tetrodotoxin, catecholamines can stimulate a slowly-rising, Ca^{2+}-dependent action potential (Figure 5.6) (Sperelakis and Shigenobu, 1972; Josephson and Sperelakis, 1978), which can also be found in some non-cultured cardiac muscle (Shigenobu and Sperelakis, 1972). This effect is mediated through β-adrenoceptors, since it is blocked by drugs such as propranolol, and it probably involves cyclic AMP, as other agents which increase the intracellular concentration of the cyclic nucleotide can also induce the slow action potential. From recent work using patch clamp techniques to record from single Ca^{2+} channels in cultured heart cells (Reuter *et al.*, 1982; Reuter, 1983), it appears that catecholamines increase the probability that a Ca^{2+} channel will open, without modifying the fundamental conductance properties of the channel.

The ability of catecholamines to increase the concentration of cyclic AMP in cultured heart muscle has been reported by several authors. Moura and Simpkins (1975) demonstrated that the increase in cyclic AMP by adrenaline and noradrenaline was inhibited by the β_1-adrenoceptor antagonist practolol, which had no effect on the stimulatory effect of glucagon. Renaud *et al.* (1978) showed that isoprenaline increased cyclic AMP levels in cultured and non-cultured heart tissue from chick embryos of different ages; the basal level of cyclic AMP in intact hearts decreased with age. Tsai and Chen (1978) found that adrenaline increased cyclic AMP levels within 2 min. and that the increase waned over a 30 min. period. The effect of adrenaline was inhibited by the β-receptor antagonist, alprenolol.

Modulation of Adrenoceptors by Hormones

Tissue culture methods are well suited to examine the long-term effects of hormones or drugs on the properties of natural receptors. There have been a few attempts to do this with cardiac muscle in culture, which serve to illustrate the possibilities of the technique.

Long-term incubation with triiodothyronine was found to increase the sensitivity of organ-cultured mouse hearts to isoprenaline but not to other positive chronotropic agents, suggesting a selective effect on β-adrenoceptors (Wildenthal, 1974) or on enzymes coupled to them. Subsequently, it was shown that triiodothyronine treatment enhanced the ability of adrenaline to increase cyclic AMP levels, and that this effect was accompanied by a two-fold increase in the number of β-adrenoceptors (Tsai and Chen, 1978). Effects of parathyroid hormone on β-adrenoceptor function in cultured cardiac cells have also been described (Larno *et al.*, 1980) and it would be interesting to compare the long-term effects of parathyroid hormone with those of other hormones.

Another effect on the development of β-adrenoceptors was found by Lipschultz *et al.* (1981). Cells cultured from hearts of 2-day chick embryos were unresponsive to adrenaline, even after several days in culture. However, if chick embryo extract was added to the culture medium, adrenaline sensitivity developed. This finding of possible 'trophic' control of adrenoceptor synthesis is extremely interesting and is obviously susceptible to further study.

6 SMOOTH MUSCLE

Soon after the development of the plasma clot technique for growing cells in culture (Carrel and Burrows, 1910) there were reports of the behaviour of smooth muscle in tissue cultures. Champy (1913/14) noted that smooth muscle cells from explants of rabbit arteries and veins could divide in culture, although the cells subsequently lost their characteristic morphological features to become 'dedifferentiated'. Laqueur (1914) and Lewis and Lewis (1917a) found that smooth muscle cells could contract spontaneously in culture. The Lewises observed that the cell cytoplasm appeared to be responsible for the movement, with the nucleus being passive. The spontaneous activity of the cells lasted for only a few days in culture. However, quiescent cells could be stimulated to contract by addition of more Ca^{2+} to the medium or by gentle prodding with a fine needle (Lewis and Lewis, 1917a).

In a later study of smooth muscle from the respiratory tracts of chicken embryos, Lewis (1924) demonstrated that the behaviour of the cells in culture depended on the age of the donor embryo and that the spontaneous activity of the muscle cells was extremely sensitive to incubation temperature. In one experiment, cells contracted rhythmically once every 10-15 s at 40°C but only once every 60 s at 37°C, and not at all at 36°C. These cultured cells maintained their ability to contract for up to two weeks, which was much longer than cells from any previous culture had retained this ability. Further details of early work on cultured smooth muscle can be found in the review by Murray (1965a), and a full bibliography was compiled by Murray and Kopech (1953). The appearance of smooth muscle cells in primary and continuous cultures is illustrated in Figures 6.1 and 6.2.

Despite the early attention given to smooth muscle, the culture system has been little used for pharmacological or electrophysiological studies. This may be partly because of the strength of traditional pharmacological techniques in the study of smooth muscle. However, since standard smooth muscle preparations are notoriously difficult to study with microelectrodes, there is a need for simpler model systems, and these could perhaps be created by the use of culture techniques.

Culture Methods

Descriptions of the various methods for culturing smooth muscle are given by Chamley-Campbell *et al.* (1979). Basically, only two major methods have been used: explants followed by subculturing, and dissociated cell cultures.

Explants and Subcultures

All early work with cultured smooth muscle was concerned with the behaviour of explants and the cells that migrated from them. Because of the disadvantages of this system, Ross (1971) described a method for the trypsinization and subsequent subculture of smooth muscle cells derived from explants of the aortas of young guinea pigs. This technique, or one of several variations of it, has been used for smooth muscle from many different sources including human vascular muscle (Gimbrone and Cotran, 1975) and human uterine muscle (Rifas *et al.*, 1979).

Although the subcultured cells retain some of the characteristic features of smooth muscle (Ross, 1971; Gimbrone and Cotran, 1975), there are significant biochemical changes (Fowler *et al.*, 1977) and the cells lose their ability to contract and to respond to noradrenaline and angiotensin (Mauger *et al.*, 1975; Chamley *et al.*, 1977). The phenotypic modification of smooth muscle culture is discussed in detail by Chamley-Campbell *et al.* (1979).

Attempts have also been made to establish continuous cell lines with the properties of smooth muscle, but these have not been altogether successful. Using material from a brain tumour induced in mice, Schubert *et al.* (1974a) produced a cell line (BC_3H1) that had some properties characteristic of smooth muscle but also had some properties consistent with skeletal muscle. Kimes and Brandt (1976a) managed to establish two cell lines which retained electrophysiological properties typical of smooth muscle, by repeated subculture of cells from embryonic rat aorta. In one other cell line from the same source, a small percentage of cells fused to form skeletal muscle-like myotubes (Kimes and Brandt, 1976a). A continuous cell line (DDT_1) was established from a smooth muscle tumour in hamsters and it was shown to contain receptors for steroid hormones and for noradrenaline (Cornett and Norris, 1982). Another line derived from human oviduct smooth muscle has remained visible for several years (Sinback and Shain, 1979, 1980). The appearance of these cells is shown in Figure 6.2. Another recent study has demonstrated that cells in a

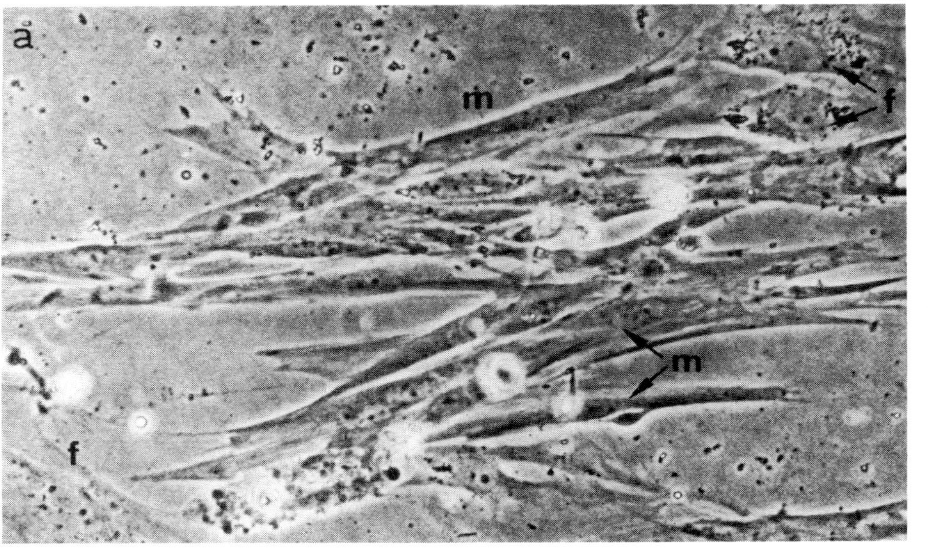

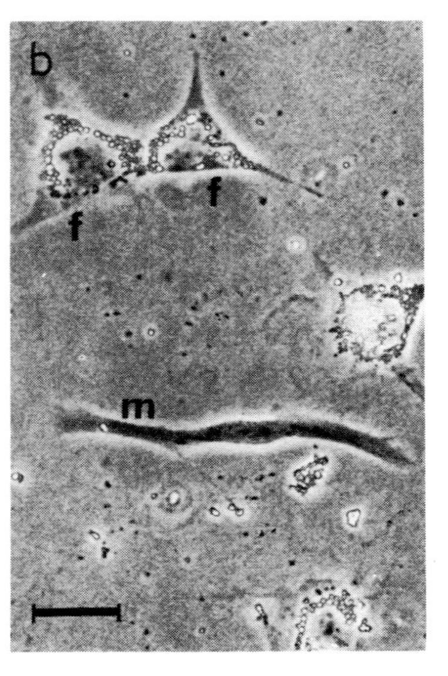

Figure 6.1: Phase Contrast Micrographs of Smooth Muscle Cells in Dissociated Cell Cultures Established from Newborn Guinea Pig Taenia Caeci

(a) A clump of muscle cells in a 5-day culture. (b) A single muscle cell in a 2-day culture. m, muscle cells; f, fibroblasts; scale marker = 25 μm. From Purves *et al*. (1973), with permission.

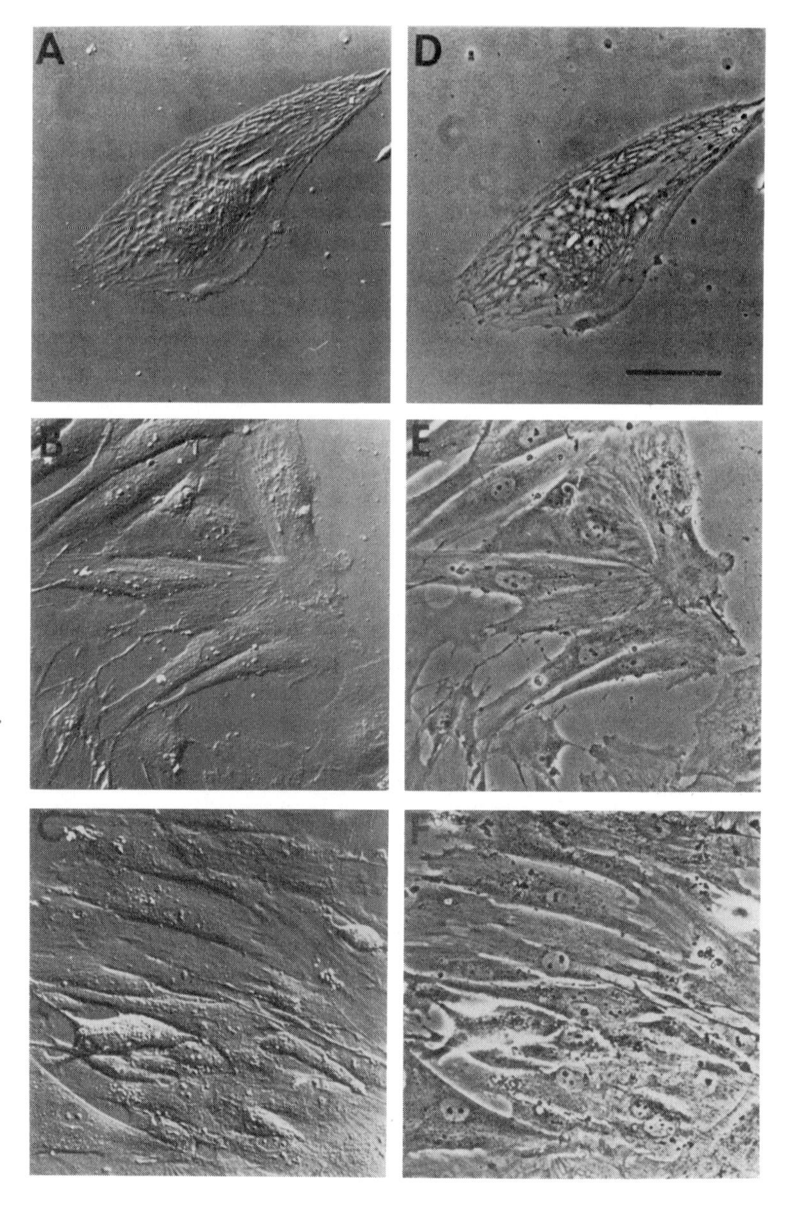

Figure 6.2: Smooth Muscle Cells of a Continuous Cell Line Derived from Human Oviduct

Left, differential interference contrast and right, phase contrast micrographs of solitary cells (A, D), semiconfluent cells (B, E) and confluent cells (C, F). Scale marker = 50 μm. From Sinback and Shain (1979), with permission.

continuous cell line from rat vascular smooth muscle retained the ability to synthesize the form of actin characteristic of smooth muscle, even after about 200 passages (Franke *et al.*, 1980). There have been no reports, however, of contractile activity in any of the putative smooth muscle cell lines.

Dissociated Cell Cultures

The use of enzymes to dissociate smooth muscle to single cells was established by Burnstock's group, who found that repeated trypsinization after pretreatment with collagenase gave a good yield of viable cells which retained properties of smooth muscle. This method has been successful with guinea pig vas deferens (Mark *et al.*, 1973) and taeniae caeci (Purves *et al.*, 1973). Other workers have used trypsin, collagenase, elastase and hyaluronidase, either alone or in various combinations. Table 6.1 contains a summary of the tissues which have been used together with the various dissociation procedures.

The main advantage of cells in primary cultures over those in subcultures is that the cells in the primary cultures retain to a much greater extent the characteristic morphological and electrical properties of smooth muscle. The morphological development of dissociated smooth muscle in culture is well reviewed by Chamley-Campbell *et al.* (1979). In primary culture the smooth muscle cells are spindle- or ribbon-shaped. At first there is little cell division, the cells have actin and myosin filaments and can contract spontaneously. After a variable period in culture, the smooth muscle cells begin to divide, eventually forming a confluent monolayer. At this stage the cells have lost both the myosin filaments and their contractility. Subsequently there can be some 're-differentiation' with the reappearance of a greater density of myosin filaments but the level of contractile activity is less than in the young cultures (Chamley *et al.*, 1977; McLean *et al.*, 1979).

Electrophysiological Studies

Analysis of the electrical properties of smooth muscle cell membranes and the nature of the ionic conductances which determine the physiological responses of the membranes is not easy for a number of reasons. Smooth muscle cells tend to be small and flat so that it is difficult to insert recording electrodes without severely damaging the membranes. Intact tissues often have spontaneous mechanical activity which makes it even harder to maintain electrode penetrations. Individual cells in a

Table 6.1: Tissues Used for Preparation of Dissociated Cell Cultures of Smooth Muscle

Tissue source	Species	Dissociation technique	Reference
Omphalomesenteric arteries and veins	chick embryo	0.0125–0.025% trypsin and/or 0.1–0.3% collagenase	Hermsmeyer *et al.* (1976)
Aorta, brachiocephalic artery, mesenteric and pulmonary vessels	chick embryo	0.05% trypsin (4 times)	McLean and Sperelakis (1977)
Ear and thoracic arteries	rabbit	0.125% trypsin + 0.1% collagenase	Chamley and Campbell (1976)
Thoracic aorta	rabbit	0.08% collagenase + 0.12% elastase	Ives *et al.* (1978)
Thoracic aorta	rabbit and monkey	0.1% collagenase + 0.05% elastase	Chamley *et al.* (1977)
Aorta	calf	1,800 U/ml collagenase + 230 U/ml elastase	Fowler *et al.* (1977)
Arteries and veins from umbilical cord	human	0.2% collagenase	Gimbrone and Cotran (1975)
Vas deferens	guinea pig	0.5% collagenase then 0.125% trypsin	Mark *et al.* (1973)
Vas deferens	guinea pig	0.125% trypsin + 0.1% collagenase	Gröschel-Stewart *et al.* (1975)
Vas deferens	guinea pig	0.125% trypsin (2 times) then 0.01% collagenase + 0.1% hyaluronidase	McLean *et al.* (1979)
Oviduct	human	collagenase + elastase (10 times)	Sinback and Shain (1979)
Taenia caeci	guinea pig	0.5% collagenase then 0.125% trypsin	Purves *et al.* (1973)

tissue are usually electrically coupled to neighbouring cells, and hence potential changes recorded in one cell may have been generated at a distant site and are modified by the electrical properties of the intervening cells. The release of neurotransmitters from the nerve varicosities that are frequently present in smooth muscle tissues may also give rise to electrical responses which are difficult to interpret.

Although electrical recordings have now been made from a number of intact smooth muscles, the inherent difficulties, especially when attempting to use voltage clamp techniques, provide a stimulus for the development of simpler preparations which can be used more easily. Single smooth muscle cells, isolated from toad stomach, can be studied successfully with microelectrodes (Singer and Walsh, 1980a, b; Walsh and Singer, 1980a, b). These cells have resting membrane potentials which are dependent on the extracellular concentrations of both Na^+ and K^+, and they have Ca^{2+}-dependent action potentials. Their passive electrical properties are consistent with data obtained from intact smooth muscle tissues (Table 6.2). As Walsh and Singer (1981) have now demonstrated that these isolated smooth muscle cells can be voltage clamped with two intracellular microelectrodes, the way would appear to be open to a detailed understanding of the ionic conductances in smooth muscle membranes. However, the muscle cells isolated from toad stomach are unusually large and it is unlikely that other smooth muscles will yield single cells that are as suitable.

There have been only a few reports of the electrophysiological properties of smooth muscle cells in culture. Visceral (Purves *et al.*, 1973) and vascular (McLean and Sperelakis, 1977) smooth muscle cells have been studied, as well as cells in smooth muscle-derived cell lines (Kimes and Brandt, 1976a; Sinback and Shain, 1979), but there has been no comprehensive attempt to use culture techniques to compare the fundamental properties of cells from different sources.

In primary cultures of enzyme-dispersed guinea pig taeniae caeci (Purves *et al.*, 1973), cells had resting potentials between –20 and –45 mV and nearly all cells displayed spontaneous electrical activity which was similar to that found in intact preparations. The spontaneous activity consisted of either regular repetitive spikes or bursts of spikes superimposed on slow waves of depolarization. Both isolated cells and cells in clusters were spontaneously active (Figure 6.3). Purves *et al.* (1973) commented that it was far easier to record from cells which were in small clumps than from isolated cells; and McLean and Sperelakis (1977) made use of this by preparing small reaggregates of dissociated smooth muscle cells, a technique which they had developed

Table 6.2: Some Electrical Properties of Cultured and Non-cultured Smooth Muscle Cells

Preparation	Membrane potential (−mV)	Input resistance (MΩ)	Specific membrane resistance (Ωcm^2)	Specific membrane capacitance ($\mu F/cm^2$)	Time constant (ms)	Action potential amplitude (mV)	Maximum rate of depolarization (V/s)	Reference
Guinea pig vas deferens								
monolayers	38 ± 1	6.9 ± 0.5	n.d.	n.d.	n.d.			McLean *et al.* (1979)
reaggregates	58 ± 2		n.d.	n.d.	n.d.	70 – 80	1 ± 4	
Chick embryo aorta								
reaggregates	50 ± 2	~8	n.d.	n.d.	n.d.	80 – 250	6 ± 1	McLean and Sperelakis (1977)
Chick embryo mesenteric blood vessel								
reaggregates	41 ± 1	~8	n.d.	n.d.	n.d.	80 – 250	3 ± 1	McLean and Sperelakis (1977)
Human oviduct cell line								
isolated cells	27 ± 2	66 ± 7	63	1.5	96 ± 5	20 – 30	0.2	Sinback and Shain (1979)
semiconfluent cells	38 ± 2	26 ± 3	n.d.	n.d.	56 ± 4	n.d.	n.d.	
confluent cells	n.d.	6 ± 0.8	n.d.	n.d.	5 ± 1	n.d.	n.d.	
Toad stomach isolated non-cultured cells	41 ± 1.6	1400 ± 120	10,000 – 100,000	1.3 ± 0.1	20 – 160			Singer and Walsh (1980a)

Note:
Where possible, values are given as means ± standard error of the means.

for cultured heart cells (see Lieberman *et al.*, 1981, and Chapter 5). The reaggregates, formed from cells dissociated by trypsinization of blood vessels from chick embryos, often had both mechanical and electrical spontaneous activity. Cultures from the great vessels near the heart had significantly greater resting potentials (about –50 mV) and larger action potentials (amplitude about 64 mV) than those from mesenteric vessels (membrane potentials about –40 mV and action potentials about 55 mV in amplitude). The maximum rate of depolarization in the cultures derived from mesenteric vessels was also lower (3 V/s compared to 6 V/s). The smooth muscle action potentials were

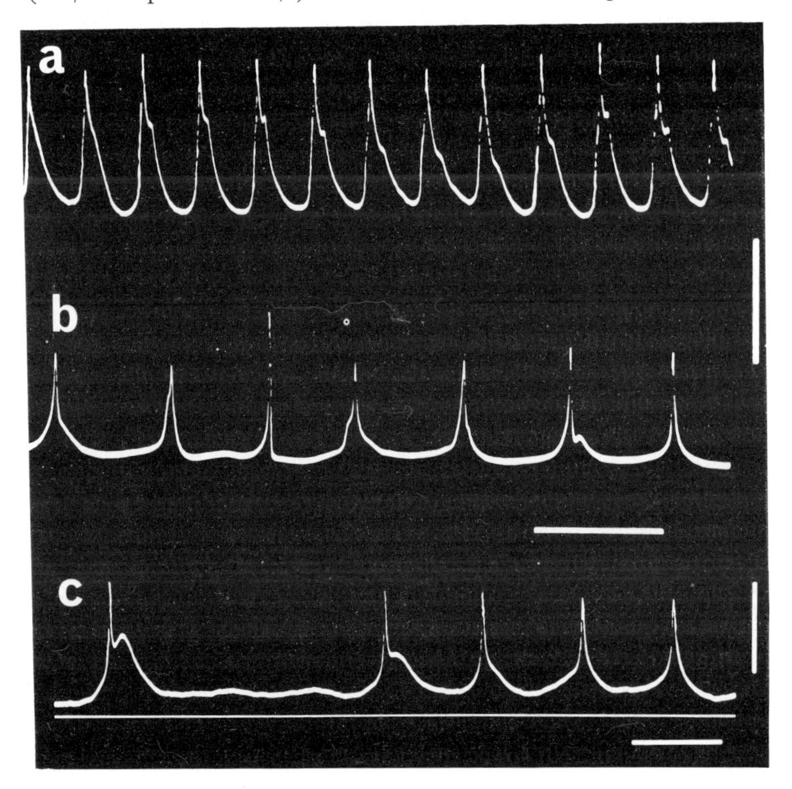

Figure 6.3: Intracellular Recording of Spontaneous Action Potentials in Cultured Taenia Caeci Smooth Muscle Cells
(a) Recording made from a clump of cells in a 3-week culture. Vertical calibration, 20 mV; horizontal calibration, 2 s. (b) Recording made from a clump of cells in a 9-day culture. Calibrations, 40 mV and 1 s. (c) Recording made from a single isolated cell in a 3-week culture. Calibrations, 20 mV and 1 s. From Purves *et al.* (1973), with permission.

insensitive to tetrodotoxin. The properties of both sets of smooth muscle cultures were significantly different from those of cardiac muscle cultures which were tested in parallel.

In a similar study, McLean *et al.* (1979) examined the properties of cells from guinea pig vas deferens, both in monolayer culture and in spontaneously-formed reaggregates. The cells in the reaggregates retained the morphological characteristics of smooth muscle for longer than the cells in the monolayers and, although fewer cells were spontaneously active compared to the chick embryo vascular cultures of McLean and Sperelakis (1977), the ability to generate action potentials was retained for up to four weeks in culture. Cells in reaggregates had higher resting membrane potentials than cells in monolayers (a mean of –58 mV compared to –38 mV), and there was a large contribution to the resting potential by a ouabain-sensitive electrogenic pump. The action potentials had positive overshoots of 10-20 mV and very slow rates of depolarization (1-4 V/s). Action potentials were unaffected by tetrodotoxin but were blocked by verapamil. It is curious that action potentials in the chick vascular smooth muscle cultures were not affected by the verapamil analogue D600 (McLean and Sperelakis, 1977). This difference between the two cell types deserves further study to determine whether it reflects a fundamental difference in channel properties or some artefact induced by the culture system.

The studies described so far have concerned the properties of primary cultures, but subcultured smooth muscle cells also can retain electrical properties similar to those of intact tissue (Kimes and Brandt, 1976a; Sinback and Shain, 1979). Although these cells have lost the ability to contract, they can generate action potentials, either spontaneously or in response to electrical stimulation. In their study of smooth muscle cells derived from human oviduct, Sinback and Shain (1979) found that some of the electrical properties of the cells depended on whether the cells were isolated or in contact with other cells (Figure 6.4). Resting potentials were similar (about –35 mV) but membrane time constant and input resistance differed according to the degree of contact. Isolated cells had time constants of about 96 ms and input resistances of about 66 MΩ, whereas cells in confluent cultures had time constants of about 5 ms and input resistances of about 6 MΩ. Similar low values for input resistances had been obtained from cells in reaggregates (McLean and Sperelakis, 1977; McLean *et al.*, 1979). The action potentials recorded in the oviduct smooth muscle cells were slow (0.2 V/s), variable in amplitude (10-50 mV) and blocked by Co^{2+} and by EGTA, suggesting that they are Ca^{2+}-dependent. The

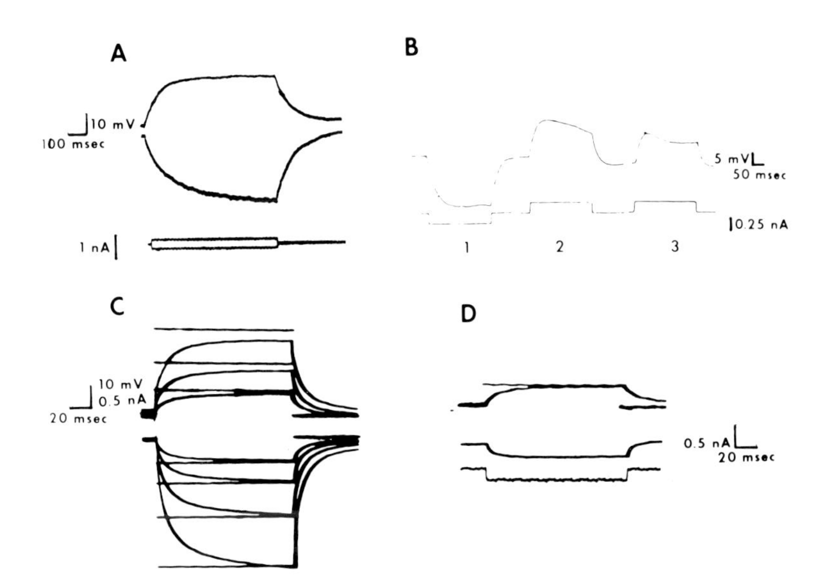

Figure 6.4: Passive Electrical Responses in Cultured Smooth Muscle Cells Derived from Human Oviduct

The responses differ in isolated cells, semiconfluent cells and confluent cells. (A) In a solitary cell with membrane potential –30 mV, there is a smaller depolarizing voltage response (20 mV) compared to 30 mV hyperpolarizing voltage responses to equal hyperpolarizing and depolarizing current pulses. This is due to delayed rectification. Input resistance is 120 MΩ and time constant is 150 ms. (B) In some solitary cells, active responses (2, 3) are superimposed on electronic potential. The active response (2) is smaller than the response to an equal hyperpolarizing current pulse (1). Repetitive depolarizing pulses increase resting membrane potential and accentuate active response. (C) In a semi-confluent cell, delayed rectification responsible for 15 per cent decrease in voltage response (3 mV/20 mV) is smaller than delayed rectification responsible for 50 per cent decrease (10 mV/20 mV) in solitary cell in A. Input resistance is 20 MΩ time constant is 30 ms and membrane potential is –30 mV. (D) In confluent cell with membrane potential –38 mV, input resistance is 10 MΩ and time constant is 9 ms. From Sinback and Shain (1979), with permission.

frequency of firing of spontaneous action potentials was increased with depolarization and was reduced by hyperpolarization (Figure 6.5).

Table 6.2 gives a summary of some of the electrical properties of various smooth muscle preparations.

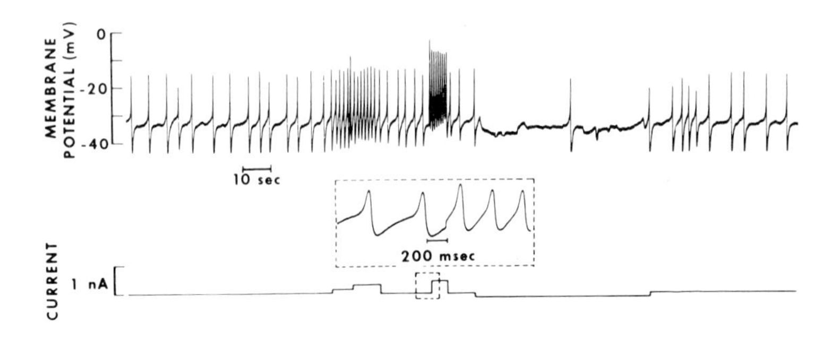

Figure 6.5: Smooth Muscle Cells in Continuous Culture Derived from Human Oviduct Can Have Spontaneous Action Potentials
Upper record, membrane potential. Lower record, current injected into the cell. Action potentials occur without injecting current. Inset shows that spontaneous action potentials fire when pacemaker depolarization reaches threshold. Depolarizing current increases the slope of the pacemaker potential and therefore the rate of firing. Hyperpolarizing current abolishes spontaneous action potentials. From Sinback and Shain (1979), with permission.

Pharmacological Studies

Smooth muscle has long fascinated pharmacologists because of the great variety of drugs that can affect its function and because of the regional variations in responses to certain drugs. Culture systems are potentially invaluable in the study of such questions as: How many different receptor types can be found on a single cell? How do the receptor mechanisms of different tissues differ? Can different receptor systems share the same final pathway to produce the biological response? In spite of the temptations to use culture methods, few detailed studies have appeared. Some attempts have been made to monitor drug action in culture by visual assessment of spontaneous mechanical activity (Hermsmeyer *et al.*, 1976; Chamley *et al.*, 1977; Ives *et al.*, 1978), and smooth muscle cells in culture were found to be extremely sensitive to noradrenaline, 5HT and acetylcholine (Hermsmeyer, 1976). However, the slowness and irregularity of the contractions makes visual monitoring a rather unattractive method. Intracellular recording techniques, despite being technically more demanding, would appear to be necessary. Biochemical studies of drug binding or enzyme activity are also possible; indeed, monolayer preparations in culture should be better than whole muscle for such work.

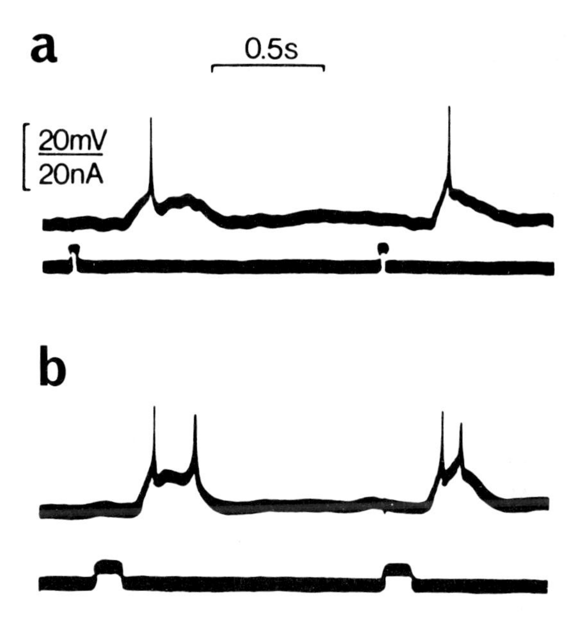

Figure 6.6: Depolarization by Acetylcholine of Cultured Smooth Muscle Cells from Guinea Pig Taenia Caeci
In both (a) and (b), upper records are membrane potential and lower records indicate the current passed through the acetylcholine-containing electrode. (a) 50 ms pulses of acetylcholine cause slow depolarizations which trigger single action potential spikes. Note the long latency of the response (about 250 ms). (b) Longer pulses (130 ms) of acetylcholine produce more rapidly developing depolarizations that trigger double action potential spikes. Latency of the response, 200 ms. From Purves (1974), with permission. Copyright 1974, Macmillan Journals Ltd.

In primary cultures of smooth muscle from guinea pig taeniae caeci, acetylcholine produced depolarization on iontophoretic application (Purves, 1974). The depolarizations were dose-dependent and began after a delay of several hundred milliseconds (Figure 6.6). This delay appears to be a characteristic of the receptors, which were presumably muscarinic, being blocked by hyoscine but not by hexamethonium or tubocurarine. The ionic basis of the conductance change produced by acetylcholine and the properties of the receptor ion channels remain to be elucidated.

The effects of acetylcholine, noradrenaline and histamine have also been studied on secondary cultures derived from human oviduct smooth muscle (Sinback and Shain, 1980). Each of the three agonists was capable of producing depolarization, hyperpolarization or biphasic responses (Figure 6.7), but a single cell could respond differently to the different drugs. This suggests that each receptor type can be linked to more than one kind of ionic channel and that individual cells can possess more than one class of receptor. Nearly 20 per cent of cells tested responded to histamine, about 60 per cent to acetylcholine and about 50 per cent to noradrenaline. The depolarization response was seen most commonly with acetylcholine and had a reversal potential of –5 to –10 mV, while histamine most often caused hyperpolarization and the null potential of this response was estimated to be –60 mV. Biphasic responses could be obtained only when the agonists were applied to certain positions on the surface of the cells: application of drug to one end of the cell could give depolarization, whereas application to the other end could cause hyperpolarization, with dual responses being recorded in a 'border region' (Figure 6.7). Since

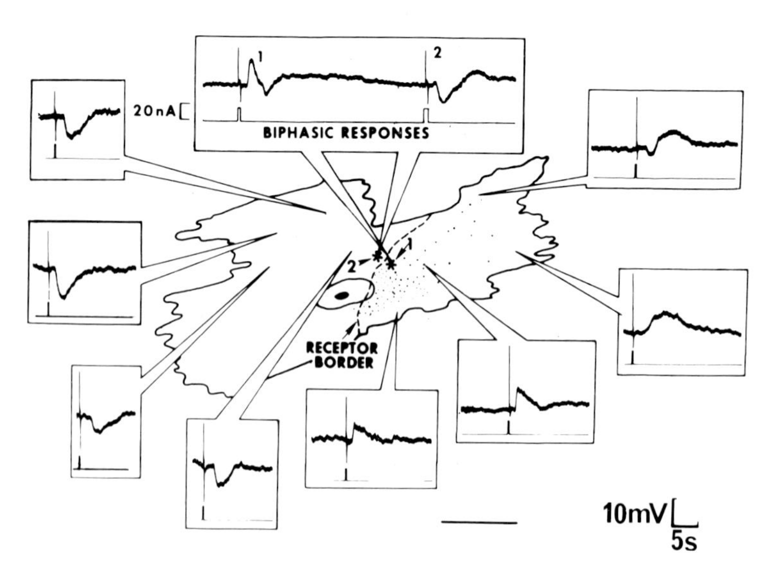

Figure 6.7: Responses of a Single Cultured Smooth Muscle Cell to Histamine Biphasic responses are due to segregation of receptors that mediate hyperpolarizing and depolarizing responses. Equal histamine pulses elicited biphasic responses at positions 1 and 2. Sequence of response polarity depended on position of histamine electrode. Smaller histamine pulses elicited depolarizing responses at position 1 and four positions on the right side of the cell. Histamine elicited hyperpolarizing responses at position 2 and at four positions on the left side of the cell. Horizontal calibration = 50 μm. From Sinback and Shain (1980), with permission.

antagonists (atropine for acetylcholine and mepyramine for histamine) blocked both phases of the biphasic response, the receptors involved are probably the same, but they must be linked to different ion channels in different regions of the cell. The mechanism controlling this process is not understood but is obviously interesting. Whether this phenomenon is shared by other smooth muscle cells or whether it is only a function of some transformation induced by the continuous subculturing must also be determined.

Adrenoceptors are also found on the smooth muscle cell line DDT_1 (Cornett and Norris, 1982). Binding of the α-adrenoceptor antagonist $[^3H]$-dihydroergocryptine was reversible, saturable and of high affinity. The density of binding sites was about 60-80/μm^2, and from competition studies with the selective α_1-adrenoceptor antagonist prazosin and the α_2-antagonist yohimbine, it was concluded that the binding sites were of the α_1-subtype.

Adenosine is a potent vasodilator substance *in vivo* and cultures of aortic smooth muscle have been used in studies on the mechanism of action of adenosine. Adenylate cyclase activity has been shown to be increased by adenosine and various analogues (Anand-Srivastava *et al.*, 1982) and depolarization-induced Ca^{2+} uptake is decreased by adenosine (Fenton *et al.*, 1982).

Angiotensin II has been reported to depolarize smooth muscle cells in culture (Zelcer and Sperelakis, 1982) and it has also been demonstrated to increase Na^+ uptake (Brock *et al.*, 1982). The Na^+/K^+ pump activity was also elevated by angiotensin but this was probably secondary to the higher internal Na^+ concentration.

7 SYNAPSES AND NEUROMUSCULAR JUNCTIONS

The primary purpose of nerve cells is to facilitate intercellular communication. Hence, a fundamental property of neurones is the ability to form functional connections with other nerve cells or with effector cells. How the nervous system develops and what controls its exquisite specificity are two of the major questions in biology. Although the answers must come largely from genetic approaches, some informative experiments can be performed utilizing tissue culture methods to provide simple models of the nervous system. For example, culture systems could be useful for investigations of synaptic development, of the limits of synaptic specificity, and of long-term regulatory effects (the so-called trophic functions of neurones). Cell cultures could also be used to determine which properties are intrinsic to the nerve and which develop only after establishment of normal environmental influences. In contrast to the potential value of tissue culture in such studies, it is probably true to say that the technique does not offer many obvious advantages for the study of synaptic pharmacology other than the chance to make available otherwise inaccessible areas of the nervous system. Moreover, methods may be devised for growing functional synapses between two defined types of nerve cell, thus allowing pharmacological experiments to be made on a precisely defined system.

There are several possible approaches to the examination of synapses in tissue culture. First, the original organization of the tissue can be preserved by use of organ or explant cultures. Secondly, the experimental system can be simplified by use of regions containing known types of cells whose functions *in vivo* are well characterized. Examples are cultures of retina and of cerebellum. A variation of this approach is to separate and grow only specific cells, such as motor neurones or cerebellar Purkinje cells. A third approach is to use cells of clonal origin if clones with the desired properties can be established. Finally, an equally valid strategy is not to attempt any simplification other than tissue dissociation. An advantage of this last method is that it does not confine the experimenter to examining those cellular relationships known to exist, but offers the possibility of discovering unsuspected interactions.

During early work on explant cultures of the central nervous system, complex spontaneous activity could be recorded with extracellular electrodes (see review by Crain, 1976). As this activity was modified by agents such as Mg^{2+} known to suppress synaptic transmission, it was concluded that functioning synapses existed in these cultures. However, in the mass of cells in the bulk of any explant, it is hard to identify the cellular connections and consequently it is difficult to analyse the working of synaptic networks. Use of roller culture techniques (see Gähwiler, 1981) to provide near monolayer preparations preserves most of the original synaptic connections and appears to provide the best system. These cultures are also suitable for cell labelling by intracellular dye or horseradish peroxidase injection to allow subsequent histological tracing of the projections of identified cells (e.g. Calvet and Calvet, 1979; Gähwiler, 1981). None the less, such preparations generally allow study of preformed synapses only; they are useless for investigations of synaptic development.

For developmental studies, some degree of cell separation is necessary. Unfortunately, the lack of adequate markers for specific types of nerve cells and the lack of distinguishing morphological features of most nerve cells when grown in culture have so far precluded proper control over the range of choices offered to the neurones in culture. Many cell cultures of neural origin have complicated synaptic activity but even in the lowest density cultures connections between cells are difficult to trace. Therefore, a rigorous analysis of synapse formation has not been possible. The ideal situation would be where a pure sample of one type of neurone was confronted with another homogeneous group of cells of known origin under conditions in which the geographical relationships were controlled. This situation has not yet been reached, although it would not be impossible. The obvious strategy of using cells of clonal origin has been hampered by the fact that many neural cell lines cannot form synapses, but some progress has been made with primary cultures in which it is possible to recognize the different elements visually. Examples are peripheral neuromuscular or neuroeffector junctions where nerves make contact with completely different types of cells, and synapses between spinal cord cells and dorsal root ganglion neurones because the two cell types have characteristically different morphologies.

Synapses in Cultures of the Central Nervous System

Brain Tissue

Explants. Spontaneous potentials can be readily recorded with extra-

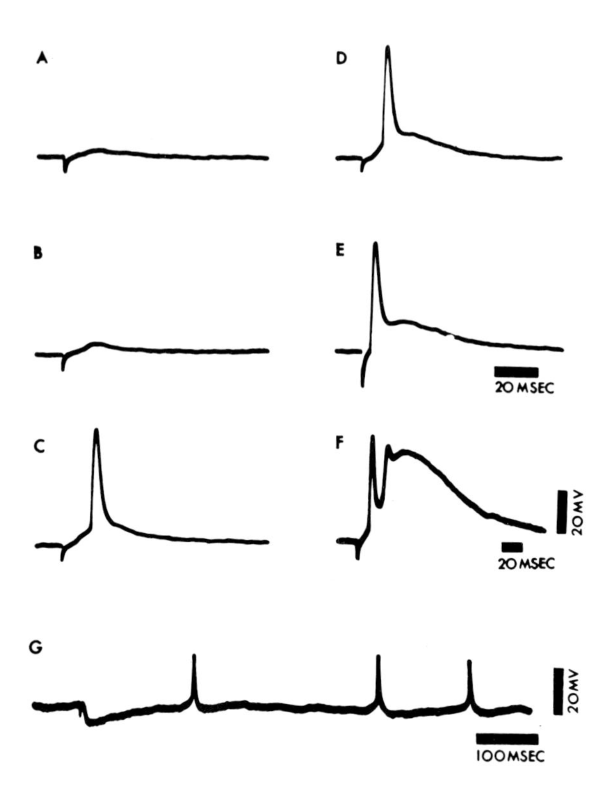

Figure 7.1: Excitatory and Inhibitory Postsynaptic Potentials Generated by Neurones in Explants of 18-day Fetal Mouse Hippocampus in Response to Presynaptic Impulses Initiated by Single Electric Stimuli to Other Neurones in Explant (24 days in vitro)

(A) Intracellular recording of an EPSP of small amplitude evoked by a weak stimulus (after introduction of strychnine at 3 X 10^{-6} M). As stimulus strength is progressively increased, the EPSP becomes larger (B), and then a spike potential appears during the rising phase of the EPSP (C), with decreasing latency after a larger stimulus (D). (E) Still stronger stimulus evokes an antidromic spike potential which occludes the synaptically-evoked spike, but a large EPSP continues after the end of the spike. (F) Depolarizing ('paroxysmal') potential with much larger amplitude (about 40 mV) and longer duration ($>$100 msec) than the usual EPSPs is evoked, at times, by stimulus of intermediate strength. (G) IPSP evoked in a neurone in another hippocampal explant (15 days *in vitro*), lasting about 100 msec and followed by repetitive spikes. Time and amplitude calibrations apply to all preceding records unless otherwise noted. From Zipser *et al*. (1973), with permission.

cellular electrodes from explant cultures established from many different brain regions, and equivalent intracellular recordings have also been made (e.g. Zipser *et al.*, 1973; Marshall *et al.*, 1980). Some of this spontaneous activity would appear to involve synaptic transmission as the level of activity can be greatly diminished by addition of high concentrations of Mg^{2+} (Gähwiler *et al.*, 1973). Both excitatory and inhibitory postsynaptic potentials can be recorded, and synaptic potentials can also be evoked by electrical stimulation of the presynaptic nerve cell body (Figure 7.1). This work has been extensively reviewed by Crain (1976) and as there has not been a great deal added since then it will only be considered briefly.

Although in a few cases, such as cerebellar Purkinje neurones, recordings can be made from known cell types, it is difficult with explant cultures to analyse the details of synaptic function. Apart from attempting to deduce which transmitter gives rise to a particular postsynaptic potential and determining the actions of drugs on overall activity, not a great deal can be established from studies on single explants. Examples of the different types of brain regions that have been used are the cerebellum (Schlapfer *et al.*, 1972; Gähwiler *et al.*, 1973; Leiman and Seil, 1973; Wojtowicz *et al.*, 1978; Marshall *et al.*, 1980; Gähwiler and Dreifuss, 1982), cerebral cortex (Calvet, 1974; Leiman *et al.*, 1975), hippocampus (Zipser *et al.*, 1973) and hypothalamus (Gähwiler and Dreifuss, 1979). Not a great deal of pharmacology has been performed on these explant preparations but there is some evidence that they retain their expected sensitivity to drugs. For example, the inhibitory responses of deep cerebellar nuclei neurones evoked by cortical stimulation were reduced by bicuculline, picrotoxin and strychnine in concentrations that blocked iontophoretic responses to GABA and glycine (Wojtowicz *et al.*, 1978).

A different, and perhaps more promising use is to place explants of different origin in close proximity, and to investigate the extent and nature of functional synaptic connections that are formed. Mostly this work has involved combination of elements of the central and peripheral nervous systems; it is described in a later section (p. 183).

Dissociated cell cultures. There is abundant evidence for the formation of functional synapses in dissociated cell cultures of the mammalian central nervous system. Most authors report that about half of the cells sampled have spontaneous postsynaptic potentials. Despite the availability of large numbers of synapses, the random and disorganized nature of the cultures has prevented them from being used frequently for detailed studies.

In cultures derived from whole brain of fetal rats, Godfrey *et al.* (1975) found that the number of cells with excitatory postsynaptic potentials was about the same as that with inhibitory postsynaptic potentials. However, only two cells of the 67 tested had both types of activity. The reversal potentials were around –50 mV for the inhibitory potentials and more positive than –10 mV for the excitatory potentials. Godfrey *et al.* (1975) reported that electrical excitability was found only in cultures that had been treated with the antimitotic agent fluorodeoxyuridine. However, such treatment is not always necessary (Dichter, 1978).

Other authors have described synapses in cultures established from particular brain regions. The region of choice has been the cerebellum because it is composed of relatively few and well-characterized cell types. In their pioneering experiments Nelson and Peacock (1973) discovered that excitatory and inhibitory postsynaptic potentials could be recorded from many cells in cultures of fetal mouse cerebellum. Stimulation of one cell could sometimes evoke an excitatory postsynaptic potential in a neighbouring cell and the synaptic potential could be large enough to initiate an action potential. One problem encountered in this study was that the isolated nerve-like cells tended to degenerate with time in culture. More recently, Nelson and his co-workers have described an improved culture method which allows prolonged survival of cerebellar Purkinje neurones (Moonen *et al.*, 1982). This technique involves dissociating the starting tissue into small clumps of cells rather than into single cells. These 'microexplants' spread in culture to form virtual monolayers that contain many well-differentiated neurones with morphological characteristics of Purkinje cells. These neurones often have spontaneous postsynaptic potentials. The inhibitory potentials are probably caused by release of GABA, whereas the excitatory potentials may be mediated by glutamate (Macdonald *et al.*, 1982). Other possible transmitter candidates, glycine, β-alanine, taurine and aspartate had little or no effect on the cultured cerebellar neurones.

Other regions of the brain that have been grown successfully as cell cultures include the cerebral cortex (Dichter, 1978, 1980) and the hippocampus (Peacock, 1979; Peacock *et al.*, 1979). In both types of culture, most cells were found to have excitatory or inhibitory postsynaptic potentials. The characteristics of the synaptic potentials were not examined in detail, although once again GABA would appear to be the inhibitory transmitter. The retina is another example of a specialized area of the nervous system that can be grown in culture. Dissociated and reaggregated cultures established from chick embryo retina do form synapses, some of which are probably cholinergic (Vogel *et al.*, 1976). However, it is not certain whether the neurones in these cultures are suitable for electrophysiological studies.

Spinal Cord

Relatively little work has been done on synaptic transmission in explant cultures of spinal cord. Results from earlier studies have been discussed by Crain (1976). More recently it has been shown that morphine and other opiates depress electrical activity of the dorsal, but not the

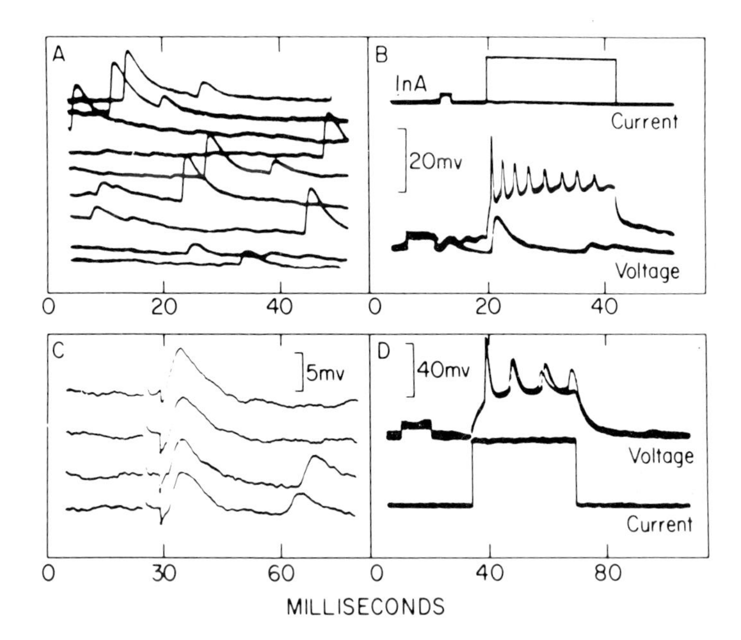

Figure 7.2: Spontaneous and Evoked Postsynaptic Excitatory Potentials in Fetal Mouse Spinal Cord Cell Cultures

(A) Successive sweeps of the oscilloscope showing spontaneous postsynaptic potentials. (B) Directly elicited action potentials from the cell in (A). Two oscilloscope sweeps are superimposed, one with and one without a depolarizing current pulse. Action potentials are evoked by the depolarizing pulse and excitatory postsynaptic potentials are seen during the other sweep. Voltage calibration in (B) also holds for (A). (C) Evoked excitatory postsynaptic potentials recorded in a spinal cord cell in response to stimulation of another cell. (D) Directly elicited action potentials from the cell in (C). From Peacock *et al.* (1973b), with permission.

ventral, regions of spinal cord explants following stimulation of attached dorsal root ganglia (Crain *et al.*, 1977). Similar effects were found with several opioid peptides (Crain *et al.*, 1978). In all cases the depressant effect was antagonized by naloxone. Enhancement of transmitter release by 4-aminopyridine reversed the opiate-induced depression, although 4-aminopyridine had no effect on binding of opiates (Crain *et al.*, 1982b). 5-Hydroxytryptamine can produce an opiate-like depression of synaptic transmission in spinal cord explant cultures (Crain *et al.*, 1982a). This effect did not appear to be mediated through 5HT receptors or opiate receptors. However, in cultures made tolerant to the actions of opiates by chronic exposure to morphine, 5HT was also less effective.

The first demonstration that synapses could form between dissociated spinal cord cells was provided by Fischbach (1970). Both spontaneous and evoked synaptic potentials were recorded. When cultured together with cells from dorsal root ganglia, almost 70 per cent of spinal cord neurones from fetal mice had spontaneous postsynaptic potentials while only about 10 per cent of the dorsal root ganglia nerves gave evidence of synaptic potentials (Peacock *et al.*, 1973b). Stimulation of one spinal cord cell could often evoke an excitatory postsynaptic potential in a nearby spinal cord cell (Figure 7.2). Some dorsal root nerves formed functional synaptic contacts with spinal cord cells. In these cultures most synaptic activity was excitatory as judged by the presence of depolarizing events at the resting membrane potential. It is possible that inhibitory responses were present but could not be distinguished easily because the resting potential (–40 to –50 mV) was too close to the null potential for the inhibitory responses.

The synaptic interactions in cultures of fetal mouse spinal cord have subsequently been studied in more detail (Nelson *et al.*, 1977; Ransom *et al.*, 1977 a, b). Because of the different morphologies of spinal cord and dorsal root ganglion nerves in culture, it proved possible to study the synaptic connections of identified pairs of cells. Given the difficulty of visualizing the contacts made by fine neurites, the rate of success was encouragingly high. Half the spinal cord to spinal cord cell pairs were functionally coupled; most pairs had excitatory potentials, although 8 per cent had inhibitory postsynaptic potentials. Dorsal root ganglion to spinal cord cell pairs had only excitatory postsynaptic potentials, and again almost 50 per cent of the pairs examined were synaptically connected. However, no evidence of functional connections between dorsal root ganglion nerves was found (Ransom *et al.*, 1977b). The size of the excitatory potentials evoked by stimulation of a presynaptic spinal cord cell was about three times the amplitude of those evoked by stimulation of dorsal root ganglion cells (6.2 mV compared to 1.9 mV). Analysis of the variance of the amplitude of the excitatory postsynaptic potentials evoked by spinal cord cell and dorsal root ganglion cell stimulation revealed that the average quantal size was 200-250 μV in both cases. Hence, the difference in the size of the postsynaptic potential must reflect a larger release of transmitter from spinal cord cells than from dorsal root ganglion cells. The average quantal contents were estimated to be about 35 and 10, respectively (Ransom *et al.*, 1977b). The small size calculated for the unitary potentials is somewhat surprising in view of the high input resistance of the cultured spinal cord cells. Direct measurements of spontaneous miniature excitatory potentials were not possible because of the high noise levels of the high resistance recording electrodes.

The reversal potential for the spinal cord to spinal cord excitatory postsynaptic potentials was determined to be about +20 mV. Because the recording electrodes were not able to pass enough current, the cells could not be reversed in polarity and the null potential had to be estimated by extrapolation. This may be unreliable for several reasons. At hyperpolarized membrane potentials the large excitatory potentials may be reduced in amplitude because of non-linear summation, at more negative potentials rectification of the membrane's voltage response may distort the excitatory potentials, and there is also the possibility that there may be differences in the open times of ion channels at different membrane potentials. Nevertheless, similar measurements with iontophoretically-elicited glutamate responses revealed consistent differences in the null potentials of the excitatory postsynaptic potentials and the glutamate responses (Ransom *et al.*, 1977a). This suggests that different conductance changes mediate the glutamate response and the synaptic potential, i.e. glutamate is not the excitatory transmitter. Voltage clamp techniques do not yet appear to have been used extensively as these should help to eliminate some of the problems and give more reliable estimates.

In contrast to the difficulty in examining the voltage dependence of excitatory postsynaptic potentials it has proved easier to study inhibitory postsynaptic potentials. When using K^+ acetate electrodes to avoid changes in cellular Cl^- concentration, these potentials reverse in direction to become depolarizing after the membrane potentials are made more negative than about –80 mV (Ransom *et al.*, 1977a, b). Responses to GABA and glycine have similar null potentials around –80 mV, so that the inhibitory transmitter at these sites has not yet been unambiguously identified. Maintenance of the cultures in a medium such as Dulbecco's Minimum Essential Medium, which contains 0.4 mM glycine, decreases the number of cells found to have inhibitory postsynaptic potentials (Nelson *et al.*, 1977). Cells in such cultures do not respond to iontophoretic application of glycine although their sensitivity to GABA is normal. Addition of GABA to medium without glycine caused the reverse effect, i.e. the cells were sensitive to iontophoretically applied glycine but not to GABA. Long-term exposure to GABA also decreased the number of excitatory potentials as well as the number of inhibitory potentials. These results were interpreted on the basis of glycine being the common inhibitory transmitter responsible for inhibitory postsynaptic potentials and GABA being able to act presynaptically to depress the release of both excitatory and inhibitory transmitters (Nelson *et al.*, 1977). However, some of the inhibitory postsynaptic potentials in these cultures are probably mediated by GABA rather than glycine because they are augmented by phenobarbitone, which has little action on glycine responses (Barker and McBurney, 1979b).

The basic properties of postsynaptic potentials have also been studied in cell cultures established from chick embryo spinal cord (Fischbach and Dichter, 1974). Most cells had spontaneous synaptic potentials which could be either inhibitory or excitatory. The reversal potential for the excitatory postsynaptic potentials was estimated to be close to 0 mV and for inhibitory postsynaptic potentials to be between –45 and –65 mV (Choi *et al.*, 1981b). In contrast to the work with mouse spinal cord cells in culture, miniature excitatory postsynaptic potentials with amplitudes of around 0.5 mV could be recorded in the presence of tetrodotoxin to block transmitter release induced by action potentials (Fischbach and Dichter, 1974). As evoked excitatory postsynaptic potentials were 15-20 mV in amplitude, the quantal content would be 30-40. The excitatory transmitter at these synapses has not yet been identified.

The inhibitory postsynaptic potentials in cultured chick spinal cord cells, as in mouse spinal cord, are mediated by an increase in Cl^- conductance (Fischbach and Dichter, 1974; Choi *et al.*, 1981b). On the basis of selective antagonism by strychnine or by bicuculline or picrotoxin, it was estimated that glycine was the inhibitory transmitter in about 70 per cent of the synapses and GABA in the remaining 30 per cent. However, it was noted that taurine and β-alanine also produced strychnine-sensitive inhibitory potentials in these cells so that glycine may not be the only alternative inhibitory transmitter to

GABA (Choi *et al.*, 1981b).

In contrast to the findings with fetal mouse spinal cord-dorsal root cultures (Ransom *et al.*, 1977b), there was evidence that nearly 20 per cent of chick embryo dorsal root ganglion cells had spontaneous inhibitory postsynaptic potentials. Since there were no synaptic potentials in pure dorsal root ganglion cultures, it implies that some of the dorsal root cells are innervated by spinal cord neurones. As the inhibitory potentials in these dorsal root ganglion nerves were blocked by bicuculline but not by strychnine, and were augmented by the benzodiazepine compound, chlordiazepoxide, it was suggested that GABA was the transmitter (Choi *et al.*, 1981b).

Attempts have also been made to culture selectively motor neurones from chick embryo spinal cord (Berg and Fischbach, 1978; Masuko *et al.*, 1979). Some of the large diameter (10-18 μm) nerve-like cells that were cultured (Berg and Fischbach, 1978) formed functional neuromuscular junctions when grown in the presence of skeletal muscle. However, these dissociation methods did not give pure populations of motor nerves. For example, the majority of the large nerve cells did not form neuromuscular junctions. Also both excitatory and inhibitory postsynaptic potentials could be recorded from the large diameter nerve cells. However, the cells had little sensitivity to iontophoretically-applied acetylcholine which would be expected to be released from the terminals of motor neurones (Berg and Fischbach, 1978). A more promising approach is to culture spinal cord fragments from a stage when only the motor neurones have withdrawn from the cell cycle. Treatment with antimitotic drugs will then kill other cell types. Cells prepared in this way do have high concentrations of choline acetyltransferase and they can form functional neuromuscular junctions with skeletal muscle cells (Masuko *et al.*, 1979), but their ability to form synapses with one another or with other neurones has not been examined.

Synapses in Cultures of the Peripheral Nervous System

Explants

Explants of peripheral autonomic or sensory ganglia have been maintained in culture usually for biochemical and morphological experiments, or to test the effects of nerve growth factor (for review, see Thoenen and Barde, 1980). In a few instances, explants have been used in order to study the specificity of synapse formation. For example, Olson and Bunge (1973) grew fetal rat superior cervical ganglia in combination with spinal cord or cerebral cortex explants, and looked for evidence of synapses with electron microscopy. When two superior cervical ganglia were grown together there was almost no sign of synapse formation, but in combined ganglia-spinal cord cultures there was abundant innervation of ganglion cells by nerves originating from the spinal cord explants. There were few synapses on superior cervical ganglion cells when they were cultured with cerebral cortex explants, and there was no evidence of adrenergic innervation of the cortex by ganglion cells. In other studies (reviewed by Crain, 1976), dorsal root ganglion explants were found to form functional synapses on medullary tissue while ignoring explants of midbrain that were present in the same cultures. Hence, some nerve cells in culture would appear to retain the ability to be selective in their choice of target cell. Whether these cells also retain their normal specificity remains to be determined.

Dissociated Cell Cultures

Sympathetic neurones. Most work has been done on sympathetic nerves obtained from spinal sympathetic chains or, more commonly, superior cervical ganglia. Despite the presumed adrenergic phenotype of these cells, culture studies have revealed that some of them can synthesize acetylcholine and form cholinergic synapses. The culture system has thus provided a model of neural plasticity and may be useful in the study of what determines the choice of transmitter in postganglionic sympathetic nerves *in vivo*.

Although Chalazonitis *et al.* (1974) reported that sympathetic neurones from chick embryos did not form synapses, other workers have found that mammalian sympathetic nerves do form synapses in cell culture. Some of the neuroeffector junctions between rat superior cervical ganglion cells and cultured heart cells have noradrenaline as their transmitter (Furshpan *et al.*, 1976), but the synapses between the nerves themselves are always cholinergic (O'Lague *et al.*, 1974; Johnson *et al.*, 1976) or electrotonic (Higgins and Burton, 1982). These sympathetic nerves can also form cholinergic junctions with skeletal and cardiac muscle in cultures (Nurse and O'Lague, 1975; Furshpan *et al.*, 1976; see also below).

Most of the studies on the properties of sympathetic nerves in culture have concentrated on the mechanisms controlling the choice of transmitter (see review by Burton and Bunge, 1981). The change from adrenergic to cholinergic properties is not simply because the cultured cells have been removed from immature animals; superior cervical ganglion cells from 2-month-old rats also form cholinergic synapses in culture although these cells would be adrenergic *in vivo* (Wakshull *et al.*, 1979a, b). Biochemical and histochemical studies have demonstrated that superior cervical ganglion nerves in culture can synthesize and store both noradrenaline and acetylcholine (Iacovitti *et al.*, 1981). The relative amounts of each transmitter can be varied by culture conditions. For example, growing sympathetic nerves in the presence of non-neural cells or in media conditioned by cardiac or skeletal muscle increases the amount of acetylcholine synthesised. Recently, it has been found that growth in a defined medium without serum maintains to a certain extent the adrenergic properties of the cells while supressing cholinergic development (Iacovitti *et al.*, 1982). Synapses formed between sympathetic nerves in such cultures have electrotonic, rather than chemical, transmission (Higgins and Burton, 1982).

The properties of synaptic transmission between superior cervical ganglion nerves in culture have been extensively investigated (O'Lague *et al.*, 1974, 1978 a, c; Ko *et al.*, 1976b; Wakshull *et al.*, 1979a, b; Higgins and Burton, 1982). Mostly, postsynaptic potentials recorded from these nerve cells were depolarizing and the amplitudes of evoked excitatory postsynaptic potentials were increased by high Ca^{2+} and decreased by high Mg^{2+} solutions (Ko *et al.*, 1976a; O'Lague *et al.*, 1974, 1978a). However, in a recent study, Higgins and Burton (1982) found that superior cervical ganglion nerves grown in defined medium formed electrotonic synapses. In these cells the spread of signals from one cell to another was not affected by changes in the concentrations of Ca^{2+} and Mg^{2+} or by various antagonist drugs (atropine, hexamethonium, propranolol, phentolamine, bicuculline and chlorpromazine).

In cultures of superior cervical ganglion cells in which chemical transmission did occur (Ko *et al.*, 1976a; O'Lague *et al.*, 1978a), miniature excitatory postsynaptic potentials could be recorded in the presence of tetrodotoxin. The largest of these was about 2 mV but there was a wide range of amplitudes and rise times, suggesting that the events were occurring at different sites. Hence, it is difficult to calculate the quantal content simply by the ratio of mean amplitude of the excitatory postsynaptic potentials to that of the miniature potentials. By use of the method of failures (see Hubbard *et al.*, 1969), quantal content was estimated to be from 1 to 4 (O'Lague *et al.*, 1978a).

The cholinergic nature of the transmission process in these cultures is supported by pharmacological evidence. The excitatory postsynaptic potentials are reduced in amplitude by drugs known to block nicotinic cholinoceptors of autonomic ganglia, e.g. hexamethonium, mecamylamine, tetraethylammonium and tubocurarine (Ko *et al.*, 1976a, b; O'Lague *et al.*, 1974). The postsynaptic potentials are not blocked by other receptor antagonists such as atropine, α-bungarotoxin, propranolol, dibenamine, phenoxybenzamine and phentolamine. Additionally, iontophoretic application of acetylcholine mimics the excitatory postsynaptic potentials and has the same sensitivity to blocking drugs (Ko *et al.*, 1976a; O'Lague *et al.*, 1978c; Wakshull *et al.*, 1979a). Localized areas of very high sensitivity to acetylcholine (greater than 1,000 mV/nC) could be found both on cell bodies and on processes (O'Lague *et al.*, 1978c). It is possible that these areas correspond to points of synaptic contact.

The superior cervical ganglion nerves in cell cultures are able to accept cholinergic innervation from other sources. When cells from dissociated fetal rat cervical ganglia were cultured with explants of rat spinal cord, synapses formed which could be activated by electrical stimulation of the spinal cord explant (Ko *et al.*, 1976a). The resulting excitatory postsynaptic potentials were decreased in size by increases in Mg^{2+} or decreases in Ca^{2+} concentrations. Transmission from spinal cord to superior cervical ganglion cell was blocked by hexamethonium, mecamylamine and tubocurarine but not by α-bungarotoxin, atropine, phentolamine or phenoxybenzamine.

Parasympathetic Neurones. In comparison to the amount of work that has been done on cultured cells from the sympathetic nervous system, there has been little on parasympathetic nerves. Partly this has been because sympathetic cells can be made to differentiate in culture by nerve growth factor, whereas no comparable trophic factor for parasympathetic cells has yet been characterized.

Parasympathetic nerves from the ciliary ganglia of chick embryos will grow in cell culture, but many of the neurones die within the first week of culture (Helland *et al.*, 1976; Nishi and Berg, 1977). Although this appears to parallel the normal pattern of development *in vivo*, it limits the usefulness of these cultures for studies on isolated parasympathetic nerves. The loss of neurones can be prevented by culturing ciliary ganglion cells in the presence of skeletal muscle or in media conditioned by cardiac or skeletal muscle cells (Nishi and Berg, 1977, 1979). The survival, although not the differentiation, of ciliary ganglion cells in culture can also be promoted by medium containing 25 mM K^+ (Bennett and White, 1979). This effect is similar to that first reported for cells from dorsal root ganglia (Scott and Fisher, 1970).

Ciliary ganglion cells were shown to form functional neuromuscular junctions when grown with skeletal muscle (Hooisma *et al.*, 1975; Nishi and Berg 1977; see p. 191) but it was not clear if they could form synapses between themselves. Crean *et al.* (1982) found morphological evidence for synapses but no sign of chemical transmission. Recently, spontaneous depolarizing postsynaptic potentials have been recorded from ciliary ganglion nerves in cell culture (Margiotta and Berg, 1982). The size of these potentials ranged from 0.5 to 15 mV and they could sometimes trigger action potentials. Although no attempts were made to evoke postsynaptic potentials by presynaptic stimulation, the spontaneous potentials were abolished by tetrodotoxin, suggesting that they were associated with spontaneously firing action potentials. The incidence of synapses was greater in combined nerve-muscle cultures, where about 90 per cent of neurones had spontaneous synaptic input, than in nerve-only cultures, where about 40 per cent of cells showed signs of synaptic activity (Margiotta and Berg, 1982). In a related study (Crean *et al.*, 1982), some evidence for functional synapses was obtained in co-cultures, and the number of nerves sensitive to acetylcholine was found to be highest in co-cultures. Whether conditioned media would have increased synapse formation in nerve-only cultures was not determined, although it does not seem likely (Crean *et al.*, 1982).

The receptors mediating the spontaneous depolarizing potentials in cultured ciliary ganglion cells would be expected to be classical ganglionic nicotinic cholinoceptors. Their pharmacology has not been studied in detail, but they are known to be blocked by tubocurarine and by a toxin from *Bungarus multicinctus* venom that has ganglion blocking properties. Alpha-bungarotoxin itself, which acts selectively on nicotinic receptors of skeletal muscle, has no effect on the excitatory postsynaptic potentials of the ciliary ganglion cells in culture (Margiotta and Berg, 1982).

Other peripheral neurones. There is little evidence for synapse formation between isolated sensory nerves from dorsal root ganglia (Peacock *et al.*, 1973b; Ransom *et al.*, 1977b; Higgins and Burton, 1982). However, cells from nodose ganglia of newborn rats can form synapses in culture (Baccaglini and Cooper, 1982a). In virtually pure nerve cultures, up to 50 per cent of cells tested had spontaneous excitatory postsynaptic potentials and it was possible to stimulate one cell and record an evoked postsynaptic potential in a neighbouring nerve. The responses were mimicked by acetylcholine and reduced by the ganglion blocking drugs, hexamethonium, pempidine, chloroisondamine and tubocurarine. Excitatory postsynaptic potentials could also be reduced by atropine, although the concentrations required (0.1-0.5 mM) would be expected to affect nicotinic as well as muscarinic cholinoceptors (Baccaglini and Cooper 1982a).

Although it is not clear whether these presumed sensory nerves can form cholinergic synapses *in vivo*, cultures of dissociated nodose ganglia provide another system for the study of the control of synapse formation. In contrast to results with sympathetic and parasympathetic nerves in cultures, co-culture of nodose ganglion cells with skeletal or cardiac muscle resulted in a decrease in the number of synapses formed between nerves (Baccaglini and Cooper, 1982b). There were also many fewer acetylcholine-sensitive cells in nerve-muscle cultures. These effects were related to the presence of muscle cells in the culture from an early stage: muscle-conditioned media had no effect on synapse formation, and neither did co-culture if it was begun one week after the initial plating of the neurones.

Neuromuscular Junctions in Culture

Skeletal Muscle

Organ and explant cultures. Coincidentally to his observations on developing nerve fibres in explants of frog embryo neural tube, Harrison (1907, 1910) noted that contracting skeletal muscle fibres could develop in culture if portions of the myotome were left attached to the spinal cord explant. Forty years later, Szepsenwol (1946, 1947) grew explant cultures of spinal cord and muscle from chick embryos. Skeletal muscle cultured alone was spontaneously active giving random contractions, but when the muscle was grown together with spinal cord tissue the muscular activity was regular. Contractions in nerve-muscle cultures, but not in cultures of muscle alone, were abolished by tubocurarine. These findings are consistent with there being functional neuromuscular transmission, but there was no direct morphological or physiological evidence. This required application of electron microscopical and electrophysiological techniques.

As reviewed by Shimada and Fischman (1973) and Crain (1976), the first type of culture in which the development of neuromuscular junctions was studied extensively consisted of entire cross-sections of spinal cord-somite regions from embryos. In these cultures, the

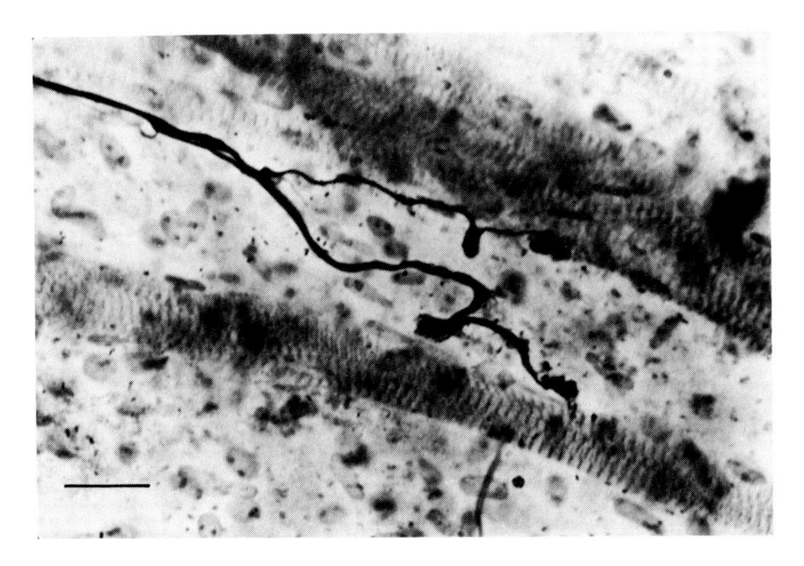

Figure 7.3: Nerve-Muscle Contacts Revealed by Silver Staining of a 14-day Cell Culture of Chick Embryo Skeletal Muscle and Spinal Cord
Calibration = 20 μm. From Shimada *et al.* (1969), with permission.

normal spatial relationships are maintained and the tissues continue to differentiate *in vitro*. Stimulation of the spinal cord results in contraction of the skeletal muscle and this is blocked by tubocurarine. Physostigmine caused repetitive muscle contractions, in response to one stimulus to the nerves (Crain, 1970). Corresponding morphological investigations revealed the presence of mature neuromuscular junctions.

Although this type of 'organotypic' culture preserves many of the normal features of nerve and muscle, it is not ideal for studying the interactions between identified cells and it is always open to the possibility that the neuromuscular junctions found in culture may have already been forming at the time of explantation. *De novo* formation of neuromuscular junctions is more convincingly demonstrated by co-culture of separate nerve and muscle explants. This approach also permits the study of selectivity between different types of nerves and between different species. When small fragments of spinal cord and muscle from fetal rats or mice are grown 0.5-1 mm apart, functional neuromuscular transmission can be demonstrated after two weeks of co-culture (Crain, 1970). Extracellular recording from spinal cord and muscle cells shows that stimulation of the cord generates a pattern of activity that is followed by electrical and mechanical responses in the muscle. Tubocurarine blocks the muscle response without affecting the electrical activity of the spinal cord. Strychnine augments nervous activity and this results in repetitive responses in the muscle explant. The ability to form functional connections in culture is not restricted to nerves and muscles taken from the same species: similar interactions occur with rat spinal cord and mouse muscle, and vice versa, and with mouse spinal cord and human muscle (Crain *et al*., 1970).

Dissociated cell cultures. The first suggestion that nerves in dissociated cell cultures could form specialized contacts with muscle cells appears to be in the paper by Shimada (1968). Silver staining of nerve-muscle cultures revealed bulbous thickening of the neurites where they were in contact with myotubes (Figure 7.3). Subsequently, electron microscopical evidence for neuromuscular junctions was provided (Shimada *et al*., 1969). It was soon confirmed that nerve-muscle contacts established between dissociated cells in culture could be physiologically competent (Fischbach, 1970; Kano and Shimada, 1971; Robbins and Yonezawa, 1971). Transmission was revealed by the presence of spontaneous depolarizing potentials in myotubes that were contacted by nerve cells. Similar potentials could be evoked by stimulation of the nerve and these endplate potentials could be large enough to trigger muscle action potentials (see Figure 4.4). As at the neuromuscular junction of mature animals, the amplitude of the endplate potential increased with hyperpolarization of the membrane potential. The reversal potential was determined to be about 0 mV (Fischbach, 1970, 1972). The endplate potential could be mimicked by application of acetylcholine (Kano and Shimada, 1971) and both responses were blocked by tubocurarine (Kano and Shimada, 1971; Fischbach, 1972). In contrast to the effects at neuromuscular junctions, the time course and amplitude of the endplate potentials in these first nerve-muscle cultures were unaffected by the anticholinesterase physostigmine (Kano and Shimada, 1971; Fischbach, 1972). Acetylcholine responses in similar cultures were also shown to be unaffected by neostigmine and dyflos (Harvey and Dryden, 1974c). Although these results imply that acetylcholinesterase does not develop at neuromuscular junctions formed in cell culture, this may not be so. In later experiments, Fischbach's group demonstrated that functional acetylcholinesterase very quickly accumulated at neuromuscular junctions in cultures that were not dissociated with proteolytic enzymes or exposed to antimitotic drugs (Frank and Fischbach, 1979; Rubin *et al*., 1979).

In cell cultures, neuromuscular junctions can be formed by a variety of nerves from several species as well as by spinal cord cells from chick embryos. Rat spinal cord has been shown to innervate rat myotubes (Robbins and Yonezawa, 1971), and cells from rat superior cervical ganglia can also form functional contacts with rat skeletal muscle cells in culture (Nurse and O'Lague, 1975; Nurse, 1981a, b). Myotubes of the rat clonal line L_6 can be innervated by rat spinal cord (Kidokoro and Heinemann, 1974). Although the mouse neuroblastoma cells C1A do influence the distribution of acetylcholine receptors on L_6 myotubes, they do not make functionally competent junctions (Harris *et al.*, 1971). However, a neuroblastoma × glioma hybrid line, NG108-15, can form functional neuromuscular junctions both with skeletal muscle cells in primary cultures (Nelson *et al.*, 1976; Puro and Nirenberg, 1976) and with a myogenic cell line, G8 (Christian *et al.*, 1977). Cells from the medulla oblongata (Obata 1977) and ciliary ganglia (Hooisma *et al.*, 1975; Betz, 1976a, b; Obata, 1977) of chick embryos can form neuromuscular junctions, whereas cells from cerebellum, cerebral cortex and dorsal root ganglia do not (Fischbach, 1972; Obata, 1977). Dissociated neural tube and myotome from *Xenopus* embryos can also form functional nerve-muscle connections in culture (Kidokoro *et al.*, 1980) and *Xenopus* nerves can innervate L_6 myotubes (Kidokoro and Yeh, 1981). Although sensory nerves from dorsal root ganglia do not make neuromuscular junctions *in vitro* (Fischbach, 1972; Obata, 1977; Cohen and Weldon, 1980), presumptive sensory neurones from the nodose ganglia of rat vagus nerves can innervate skeletal muscle in culture (Baccaglini and Cooper, 1982b).

In terms of detailed examination of the early stages of the formation of neuromuscular contacts, most experiments have been carried out on chicken or *Xenopus* embryo nerve and muscle co-cultures. The time course of junction formation differs considerably in different species. In *Xenopus* cultures functional junctions form very rapidly and morphologically-mature neuromuscular junctions are seen within a few hours (Peng and Nakajima, 1978), but only after several days in rat or chicken cultures (Nakajima *et al.*, 1980). The different stages of junction formation are summarized diagrammatically in Figure 7.4. It is apparent that growth cones of motor nerves can release acetylcholine although they contain few of the structural specializations of mature terminals (Cohen, 1980; Kidokoro and Yeh, 1982).

Additionally, the efficiency with which nerves in culture form neuromuscular junctions appears to vary with species and with the state of dissociation of the neural tissue. Junction formation is a

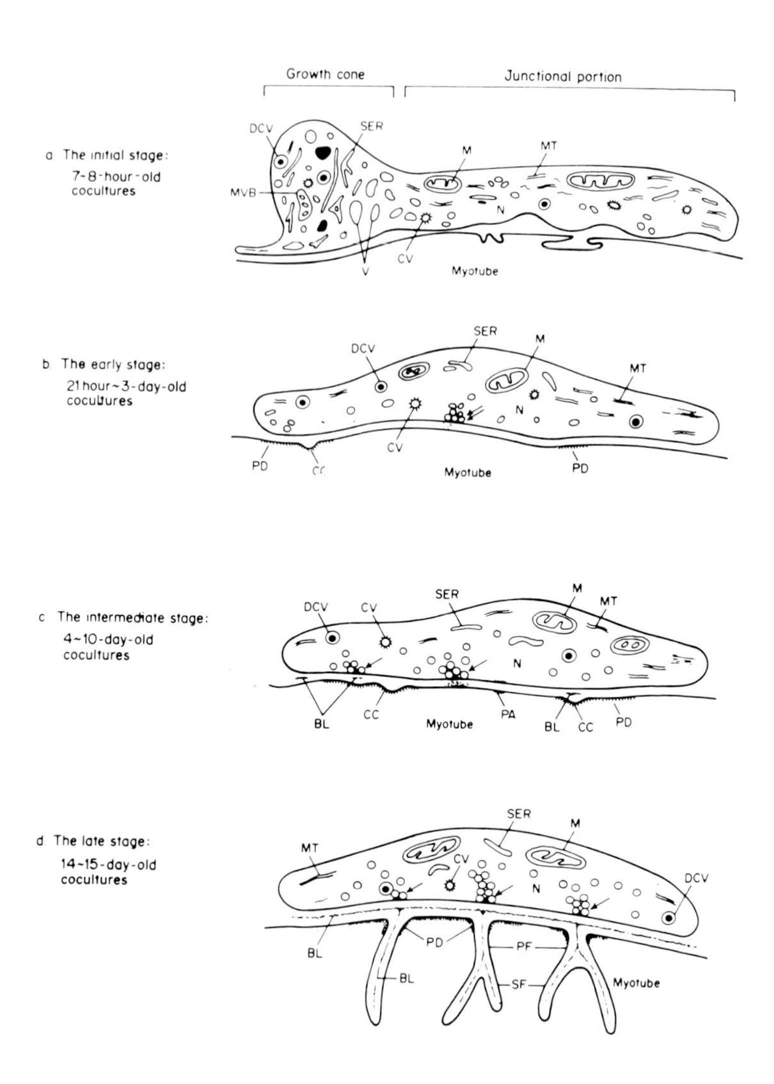

Figure 7.4: Schematic Diagram of Functional Neuromuscular Junctions Observed in Rat Cell Cultures

Arrows, active zone structure; double arrows, vesicle accumulation; BL, basal lamina; CC, coated caveolae; CV, coated vesicles; DCV, dense-cored vesicles; M, mitochondria; MT, microtubules; MVB, multivesicular body; N, nerve endings; PA, punctum adherens; PD, postjunctional densities; ER, smooth endoplasmic reticulum; PF, primary foldings; SF, secondary foldings; V, vacuoles. From Nakajima *et al*. (1980), with permission.

difficult parameter to quantify because of the possibilities of multiple innervation and of 'silent' release sites, i.e. where the nerve releases acetylcholine at a region where the density of cholinoceptors is too low to allow a response to be measured (see Cohen, 1980). When dissociated spinal cord cells are added to muscle cultures, only about 5-10 per cent of the muscle cells have evidence of neuromuscular junctions (Fischbach, 1972) Higher incidences have been reported with cells of the neuroblastoma X glioma hybrid clone NG108-15 (Puro and Nirenberg, 1976) and with explants of spinal cord (Cohen and Fischbach, 1977) or ciliary ganglia (Hooisma *et al.*, 1975). The extent of innervation can be modified by chronic depolarization or electrical stimulation (Fishman and Nelson, 1981; Magchielse and Meeter, 1982).

Since the establishment of functional neuromuscular transmission appears to be a major factor controlling the properties of muscle cells *in vivo*, it is tempting to use the culture system to investigate neurotrophic interactions. The effects that can be studied include the redistribution of acetylcholine receptors, the numbers of acetylcholine receptors and the nature of the ion channels that contribute to the muscle action potential. Because of the ready accessibility and good visibility of neuromuscular junctions in cell cultures, and because of the ease with which long-term environmental changes can be made in such cultures, the culture approach has much to recommend itself.

In vivo, acetylcholine sensitivity is restricted to the area of muscle membrane at the neuromuscular junction. Before innervation takes place, sensitivity is widespread and denervation leads to the reappearance of widespread sensitivity to acetylcholine. The mechanism that controls receptor localization is not known but, in general, there are two possibilities. The first is that nerve-induced muscle activity is responsible, the second is that motor nerves release a 'neurotrophic' chemical which acts on the muscle (see reviews by Gutmann, 1976; Fambrough, 1979; Dennis, 1981). Use of nerve-muscle cultures allows these hypotheses to be tested and also provides the opportunity to study the mechanism whereby receptor distribution may be altered.

Kano and Shimada (1971) reported that innervated chick embryo myotubes had high sensitivity only at nerve-muscle contacts, whereas non-innervated cells had widespread sensitivity. These experiments were performed on rather high density cultures and the acetylcholine responses were very slow in onset, suggesting that the drug had to pass some diffusion barriers before reaching the receptors. In an extensive study on low density cultures which had been prepared in order to reduce the contamination by fibroblasts, Fischbach and Cohen

(1973) also found that acetylcholine sensitivity was high at points of nerve-muscle contacts. However, the extrajunctional sensitivity remained as high as on non-innervated fibres. As discussed in Chapter 4, non-innervated myotubes can also have localized areas of high receptor density, or 'hot spots' (Vogel *et al.*, 1972; Fischbach and Cohen, 1973; Hartzell and Fambrough, 1973; Sytkowski *et al.*, 1973). These findings suggest a different form of nerve-muscle trophic interaction, i.e. the muscle attracts the nerve to pre-formed areas of high receptor density, rather than vice versa. Subsequent results, however, have demonstrated that this is not so. It is possible to stimulate small portions of neurites locally with an extracellular electrode if action potential propagation is blocked by tetrodotoxin. With this method the distribution of acetylcholine sensitivity of myotubes can be correlated with release sites. At nerve-muscle contacts that are only a few hours old, acetylcholine sensitivity at release sites is the same as at extrajunctional sites (Cohen, 1980). With time, high acetylcholine sensitivity develops at the release sites (Cohen and Fischbach, 1977; Frank and Fischbach, 1979). Nerves, therefore, do not seek out existing 'hot spots'.

Neuromuscular junctions with high acetylcholine sensitivity could result either from a redistribution of existing receptors from elsewhere in the membrane or from preferential insertion of receptors at the junction. There is evidence that both mechanisms may occur, although perhaps to different extents in different species. Work on chick embryo muscle cells indicates that there is little or no decrease in extrajunctional sensitivity after innervation. In fact, innervated myotubes may have generally higher sensitivity (Betz and Osborne, 1977; Cohen and Fischbach, 1977). However, labelling of receptors in *Xenopus* muscle cells with fluorescent derivatives of α-bungarotoxin reveals redistribution of receptors into clusters at nerve-muscle contacts (Anderson and Cohen, 1977; Anderson *et al.*, 1977). Spontaneous redistribution can also occur in rat myotubes (Stya and Axelrod, 1983). The fluorescent microscopy method has lower sensitivity than autoradiography (Anderson *et al.*, 1977) so that it is not known in these studies on *Xenopus* cells whether overall extrajunctional receptor density was reduced after innervation. Subsequently, it has been shown that the background sensitivity on innervated and non-innervated *Xenopus* muscle cells is similar (Gruener and Kidokoro, 1982; Kidokoro and Gruener, 1982).

It would appear, therefore, that nerves in culture, as *in vivo*, can induce the appearance of local regions of high cholinoceptor density although the mechanism is by no means fully understood. As junctional 'hot spots' can be found in cells cultured in the presence of neuromuscular blocking agents or treated to prevent muscle action potentials (Steinbach, 1974; Anderson and Cohen, 1977; Anderson *et al.*, 1977; Cohen and Fischbach, 1977), functional transmission or muscle activity is not necessary for localization of receptors at neuromuscular junctions.

Activity does, however, affect other parameters. As described in Chapter 4, electrical stimulation of cultured myotubes results in a decrease in acetylcholine sensitivity and in the number of toxin binding sites. These changes can be blocked by incubation of the cultures with local anaesthetics or tetrodotoxin (Cohen and Fischbach, 1973; Prives *et al.*, 1976; Shainberg *et al.*, 1976; Cohen and Pumplin, 1979). Muscle activity has also been implicated in the appearance of acetylcholinesterase at the neuromuscular junction (Rubin *et al.*, 1980). When chick embryo muscle cells and spinal cord explants were co-cultured in the presence of tubocurarine, neuromuscular junctions still formed, but the time course of endplate potentials (measured after removal of the tubocurarine) was slow and unaffected by anticholinesterase drugs, and fewer focal areas of cholinesterase staining could be detected. Also, less of the form of acetylcholinesterase specific to neuromuscular junctions was found in tubocurarine-treated cultures. The prevention of the normal developmental changes was not a consequence simply of receptor blockade as similar effects were produced by treatment with proadifen (SKF525A; thought to be blocking receptor ion channels) or with tetrodotoxin. Direct muscle stimulation or incubation with dibutyrylcyclic GMP reversed the effect of tubocurarine on acetylcholinesterase localization (Rubin *et al.*, 1980). In contrast to these results obtained with cultures of chick embryo cells, Moody-Corbett *et al.* (1982) found that acetylcholinesterase activity became localized at neuromuscular junctions in *Xenopus* cultures even in the presence of tubocurarine.

During development *in vivo*, it is not only receptor distribution that alters but there are also changes in receptor channel properties and in receptor metabolism (for review, see Harvey, 1980). These characteristic changes do not seem to be reproduced in cultures of nerve and muscle, indicating some deficiency in the culture system. In rat and chicken muscle, receptor turnover slows dramatically between fetal and mature stages *in vivo*. Half lives of 15-24 h have been estimated for receptors of immature animals while half lives of the order of several days appear characteristic of cholinoceptors in older muscles. In rats the changeover occurs in the first few days after birth (Steinbach *et al.*, 1979), whereas in the chick this change is only complete a few weeks after hatching (Burden, 1977b). Because the turnover rate of extrajunctional receptors on denervated frog and rat muscle is also fast, there is a tendency to assume that the change in degradation rates is characteristic of cholinoceptors of all skeletal muscle. However, the changes brought about by innervation and by denervation have not been systematically compared. A detailed assessment of the effects of innervation on receptor metabolism in cultured muscle has not yet been made, but there is no evidence for changes during muscle development in culture or consequent to innervation. In cultured chick embryo muscle, receptor half life appears to remain short even when receptors in 'hot spots' on innervated fibres were studied (Schuetze *et al.*, 1978). However, as mentioned earlier, the change from rapid to slow receptor degradation occurs more slowly in chickens than in rats, and hence it may be more difficult to define this effect with cultured chick muscle. There are no reports of the effects of innervation on rat muscle in culture.

The time that acetylcholine-activated channels remain open also changes with development in most species. In rats, *Xenopus* and human muscle, open times are long at fetal stages and subsequently undergo a transition to shorter values (Bevan *et al.*, 1978; Sakmann and Brenner, 1978; Cull-Candy *et al.*, 1979; Fischbach and Schuetze, 1980; Kullberg *et al.*, 1981). In chick muscle, there is no developmental change in open time (Schuetze, 1980; Harvey and van Helden, 1981). Because of this, perhaps it is not surprising that studies on cholinoceptors of cultured chick embryo muscle did not reveal any changes in channel lifetime following innervation (Schuetze *et al.*, 1978; Schuetze, 1980). There is no information about what happens after innervation of rat muscle in culture. Fischbach and Schuetze (1980) reported that neuromuscular junctions in rat cell cultures could not be precisely located to allow assessment of the properties of junctional cholinoceptors. However, in *Xenopus* nerve-muscle cultures there was a reduction in the number of receptors with short open times after innervation, which is the opposite effect to that seen *in vivo* (Brehm *et al.*, 1982).

Neurotrophic factors. Although muscle activity can account for some of the changes seen during development, there is increasing evidence for the existence of additional neurotrophic factors. Much of this evidence comes from studies on cultured muscle. Although some of the active factors have been partially purified, it is not known whether any are, in fact, released by nerves *in vivo*. No doubt antibodies against these factors will soon be prepared and then used to test for their presence *in vivo*.

There are several reports that indicate that the number and distribution of acetylcholine receptors can be influenced by neurotrophic factors. This possibility was suggested by the finding of higher sensitivities to acetylcholine on myotubes nearest to spinal cord explants (Cohen and Fischbach, 1977; Podleski *et al.*, 1978). Extracts from fetal rat brain or spinal cord produced a similar increase in receptor numbers (Podleski *et al.*, 1978), and such a factor appears to be present in chick embryo brain (Jessell *et al.*, 1979). The active factor from rat nerves appeared to be a protein of around 100,000 molecular weight. It also had the ability to induce redistribution of receptors from the underside of myotubes to the more exposed surface (Salpeter *et al.*, 1982). An increase in receptor aggregation was also caused by a factor secreted by NG108-15 neuroblastoma X glioma cells (Christian *et al.*, 1978b). This effect was not associated with an overall increase in receptor numbers. The active factor is probably a protein with a molecular weight of 150,000-250,000 and it is found in embryonic nerves as well as in NG108-15 cells (Bauer *et al.*, 1981).

Other neurotrophic factors have been associated with increases in acetylcholinesterase activity (Oh, 1975, 1976), with changes in passive membrane properties (Engelhardt *et al.*, 1976, 1977) and with alterations in action potential channels (Kano *et al.*, 1979; Kuromi *et al.*, 1981).

Autonomic Neuromuscular Junctions

Cardiac muscle. Despite the great interest in the development of cardiac muscle in culture (see Chapter 5), there have been remarkably few attempts to determine if innervation of cultured heart cells is possible. This is surprising because of the difficulty of examining the growth of autonomic neuroeffector junctions *in vivo*.

There have been preliminary reports that innervation of explants of heart tissue can occur when co-cultured with sympathetic ganglia (Masurovsky and Benitez, 1967; Crain, 1968) and dissociated cardiac cells appear to be preferentially contacted by nerve fibres growing from explants of sympathetic ganglia (Mark *et al.*, 1973). The nerve-muscle contacts that are formed are stable (for at least six days) and can be formed in the presence of muscarinic and β-adrenoceptor blocking drugs (Campbell *et al.*, 1978). This implies that functional synaptic contacts may be made despite blockade of synaptic activity; similar results were found with skeletal muscle (e.g. Obata, 1977).

Functional neuromuscular contacts are established in cultures of newborn rat heart cells and explants of sympathetic ganglia (Purves *et al.*, 1974). Stimulation of the ganglia resulted either in an expected

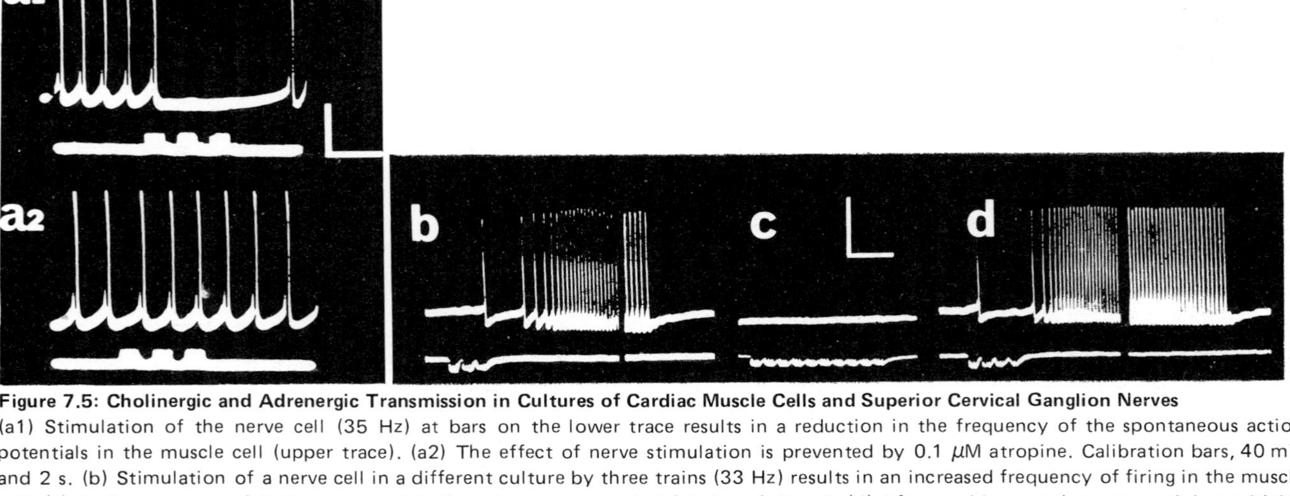

Figure 7.5: Cholinergic and Adrenergic Transmission in Cultures of Cardiac Muscle Cells and Superior Cervical Ganglion Nerves
(a1) Stimulation of the nerve cell (35 Hz) at bars on the lower trace results in a reduction in the frequency of the spontaneous action potentials in the muscle cell (upper trace). (a2) The effect of nerve stimulation is prevented by 0.1 μM atropine. Calibration bars, 40 mV and 2 s. (b) Stimulation of a nerve cell in a different culture by three trains (33 Hz) results in an increased frequency of firing in the muscle cell. (c) In the presence of 1 μM propranolol, there is no response to 10 trains of stimuli. (d) After washing out the propranolol, sensitivity to stimulation returns. Calibration bars for (b)-(d), 50 mV and 5 s. From Furshpan *et al.* (1976), with permission.

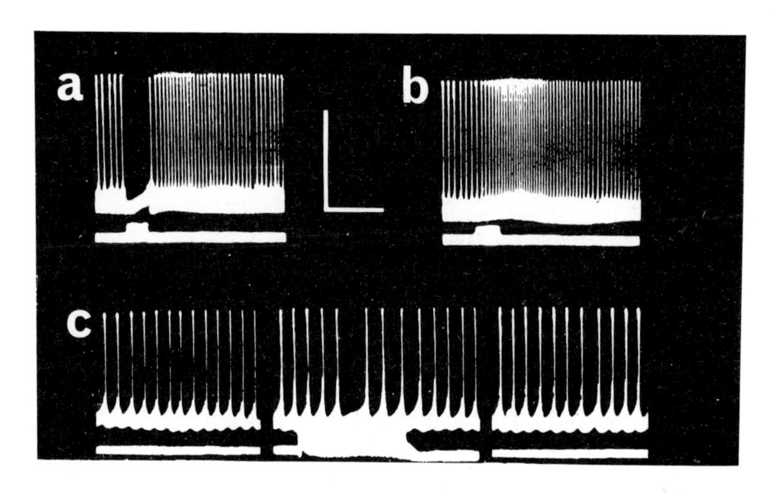

Figure 7.6: Cholinergic and Adrenergic Transmission in a Single Nerve Cell from a Superior Cervical Ganglion Cultured with Dissociated Heart Cells
(a) Stimulation of the nerve first inhibits and then stimulates the spontaneous activity of the muscle cell. (b) In the presence of 0.1 μM atropine, inhibition is blocked but stimulation still occurs. (c) After addition of 0.6 μM propranolol, the stimulatory effect of nerve stimulation (middle records) is prevented. Calibration bar, 100 mV and 12.5 s for (a) and (b), and 100 mV and 5 s for (c). From Furshpan *et al*. (1976), with permission.

increase in the frequency of the spontaneous contractions of the heart cells or, unexpectedly, in a decrease. The decrease was prevented by the muscarinic antagonist hyoscine. Although the finding of cholinergic transmission in these cultures was explained by the presence of some sympathetic cholinergic nerves (Purves *et al*., 1974), it is possible that the results arose because of a shift from adrenergic to cholinergic properties. Such neural plasticity has been found with nerves from superior cervical ganglia, as described earlier (p. 184). When dissociated superior cervical ganglion cells are grown with dissociated rat heart cells, the nerves may be adrenergic, cholinergic, or both (Furshpan *et al*., 1976). Stimulation of some nerves causes hyperpolarization and a reduction in the frequency of action potentials in cardiac cells (Figure 7.5). These effects can be mimicked by acetylcholine and blocked by atropine (Figure 7.5). Stimulation of other nerves results in depolarization and an increased frequency of action potentials; these effects are mimicked by noradrenaline and blocked by propranodol (Figure 7.5). In some cultures, stimulation of the nerve first decreased and then increased the frequency of firing. As the two effects could be blocked selectively

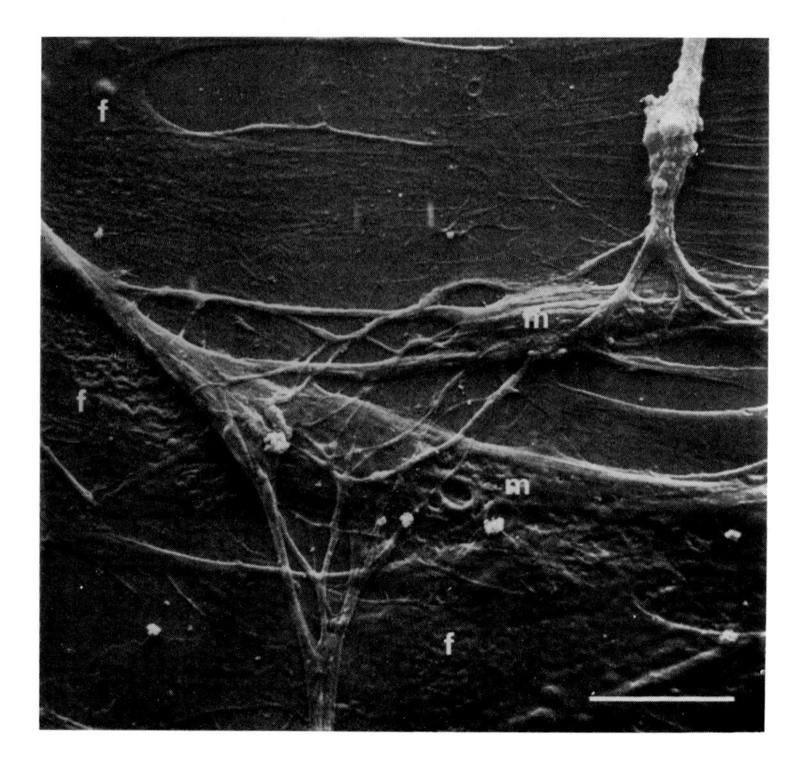

Figure 7.7: Nerve-Muscle Contacts in a Culture of Newborn Guinea Pig Sympathetic Ganglia and Dissociated Vas Deferens Cells
The nerve fibres contact muscle cells (m) but not fibroblasts (f). Calibration bar = 25 μm. From Chamley *et al.* (1973), with permission.

with atropine and propranolol, respectively (Figure 7.6), it was concluded that some nerves could release both acetylcholine and noradrenaline (Furshpan *et al.*, 1976). Adrenergic transmission was demonstrated indirectly by the finding that tyramine could increase the frequency of beating of cardiac cells grown with cells from superior cervical ganglia (King *et al.*, 1978).

Smooth muscle. Mark *et al.* (1973) demonstrated that neural processes from explants of sympathetic ganglia grew towards smooth muscle cells cultured from guinea pig vas deferens. Neural processes form extensive networks on muscle cells (Figure 7.7). Stable contacts formed between nerve and muscle cells but not between nerves and fibroblasts. Time-lapse photography revealed that the nerve growth cone was attracted to clumps of smooth muscle cells, but not specifically to isolated individual cells. Random contacts appeared to be responsible for providing the opportunity for nerve-muscle interactions (Chamley *et al*., 1973). Many of the smooth muscle cells contracted spontaneously in culture and it was common to find that the frequency of contractions increased after the initial contact by a growth cone. However, it was not determined whether this was due to release of a neurotransmitter or to some other mechanism (Chamley *et al*., 1973).

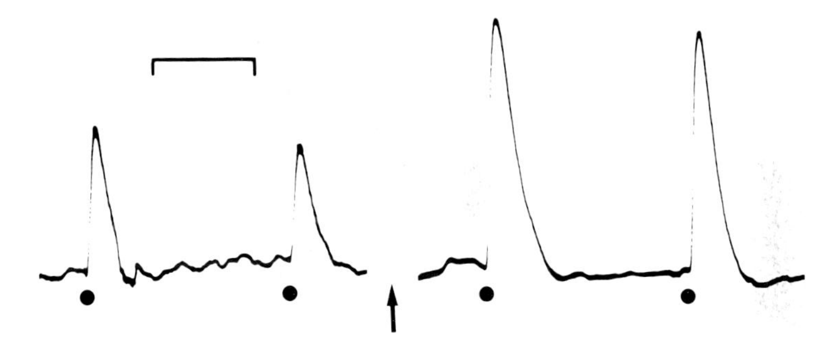

Figure 7.8: Contractions of Cultured Taenia Caeci Muscle in Response to Stimulation of a Ciliary Ganglion Explant, and their Augmentation by an Anticholinesterase

Contractions (recorded photometrically) were elicited by short trains of stimuli (7 pulses in 0.6 s) applied at the times marked by the dots. At the arrow, neostigmine (10 μg/ml) was added to the culture and responses were tested 3 min. later. Calibration bar, 20 s. From Purves *et al*. (1974), with permission.

The first demonstration that functional neurotransmitter junctions can be formed with smooth muscle cells in culture appears to be that of Purves *et al* (1974). Smooth muscle from three sources (constrictor pupillae, vas deferens and taenia caeci) were cultured with sympathetic or parasympathetic ganglia. Stimulation of the ganglia after 3-14 days in culture led to contractions of the smooth muscle cells, and the contractile activity was inhibited by hyoscine and augmented by neostigmine (Figure 7.8), indicating that the junctions were cholinergic. Similar results were obtained when sympathetic nerves were cultured with explants of the smooth muscle of rat irises (Hill *et al*., 1976). Although some nerves appeared to be adrenergic on the basis of their positive catecholamine fluorescence, there was evidence that only cholinergic transmission was functional. Again, these may be examples of the plasticity of sympathetic neurones.

REFERENCES

Adams, P.R., Constanti, A., Brown, D.A. and Clark, R.B. (1982) Intracellular Ca^{2+} activates a fast voltage-sensitive K^+ current in vertebrate neurones, *Nature, 296,* 746-9.

Albuquerque, E.X. and McIsaac, R.J. (1970) Fast and slow mammalian muscles after denervation, *Exp. Neurol., 26,* 183-202.

Amano, T. Richelson, E. and Nirenberg, M. (1972) Neurotransmitter synthesis by neuroblastoma clones, *Proc. Nat. Acad. Sci.* USA *69,* 258-63.

Amano, T., Hamprecht, B. and Kemper, W. (1974) High activity of choline acetyltransferase induced in neuroblastoma × glia hybrid cells. *Exp. Cell Res. 85*, 339-408.

Amy, C. and Kirshner, N. (1982) $^{22}Na^+$ uptake and catecholamine secretion by primary cultures of adrenal medulla cells. *J. Neurochem. 39*, 132-42.

Anand-Srivastava, M.B., Franks, D.J., Cantin, M. and Genest, J. (1982) Presence of 'Ra' and 'P' -site receptors for adenosine coupled to adenylate cyclase in cultural vascular smooth muscle cells. *Biochem. Biophys. Res. Comm. 108*, 213-19.

Anderson, M.J. and Cohen, M.W. (1977) Nerve-induced and spontaneous redistribution of acetylcholine receptors on cultured muscle cells. *J. Physiol. 268*, 757-73.

Anderson, M.J., Cohen, M.W. and Zorychta, E. (1977) Effects of innervation on the distribution of acetylcholine receptors on cultured muscle cells. *J. Physiol. 268*, 731-56.

Appel, S.H., Anwyl, R., McAdams, M.W. and Elias, S. (1977) Accelerated degradation of acetylcholine receptor from cultured rat myotubes with myasthenia gravis sera and globulins. *Proc. Nat. Acad. Sci.* USA *74,* 2130-4.

Arner, L.S. and Stallcup, W.B. (1981) Rubidium efflux from neural cell lines through voltage-dependent potassium channels. *Dev. Biol. 83*, 138-45.

Askanas, V. Engel, W.K., Ringel, S.P. and Bender, A.N. (1977) Acetylcholine receptors of aneurally cultured human and animal muscle. *Neurology 27*, 1019-22.

Athias, P., Frelin, C., Groz, B., Dumas, J.P., Klepping, J. and Padieu, P. (1979) Myocardial electrophysiology; intracellular studies on heart cell cultures from newborn rats. *Path. Biol. 27*, 13-19.

Atlas, D. and Adler, M. (1981) α-Adrenergic antagonists as possible calcium channel inhibitors. *Proc. Nat. Acad. Sci.* USA *78*, 1237-41.

Atlas, D. and Sabol, S.L. (1981) Interaction of clonidine and clonidine analogues with α-adrenergic receptors of neuroblastoma × glioma hybrid cells and rat brain. Comparison of ligand binding with inhibition of adenylate cyclase. *Eur. J. Biochem. 113,* 521-9.

Augusti-Tocco, G. and Sato, G. (1969) Establishment of functional clonal lines of neurons from mouse neuroblastoma. *Proc. Nat. Acad. Sci.* USA *64,* 311-15.

Axelrod, D., Ravdin, P., Koppel, D.E., Schlessinger, J., Webb, W.W., Elson, E.L. and Poldleski, T.R. (1976) Lateral motion of fluorescently labeled acetylcholine receptors in membranes of developing muscle fibers. *Proc. Nat. Acad. Sci.* USA *73,* 4594-8.

Axelrod, D., Ravdin, P.M. and Podleski, T.R. (1978a) Control of acetylcholine receptor mobility and distribution in cultured muscle membrane. A fluorescence study. *Biochim. Biophys. Acta 511,* 23-38.

Axelrod, D., Wight, A., Webb, W. and Horwitz, A. (1978b) Influence of membrane lipids on acetylcholine receptor and lipid probe diffusion in cultured myotube membrane. *Biochemistry 17,* 3604-9.

Azuma, J., Sawamura, A., Harada, H., Tanimoto, T., Ishiyama, T., Morita, Y., Yamamura, Y. and Sperelakis, N. (1981) Cyclic adenosine monophosphate modulation of contractility via slow Ca^{2+} channels in chick heart. *J. Mol. Cell Cardiol. 13,* 577-87.

Baccaglini, P.I. and Cooper, E. (1982a) Electrophysiological studies of new-born rat nodose neurones in cell culture. *J. Physiol. 324,* 429-39.

Baccaglini, P.I. and Cooper, E. (1982b) Influences on the expression of acetylcholine receptors on rat nodose neurones in cell culture. *J. Physiol. 324,* 441-51.

Baccaglini, P.I. and Hogan, P.G. (1983) Some rat sensory neurons in culture express characteristics of differentiated pain sensory cells. *Proc. Nat. Acad. Sci.* USA *80,* 594-8.

Bader, C.R., Bertrand, D. and Kato, A.C. (1982) Chick ciliary ganglion in dissociated cell culture. II. Electrophysiological properties. *Dev. Biol. 94,* 131-41.

Baram, D. and Simantov, R. (1983) Enkephalins and opiate antagonists control calmodulin distribution in neuroblastoma-glioma cells. *J. Neurochem. 40,* 55-63.

Barker, J.L. and Mathers, D.A. (1981) GABA analogues activate channels of different duration in cultured mouse spinal neurons. *Science 212,* 358-61.

Barker, J.L. and McBurney, R.N. (1979a) GABA and glycine may share the same conductance channel on cultured mammalian neurones. *Nature, 277,* 234-6.

Barker, J.L. and McBurney, R.N. (1979b) Phenobarbitone modulation of postsynaptic GABA receptor function on cultured mammalian neurones. *Proc. Roy. Soc. B 206,* 319-27.

Barker, J.L., McBurney, R.N. and MacDonald, J.F. (1982) Fluctuation analysis of neutral amino acid responses in cultured mouse spinal neurones. *J. Physiol. 322,* 365-87.

Barker, J.L., MacDonald, R.L. and Smith, T.G. (1977) Voltage clamp analysis of amino acid currents in cultured mammalian neurons. *J. Gen. Physiol. 70,* 1a.

Barker, J.L., Neale, J.H., Smith, T.G. and MacDonald, R.L. (1978a) Opiate peptide modulation of amino acid responses suggests novel form of neuronal communication. *Science 199,* 1451-3.

Barker, J.L. and Ransom, B.R. (1978a) Amino acid pharmacology of mammalian central neurones grown in tissue culture. *J. Physiol. 280*, 331-54.

Barker, J.L. and Ransom, B.R. (1978b) Pentobarbitone pharmacology of mammalian central neurones grown in tissue culture. *J. Physiol. 280*, 355-72.

Barker, J.L., Smith, T.G. and Neale, J.N. (1978b) Multiple membrane actions of enkephalin revealed using cultured spinal neurons. *Brain Res. 154*, 153-8.

Barrett, J.N., Barrett, E.F. and Dribin, L.B. (1981) Calcium-dependent slow potassium conductance in rat skeletal myotubes. *Dev. Biol. 82*, 258-66.

Barrett, J.N., Magleby, K.L. and Pallotta, B.S. (1982) Properties of single calcium-activated potassium channels in cultured rat muscle. *J. Physiol. 331*, 211-30.

Barry, W.H. and Smith, T.W. (1982) Mechanisms of transmembrane calcium movement in cultured chick embryo ventricular cells. *J. Physiol. 325*, 243-60.

Bartfai, T. Breakefield, X.O. and Greengard, P. (1978) Regulation of synthesis of guanosine 3′:5′ - cyclic monophosphate in neuroblastoma cells. *Biochem. J. 176*, 119-27.

Bauer, H.C., Daniels, M.P., Pudimat, P.A., Jacques, J., Sugiyama, H. and Christian, C.N. (1981) Characterization and partial purification of a neuronal factor which increases acetylcholine receptor aggregation on cultured muscle cells. *Brain Res. 209*, 395-404.

Beale, R., Dutton, G.R. and Currie, D.N. (1980) An ion flux assay of action potential sodium channels in neuron- and glia-enriched cultures of cells dissociated from rat cerebellum. *Brain Res. 183*, 241-6.

Bennett, M.R. and White, W. (1979) The survival and development of cholinergic neurons in potassium-enriched media. *Brain Res. 173*, 549-53.

Beranek, R. and Vyskočil, F. (1967) The action of d-tubocurarine and atropine on the normal and denervated rat diaphragm. *J. Physiol. 188*, 53-66.

Beranek, R. and Vyskočil, F. (1968) The effect of atropine on the frog sartorius neuromuscular junction. *J. Physiol. 195*, 493-503.

Berg, D.K. and Fischbach, G.D. (1978) Enrichment of spinal cord cell cultures with motoneurons. *J. Cell Biol. 77*, 83-98.

Berg, D.K. and Hall, Z.W. (1975) Loss of α-bungarotoxin from junctional and extrajunctional acetylcholine receptors in rate diaphragm muscle *in vivo* and in organ culture. *J. Physiol. 252*, 771-89.

Bernard, P. and Couraud, F. (1979) Electrophysiological studies on embryonic heart cells in culture. Scorpion toxin as a tool to reveal latent fast sodium channel. *Biochim. Biophys. Acta 553*, 154-68.

Berwald-Netter, Y., Martin-Moutot, N., Koulakoff, A. and Couraud, F. (1981) Na^+-channel-associated scorpion toxin receptor sites as probes for neuronal evolution *in vivo* and *in vitro*. *Proc. Nat. Acad. Sci.* USA *78*, 1245-9.

Betz, H. (1981) Characterization of the α-bungarotoxin receptor in chick embryo retina. *Eur. J. Biochem. 117*, 131-9.

Betz, H. (1982) Differential down-regulation by carbamylcholine of putative nicotinic and muscarinic acetylcholine receptor sites in chick retina cultures. *Neurosci. Lett. 28*, 265-8.

Betz, H. and Changeux, J.P. (1979) Regulation of muscle acetylcholine receptor synthesis *in vitro* by cyclic nucleotide derivatives. *Nature 278*, 749-52.

Betz, W. (1976a) The formation of synapses between chick embryo skeletal muscle and ciliary ganglia grown *in vitro*. *J. Physiol. 254*, 63-73.

Betz, W. (1976b) Functional and non-functional contacts between ciliary neurones and muscle grown *in vitro*. *J. Physiol. 254*, 75-86.

Betz, W. and Osborne, M. (1977) Effects of innervation on acetylcholine sensitivity of developing muscle *in vitro*. *J. Physiol. 270*, 75-88.

Bevan, S. and Steinbach, J.H. (1977) The distribution of α-bungarotoxin binding sites on mammalian skeletal muscle *in vivo*. *J. Physiol. 267*, 195-213.

Bevan, S., Kullberg, R.W. and Rice, J. (1978) Acetylcholine-induced conductance fluctuations in cultured human myotubes. *Nature 273*, 469-71.

Biales, B., Dichter, M. and Tischler, A. (1976) Electrical excitability of cultured adrenal chromaffin cells. *J. Physiol. 262*, 743-53.

Birnbaum, M., Reis, M.A. and Shainberg, A. (1980) Role of calcium in the regulation of acetylcholine receptor synthesis in cultured muscle cells. *Pflügers Arch. 385*, 37-43.

Blair, I.A., Hensby, C.N. and MacDermot, J. (1980) Prostacyclin-dependent activation of adenylate cyclase in a neuronal somatic cell hybrid: prostanoid structure-activity relationships. *Br. J. Pharmac.* 69, 519-25.

Blair, I.A. and MacDermot, J. (1981) The binding of [^{3}H]-prostacyclin to membranes of a neuronal somatic hybrid. *Br. J. Pharmac. 72*, 435-41.

Blau, H.M. and Webster, C. (1981) Isolation and characterization of human muscle cells. *Proc. Nat. Acad. Sci.* USA. *78*, 5623-7.

Blondel, B., Roijen, I. and Cheneval, J.P. (1971) Heart cells in culture: a simple method for increasing the proportion of myoblasts. *Experientia 27*, 356-8.

Blosser, J., Abbott, J. and Shain, W. (1976) Sympathetic ganglion cell × neuroblastoma hybrids with opiate receptors. *Biochem. Pharmac. 25*, 2395-9.

Blosser, J.C. and Appel, S.H. (1980) Regulation of acetylcholine receptor by cyclic AMP. *J. Biol. Chem. 255*, 1235-8.

Blosser, J.C., Myers, P.M. and Shain, W. (1978) Neurotransmitter modulation of prostaglandin in E_1-stimulated increases in cyclic AMP. I. Characterization of a cultured neuronal cell line in exponential growth phase. *Biochem. Pharmac. 27*, 1167-72.

Blume, A.J., Chen, C. and Foster, C.J. (1977) Muscarinic regulation of cAMP in mouse neuroblastoma. *J. Neurochem. 29*, 625-32.

Blume, A.J. and Foster, C.J. (1976) Mouse neuroblastoma cell adenylate cyclase: regulation by 2-chloroadenosine, prostaglandin E_1 and the cations Mg^{2+}, Ca^{2+}, and Mn^{2+}. *J. Neurochem. 26*, 305-11.

Blume, A.J., Lichtshtein, D. and Boone, G. (1979) Coupling of opiate receptors to adenylate cyclase: requirement for Na^+ and GTP. *Proc. Nat. Acad. Sci.* USA *76*, 5626-30.

Bonkowski, L. and Dryden, W.F. (1976) The effects of putative neurotransmitters on the resting membrane potential of dissociated brain neurones in culture. *Brain Res. 107*, 69-84.

Bonkowski, L. and Dryden, W.F. (1979) Effects of iontophoretically applied neurotransmitters on mouse brain neurones in culture. *Neuropharmac., 16*, 89-97.

Bottenstein, J.E. and Sato, G.H. (1979) Growth of a rat neuroblastoma cell line in serum-free supplemented medium. *Proc. Nat. Acad. Sci.* USA *76*, 514-17.

Bottenstein, J.E., Skaper, S.D., Varon, S.S. and Sato, G.H. (1980) Selective survival of neurons from chick embryo sensory ganglionic dissociates utilizing serum-free supplemented medium, *Exp. Cell Res. 125*, 183-90.

Brandt, M., Buchen, C. and Hamprecht, B. (1977) Endorphins exert opiate-like action on neuroblastoma × glioma hybrid cells. *FEBS Lett. 80*, 251-4.

Brandt, B.L., Hagiwara, S., Kidokoro, Y. and Miyazaki, S. (1976) Action potentials in the rat chromaffin cell and effects of acetylcholine. *J. Physiol. 263*, 417-39.

Brehm, P., Steinbach, J.H. and Kidokoro, Y. (1982) Channel open time of acetylcholine receptors on *Xenopus* muscle cells in dissociated cell culture. *Dev. Biol. 91*, 93-102.

Brock, T.A., Lewis, L.J. and Smith, J.B. (1982) Angiotensin increases Na^+ entry and Na^+/K^+ pump activity in cultures of smooth muscle from rat aorta. *Proc. Nat. Acad. Sci.* USA *79*, 1438-42.

Brookes, N. (1978) Actions of glutamate on dissociated mammalian spinal neurons *in vitro*. *Dev. Neurosci. 1*, 203-15.

Brookes, N. and Burt, D.R. (1980) Development of muscarinic receptor binding in spinal cord cell cultures and its reduction by glutamic and kainic acids. *Dev. Neurosci. 3*, 118-27.

Brown, A.M., Camerer, H., Kunze, D.L. and Lux, H.D. (1982) Similarity of unitary Ca^{2+} currents in three different species. *Nature 299*, 156-8.

Brunner, G. and Tschank, G. (1982) Contracting striated muscle fibres differentiated from primary rat pituitary cultures. *Cell Tiss. Res. 224*, 655-62.

Bulloch, K., Stallcup, W.B. and Cohen, M. (1977) The derivation and characterization of neuronal cell lines from rat and mouse brain. *Brain Res. 135*, 25-36.

Burden, S. (1977a) Development of the neuromuscular junction in the chick embryo: the number, distribution and stability of acetylcholine receptors. *Dev. Biol. 57*, 317-29.

Burden, S. (1977b) Acetylcholine receptors at the neuromuscular

junction: developmental change in receptor turnover. *Dev. Biol. 61,* 79-85.
Burgermeister, W., Klein, W.L., Nirenberg, M. and Witkop, B. (1978) Comparative binding studies with cholinergic ligands and histrionicotoxin at muscarinic receptors of neural cell lines. *Mol. Pharmac. 14*, 751-67.
Burgoyne, R.D. and Pearce, B. (1982) Muscarinic acetylcholine receptor regulation and protein phosphorylation in primary cultures of rat cerebellum. *Dev. Brain Res. 2,* 55-63.
Burrows, M.T. (1910) The cultivation of tissues of the chick embryo outside the body. *J. Amer. Med. Assoc. 55*, 2057-8.
Bursztajn, S. and Gershon, M.D. (1977) Discrimination between nicotinic receptors in vertebrate ganglia and skeletal muscle by alpha-bungarotoxins and cobra venoms. *J. Physiol. 269,* 17-31.
Burton, H. and Bunge, R.P. (1981) The expression of cholinergic and adrenergic properties by automatic neurons in tissue culture, in *Excitable Cells in Tissue Culture*, P.G. Nelson and M. Lieberman (eds.), Plenum Press, New York, pp. 1-37.
Calvet, M.-C. (1974) Patterns of spontaneous electrical activity in tissue cultures of mammalian cerebral cortex *vs.* cerebellum. *Brain Res. 69*, 281-95.
Calvet, M.-C. and Calvet, J. (1979) Horseradish peroxidase iontophoretic intracellular labelling of cultured Purkinje cells. *Brain Res. 173*, 527-31.
Campbell, G.R., Chamley, J.H. and Burnstock, G. (1978) Lack of effect of receptor blockers on the formation of long-lasting associations between sympathetic nerves and cardiac muscle cells in vitro. *Cell Tiss. Res. 187*, 551-3.
Carbone, E., Wanke, E., Prestipino, G., Possani, L.D. and Maelicke, A. (1982) Selective blockage of voltage-dependent K^+ channels by a novel scorpion toxin. *Nature 296*, 90-1.
Carbonetto, S.T., Fambrough, D.M. and Muller, K.J. (1978) Nonequivalence of α-bungarotoxin receptors and acetylcholine receptors in chick sympathetic neurons. *Proc. Nat. Acad. Sci.* USA *75*, 1016-20.
Carrel, A. and Burrows, M.T. (1910) Cultivation of adult tissues and organs outside of the body. *J. Amer. Med. Assoc. 55*, 1379-81.
Catterall, W.A. (1975a) Cooperative activation of action potential Na^+ ionophore by neurotoxins. *Proc. Nat. Acad. Sci.* USA *72,* 1782-6.
Catterall, W.A. (1975b) Sodium transport by acetylcholine receptor of cultured muscle cells. *J. Biol. Chem. 250*, 1776-81.
Catterall, W.A. (1976) Activation and inhibition of the action potential Na^+ ionophore of cultured rat muscle cells by neurotoxins. *Biochem. Biophys. Res. Comm. 68*, 136-42.
Catterall, W.A. (1977) Activation of the action potential Na^+ ionophore by neurotoxins. An allosteric model. *J. Biol. Chem. 252*, 8669-76.
Catterall, W.A. (1980) Pharmacologic properties of voltage sensitive sodium channels in chick muscle fibers developing *in vitro. Dev. Biol. 78*, 222-30.
Catterall, W.A. (1981a) Studies of voltage-sensitive sodium channels in

cultured cells using ion-flux and ligand-binding methods. In *Excitable Cells in Tissue Culture*, P.G. Nelson and M. Lieberman (eds.) Plenum Press, New York, pp. 279-317.

Catterall, W.A. (1981b) Inhibition of voltage-sensitive sodium channels in neuroblastoma cells by antiarrhythmic drugs. *Mol. Pharmac. 20,* 356-62.

Catterall, W.A. (1981c) Localization of sodium channels in cultured neural cells. *J. Neurosci. 1*, 777-83.

Catterall, W.A. and Beress, L. (1978) Sea anemone toxin and scorpion toxin share a common receptor site associated with the action potential sodium ionophore. *J. Biol. Chem. 253*, 7393-6.

Catterall, W.A. and Coppersmith, J. (1981) Pharmacological properties of sodium channels in cultured rat heart cells. *Mol. Pharmac. 20*, 533-42.

Catterall, W.A. and Nirenberg, M. (1973) Sodium uptake associated with activation of action potential ionophores of cultured neuroblastoma and muscle cells. *Proc. Nat. Acad. Sci.* USA *70*, 3759-63.

Cavanaugh, M.W. and Cavanaugh, D.J.C. (1957) Studies on the pharmacology of tissue cultures. I. The action of quinidine on cultures of dissociated chick embryo heart cells. *Arch. Int. Pharmacodyn. Ther. 110*, 43-55.

Chalazonitis, A. and Greene, L.A. (1974) Enhancement in excitability properties of mouse neuroblastoma cells cultured in the presence of dibutyryl cyclic AMP. *Brain Res. 72*, 340-5.

Chalazonitis, A., Greene, L.A. and Nirenberg, M. (1974) Electrophysiological characteristics of chick embryo sympathetic neurons in dissociated cell cultures. *Brain Res. 68*, 235-52.

Chalazonitis, A., Greene, L.A. and Shain, W. (1975) Excitability and chemosensitivity properties of a somatic cell hybrid between mouse neuroblastoma and sympathetic ganglion cells. *Exp. Cell Res. 96*, 225-38.

Chalazonitis, A., Minna, J.D. and Nirenberg, M. (1977) Expression and properties of acetylcholine receptors in several clones of mouse neuroblastoma × L cell somatic hybrids. *Exp. Cell Res. 105*, 269-80.

Chamley, J.H. and Campbell, G.R. (1976) Tissue culture: interaction between sympathetic nerves and vascular smooth muscle. In *Vascular Neuroeffector Mechanisms*, J.A. Bevan, G. Burnstock, B. Johansson, R.A. Maxwell and O.A. Nedergaard (eds.), Karger, Basel, pp. 10-18.

Chamley, J.H., Campbell, G.R. and Burnstock, G. (1973) An analysis of the interactions between sympathetic nerve fibers and smooth muscle cells in tissue culture. *Dev. Biol. 33*, 344-61.

Chamley, J.H., Campbell, G.R., McConnell, J.D. and Gröschel-Stewart, V. (1977) Comparison of vascular smooth muscle cells from adult human, monkey and rabbit in primary culture and in subculture. *Cell Tiss. Res. 177*, 503-22.

Chamley-Campbell, J., Campbell, G.R. and Ross, R. (1979) The smooth muscle cell in culture. *Physiol. Rev. 59*, 1-61.

Champy, C. (1913/14) Quelques résultats de la méthode de culture des

tissus. *Arch. Zool. Exp. Gén*, *53*, 42-51.
Chang, C.C. and Huang, M.C. (1975) Turnover of junctional and extra-junctional acetylcholine receptors of the rat diaphragm. *Nature*, *253*, 643-4.
Chang, K.-J. and Cuatrecasas, P. (1979) Multiple opiate receptors. Enkephalins and morphine bind to receptors of different specificity. *J. Biol. Chem.* *254*, 2610-18.
Chang, K.-J., Miller, R.J. and Cuatrecasas, P. (1978) Interaction of enkephalin with opiate receptors in intact cultured cells. *Mol. Pharmac.* *14*, 961-70.
Changeux, J.-P. and Danchin, A. (1976) Selective stabilisation of developing synapses as a mechanism for the specification of neuronal networks. *Nature 264*, 705-12.
Christian, C.N., Nelson, P.G., Peacock, J. and Nirenberg, M. (1977) Synapse formation between two clonal cell lines. *Science 196*, 995-8.
Christian, C.N., Nelson, P.G., Bullock, P., Mullinax, D. and Nirenberg, M. (1978a) Pharmacologic responses of cells of a neuroblastoma X glioma hybrid clone and modulation of synapses between hybrid cells and mouse myotubes. *Brain Res. 147*, 261-76.
Christian, C.N., Daniels, M.P., Sugiyama, H., Vogel, Z., Jacques, L. and Nelson, P.G. (1978b) A factor from neurons increases number of acetylcholine receptor aggregates on cultured muscle cells. *Proc. Nat. Acad. Sci.* USA *75*, 4011-15.
Choi, D.W. and Fischbach, G.D. (1981) GABA conductance of chick spinal cord and dorsal root ganglia neurons in cell culture. *J. Neurophysiol. 145*, 605-20.
Choi, D.W., Farb, D.H. and Fischbach, G.D. (1981a) Chlordiazepoxide selectively potentiates GABA conductance of spinal cord and sensory neurons in cell culture. *J. Neurophysiol. 45*, 621-31.
Choi, D.W., Farb, D.H. and Fischbach, G.D. (1981b) GABA-mediated synaptic potentials in chick spinal cord and sensory neurons. *J. Neurophysiol. 45*, 632-43.
Clay, J.R., DeFelice, L.J. and DeHaan, R.L. (1979) Current noise parameters derived from voltage noise and impedance in embryonic heart cell aggregates. *Biophys. J. 28*, 169-84.
Clay, J.R. and Shrier, A. (1981) Developmental changes in subthreshold pacemaker currents in chick embryonic heart cells. *J. Physiol. 312*, 491-504.
Clusin, W.T. (1983) Caffeine induces a transient inward current in cultured cardiac cells. *Nature 301*, 248-50.
Coetzee, G.A., Van Der Westhuyzen, D.R. and Gevers, W. (1977) Effects of 5-bromo-2'-deoxyuridine on beating heart cell cultures from neonatal hamsters. *Biochem. J. 164*, 635-43.
Cohen, M.W. and Weldon, P.R. (1980) Localization of acetylcholine receptors and synaptic ultrastructure at nerve-muscle contacts in culture: dependence on nerve type. *J. Cell Biol. 86*, 388-401.
Cohen, S.A. (1980) Early nerve-muscle synapses *in vitro* release transmitter over postsynaptic membrane having low acetylcholine

sensitivity. *Proc. Nat. Acad. Sci.* USA *77*, 644-8.
Cohen, S.A. and Fischbach, G.D. (1973) Regulation of muscle acetylcholine sensitivity by muscle activity in cell culture. *Science 181,* 76-8.
Cohen, S.A. and Fischbach, G.D. (1977) Clusters of acetylcholine receptors located at identified nerve-muscle synapses *in vitro*. *Dev. Biol. 59*, 24-35.
Cohen, S.A. and Pumplin, D.W. (1979) Clusters of intramembrane particles associated with binding sites for α-bungarotoxin in cultured chick myotubes. *J. Cell Biol. 82,* 494-516.
Colquhoun, D.. Large, W.A. and Rang, H.P. (1977) An analysis of the action of a false transmitter at the neuromuscular junction. *J. Physiol. 266*, 361-95.
Colquhoun, D., Neher, E., Reuter, H. and Stevens, C.F. (1981) Inward current channels activated by intracellular Ca in cultured cardiac cells. *Nature 294,* 752-4.
Colquhoun, D. and Sakmann, B. (1981) Fluctuations in the microsecond time range of the current through single acetylcholine receptor ion channels. *Nature 294,* 464-6.
Corner, M.A. and Crain, S.M. (1969) The development of spontaneous bioelectric activities and strychnine sensitivity during maturation in culture of embryonic chick and rodent nervous tissues. *Arch. Int. Pharmacodyn. Ther. 182,* 404-6.
Cornett, L.E. and Norris, J.S. (1982) Characterization of the α_1-adrenergic subtype in a smooth muscle cell line. *J. Biol. Chem. 257*, 694-7.
Costero, I. and Pomerat, Ç.M. (1951) Cultivation of neurons from the adult human cerebral and cerebellar cortex. *Am. J. Anat. 89*, 405-67.
Couraud, G., Rochat, H. and Lissitzky, S. (1976) Stimulation of sodium and calcium uptake by scorpion toxin in chick embryo heart cells. *Biochim. Biophys. Acta 433*, 90-100.
Crain, S.M. (1954) Action potentials in tissue cultures of chick embryo spinal ganglia. *Anat. Rec. 118*, 292.
Crain, S.M. (1956) Resting and action potentials of cultured chick embryo spinal ganglion cells. *J. Comp. Neurol. 104*, 285-329.
Crain, S.M. (1968) Development of functional neuromuscular connections between separate explants of fetal mammalian tissues after maturation in culture. *Anat. Rec. 160*, 466.
Crain, S.M. (1970) Bioelectric interactions between cultured fetal rodent spinal cord and skeletal muscle after innervation *in vitro*. *J. Exp. Zool. 173*, 353-70.
Crain, S.M. (1974) Selective depression of organotypic bioelectric activities of CNS tissue cultures by pharmacologic and metabolic agents, in *Drugs and the Developing Brain*, A. Vernadakis and N. Weiner (eds.), Plenum Press, New York, pp. 29-57.
Crain, S.M. (1976) *Neurophysiologic Studies in Tissue Culture.*, Raven Press, New York.
Crain, S.M., Alfei, L. and Peterson, E.R. (1970) Neuromuscular

transmission in cultures of adult human and rodent skeletal muscle after innervation *in vitro* by fetal rodent spinal cord. *J. Neurobiol.* *1*, 471-89.

Crain, S.M. and Bornstein, M.B. (1964) Bioelectric activity of neonatal mouse cerebral cortex during growth and differentiation in tissue culture. *Exp. Neurol. 10*, 425-50.

Crain, S.M. and Bornstein, M.B. (1974) Early onset in inhibitory functions during synaptogenesis in fetal mouse brain cultures. *Brain Res.* *68*, 351-7.

Crain, S.M., Crain, B., Peterson, E.R. and Simon, E.J. (1978) Selective depression by opioid peptides of sensory evoked dorsal-horn network responses in organized spinal cord cultures. *Brain Res. 157*, 196-201.

Crain, S.M., Crain, B. and Peterson, E.R. (1982a) Development of cross-tolerance to 5-hydroxytryptamine in organotypic cultures of mouse spinal cord-ganglia during chronic exposure to morphine. *Life Sci.* *31*, 241-7.

Crain, S.M., Crain, B., Peterson, E.R., Hiller, J.M. and Simon, E.J. (1982b) Exposure to 4-aminopyridine prevents depressant effects of opiates on sensory-evoked dorsal-horn network responses in spinal cord cultures. *Life Sci. 31*, 235-40.

Crain, S.M., Peterson, E.R., Crain, B. and Simon, E.J. (1977) Selective depression of sensory-evoked synaptic networks in dorsal horn regions of spinal cord cultures. *Brain Res. 133*, 162-6.

Crean, G., Pilar, G., Tuttle, J.B. and Vaca, K. (1982) Enhanced chemosensitivity of chick parasympathetic neurones in co-culture with myotubes. *J. Physiol. 331*, 87-104.

Crill, W.E., Rumery, R.E. and Woodbury, J.W. (1959) Effects of membrane current on transmembrane potentials of cultured chick embryo heart cells. *Am. J. Physiol. 97*, 733-5.

Cull-Candy, S.G., Miledi, R. and Uchitel, O.D. (1979) Acetylcholine receptors in organ-cultured human muscle fibres. *Nature 277*, 236-8.

Dawson, G., McLawhon, R. and Miller, R.J. (1979) Opiates and enkephalins inhibit synthesis of gangliosides and membrane glycoproteins in mouse neuroblastoma cell line N4TG1. *Proc. Nat. Acad. Sci.* USA *76*, 605-9.

De Barry, J., Fosset, M. and Lazdunski, M. (1977) Molecular mechanism of the cardiotoxic action of a polypeptide neurotoxin from sea anemone on cultured embryonic cardiac cells. *Biochemistry 16*, 3850-5.

DeFelice, L.J. and Clapham, D.E. (1981) Single channel currents from embryonic heart. *Biophys. J. 33*, 265a.

DeFeudis, F.V., Ossola, L., Schmitt, G. and Mandel, P. (1980) Substrate specificity of [^{3}H] muscimol binding to a particulate fraction of a neuron-enriched culture of embryonic rat brain. *J. Neurochem.* *34*, 845-9.

DeHaan, R.L. (1967) Regulation of spontaneous activity and growth of embryonic chick heart cells in tissue culture. *Dev. Biol. 16*, 216-49.

DeHaan, R.L. (1980) Differentiation of excitable membranes. *Curr. Top. Dev. Biol. 16*, 117-64.

DeHaan, R.L. and Fozzard, H.A. (1975) Membrane responses to current pulses in spheroidal aggregates of embryonic heart cells. *J. Gen. Physiol. 65*, 207-22.

De Mello, M.C.F., Ventura, A.L.M., Paes de Carvalho, R., Klein, W.L. and De Mello, F.G. (1982) Regulation of dopamine- and adenosine-dependent adenylate cyclase systems of chicken retina cells in culture. *Proc. Nat. Acad. Sci.* USA *79*, 5708-12.

Dennis, M.J. (1981) Development of the neuromuscular junction: inductive interactions between cells. *Ann. Rev. Neurosci. 4*, 43-68.

Devreotes, P.N. and Fambrough, D.M. (1975) Acetylcholine receptor turnover in membranes of developing muscle fibers. *J. Cell Biol. 65*, 335-58.

Dhillon, D.S. and Harvey, A.L. (1982) Properties of acetylcholine receptors on myotubes cultured from thymus and skeletal muscle of neonatal rats. *Br. J. Pharmac. 76*, 275P.

Dichter, M.A. (1978) Rat cortical neurons in cell culture: culture methods, cell morphology, electrophysiology, and synapse formation. *Brain Res. 149*, 279-93.

Dichter, M.A. (1980) Physiological identification of GABA as the inhibitory transmitter for mammalian cortical neurons in cell culture. *Brain Res. 190*, 111-21.

Dichter, M.A. and Fischbach, G.D. (1977) The action potential of chick dorsal root ganglion neurones maintained in cell culture. *J. Physiol. 267*, 281-98.

Dichter, M.A., Tischler, A.S. and Greene, L.A. (1977) Nerve growth factor-induced increase in electrical excitability and acetylcholine sensitivity of a rat pheochromocytoma cell line. *Nature 268*, 501-4.

Dryden, W.F. (1970) Development of acetylcholine sensitivity in cultured skeletal muscle. *Experientia 26*, 984-86.

Dryden, W.F., Erulkar, S.D. and de La Haba, G. (1974) Properties of the cell membrane of developing skeletal muscle fibres in culture and its sensitivity to acetylcholine. *Clin. Exp. Pharmac. Physiol. 1*, 369-87.

Dudai, Y. and Yavin, E. (1978) Ontogenesis of muscarinic receptors and acetylcholinesterase in differentiating rat cerebral cells in culture. *Brain Res. 155*, 368-73.

Dufton, M.J. and Hider, R.C. (1983) Conformational properties of the neurotoxins and cytotoxins isolated from elapid snake venoms. *CRC Crit. Rev. Biochem. 14*, 113-71.

Dunlap, K. (1981) Two types of γ-aminobutyric acid receptor on embryonic sensory neurones. *Br. J. Pharmac. 74*, 579-85.

Dunlap, K. and Fischbach, G.D. (1978) Neurotransmitters decrease the calcium component of sensory neurone action potentials *Nature 276*, 837-9.

Dunlap, K. and Fischbach, G.D. (1981) Neurotransmitters decrease the calcium conductance activated by depolarization of embryonic chick sensory neurones. *J. Physiol. 317*, 519-35.

Dvorak, D., Gipps, E., Leah, J. and Kidson, C. (1978) Development of receptors for α-bungarotoxin in chick embryo sympathetic ganglion neurons *in vitro*. *Life Sci.* *22*, 407-14.
Ebihara, L., Shigeto, N., Lieberman, M. and Johnson, E.A. (1980) The initial inward current in spherical clusters of chick embryonic heart cells. *J. Gen. Physiol.* *75*, 437-56.
El-Fakahany, E. and Richelson, E. (1983) Antagonism by antidepressants of muscarinic acetylcholine receptors of human brain. *Br. J. Pharmac.* *78*, 97-102.
Elmqvist, D., Johns, T.R. and Thesleff, S. (1960) A study of some electrophysiological properties of human intercostal muscle. *J. Physiol.* *154*, 602-7.
Elsas, L.J., Wheeler, F.B., Danner, D.J. and DeHaan, R.L. (1975) Amino acid transport by aggregates of cultured chicken heart cells. Effect of insulin. *J. Biol. Chem.* *250*, 9381-90.
Engelhardt, J.K., Ishikawa, K., Lisbin, S.J. and Mori, J. (1976) Neurotrophic effects on passive electrical properties of cultured chick skeletal muscle. *Brain Res.* *110*, 170-4.
Engelhardt, J.K., Ishikawa, K., Mori, J. and Shima-Bukuro, Y. (1977) Neurotrophic effects on the electrical properties of cultured muscle produced by conditioned medium from spinal cord explants. *Brain Res.* *128*, 243-8.
Ertel, R.J., Clarke, D.E., Chao, J.C. and Franke, F.R. (1971) Autonomic receptor mechanisms in embryonic chick myocardial cell cultures. *J. Pharmac. Exp. Ther.* *178*, 73-80.
Fambrough, D.M. (1974) Acetylcholine receptors. Revised estimates of extrajunctional receptor density in denervated rat diaphragm. *J. Gen. Physiol.* *64*, 468-72.
Fambrough, D.M. (1979) Control of acetylcholine receptors in skeletal muscle. *Physiol Rev.* *59*, 165-227.
Fambrough, D.M. and Devreotes, P.N. (1978) Newly synthesized acetylcholine receptors are located in the Golgi apparatus. *J. Cell Biol.* *76*, 237-44.
Fambrough, D.M. and Hartzell, H.C. (1972) Acetylcholine receptors: number and distribution at neuromuscular junctions in rat diaphragm. *Science* *176*, 189-91.
Fambrough, D.M. and Rash, J.E. (1971) Development of acetylcholine sensitivity during myogenesis. *Dev. Biol.* *26*, 55-68.
Fänge, R., Persson, H. and Thesleff, S. (1956) Electrophysiologic and pharmacological observations on trypsin-disintegrated embryonic chick heart cultured *in vitro*. *Acta Physiol. Scand.* *38*, 173-83.
Farley, J.M. and Narahashi, T. (1983) Effects of drugs on acetylcholine-activated ionic channels of internally-perfused chick myoballs. *J. Physiol.* *337*, 753-68.
Fayet, G., Couraud, F., Miranda, F. and Lissitzky, S. (1974) Electro-optical system for monitoring activity of heart cells in culture: application to the study of several drugs and scorpion toxins. *Eur. J. Pharmac.* *27*, 165-74.
Fedde, M.R. (1969) Electrical properties and acetylcholine sensitivity

of singly and multiply innervated avian muscle fibers. *J. Gen. Physiol. 53*, 624-37.

Fenton, R.A., Bruttig, S.P., Rubio, R. and Berne, R.M. (1982) Effect of adenosine on calcium uptake by intact and cultured vascular smooth muscle. *Am. J. Physiol. 242*, H797-804.

Fenwick, E.M., Marty, A. and Neher, E. (1982a) A patch-clamp study of bovine chromaffin cells and of their sensitivity to acetylcholine. *J. Physiol. 331*, 577-97.

Fenwick, E.M., Marty, A. and Neher, E. (1982b) Sodium and calcium channels in bovine chromaffin cells. *J. Physiol. 331*, 599-635.

Fertuk, H.C. and Salpeter, M.M. (1976) Quantitation of junctional and extrajunctional acetylcholine receptors by electron microscope autoradiography after ^{125}I-α-bungarotoxin binding at mouse neuromuscular junctions. *J. Cell Biol. 69*, 144-58.

Fischbach, G.D. (1970) Synaptic potentials recorded in cell cultures of nerve and muscle. *Science 169*, 1331-3.

Fischbach, G.D. (1972) Synapse formation between dissociated nerve and muscle cells in low density cell cultures. *Dev. Biol. 28*, 407-29.

Fischbach, G.D. and Cohen, S.A. (1973) The distribution of acetylcholine sensitivity over uninnervated and innervated muscle fibers grown in cell culture. *Dev. Biol. 31*, 147-62.

Fischbach, G.D. and Dichter, M.A. (1974) Electrophysiologic and morphologic properties of neurons in dissociated chick spinal cord cell cultures. *Dev. Biol. 37*, 100-66.

Fischbach, G.D. and Lass, Y. (1978a) Acetylcholine noise in cultured chick myoballs: a voltage clamp analysis. *J. Physiol. 280*, 515-26.

Fischbach, G.D. and Lass, Y. (1978b) A transition temperature for acetylcholine channel conductance in chick myoballs. *J. Physiol. 280*, 527-36.

Fischbach, G.D., Nameroff, M. and Nelson, P.G. (1971) Electrical properties of chick skeletal muscle fibers developing in cell culture. *J. Cell. Physiol. 78*, 289-300.

Fischbach, G.D. and Nelson, P.G. (1977) Cell culture in neurobiology in *Handbook of Physiology*, Section 1, Vol. 1. Kandel, E.R. (ed.), American Physiol. Soc., Bethesda, Maryland, pp. 719-74.

Fischbach, G.D. and Schuetze, S.M. (1980) A post-natal decrease in acetylcholine open time at rat end-plates. *J. Physiol. 303*, 125-37.

Fisher, S.K., Holz, R.W. and Agranoff, B.W. (1981) Muscarinic receptors in chromaffin cell cultures mediate enhanced phospholipid labelling but not catecholamine secretion. *J. Neurochem. 37*, 491-7.

Fishman, M.C. and Nelson, P.G. (1981) Depolarization-induced synaptic plasticity at cholinergic synapses in tissue culture. *J. Neurosci. 1*, 1043-51.

Fishman, M.C. and Spector, I. (1981) Potassium current suppression by quinidine reveals additional calcium currents in neuroblastoma cells. *Proc. Nat. Acad. Sci.* USA *78*, 5245-9.

Fosset, M., De Barry, J., Lenoir, M.C. and Lazdunski, M. (1977) Analysis of molecular aspects of Na^+ and Ca^{2+} uptakes by embryonic cardiac cells in culture. *J. Biol. Chem. 252*, 6112-17.

Fowler, S., Shio, H. and Wolinsky, H. (1977) Subcellular fractionation and morphology of calf aortic smooth muscle cells. Studies on whole aorta, aortic explants and subcultures grown under different conditions. *J. Cell Biol. 75*, 166-84.
Frank, E. and Fischbach, G.D. (1979) Early events in neuromuscular junction formation *in vitro*. Induction of acetylcholine receptor clusters in the postsynaptic membrane and morphology of newly-formed synapses. *J. Cell Biol. 83*, 143-58.
Frank, J.S., Langer, G.A., Nudd, L.M. and Seraydarian, K. (1977) The myocardial cell surface, its histochemistry and the effect of sialic acid and calcium removal on its structure and cellular ionic exchange. *Circulation Res. 41*, 702-14.
Franke, W.W., Schmid, E., Vandekerckhove, J. and Weber, W. (1980) A permanently proliferating rat vascular smooth muscle cell with maintained expression of smooth muscle characteristics, including actin of the vascular smooth muscle type. *J. Cell Biol. 87*, 594-600.
Freedman, S.B., Dawson, G., Miller, R.J., Villereal, M.L. and West, R.E. (1983) Identification and pharmacology of voltage sensitive calcium channels in clonal cell lines. *Br. J. Pharmac. 78*, 80P.
Frelin, C., Lombet, A., Vigne, P., Romey, G. and Lazdunski, M. (1981) The appearance of voltage-sensitive Na^+ channels during the *in vitro* differentiation of embryonic chick skeletal muscle cells. *J. Biol. Chem. 256*, 12355-61.
Frere, R.C., MacDonald, R.L. and Young, A.B. (1982) GABA binding and bicuculline in spinal cord and cortical membranes from adult rat and from mouse neurons in cell culture. *Brain Res. 244*, 145-53.
Freschi, J.E. (1982) Effect of serum-free medium on growth and differentiation of sympathetic neurons in culture. *Dev. Brain Res. 4*, 455-64.
Freschi, J.E., Parfitt, A.G. and Shain, W.G. (1979) Electrophysiology and pharmacology of striated muscle fibres cultured from dissociated neonatal rat pineal glands. *J. Physiol. 293*, 1-10.
Freschi, J.E. and Shain, W.G. (1980) Slow muscarinic depolarization in neurons of dissociated rat superior cervical ganglia can be evoked by iontophoresis of acetylcholine. *Brain Res. 185*, 429-34.
Freschi, J.E. and Shain, W.G. (1982) Electrophysiological and pharmacological characteristics of the serotonin response on a vertebrate neuronal somatic cell hybrid. *J. Neurosci. 2*, 106-12.
Fukada, J. and Kameyama, M. (1979) Enhancement of Ca spikes in nerve cells of adult mammals during neurite growth in tissue culture. *Nature 279*, 546-8.
Fukada, J., Fischbach, G.D. and Smith, T.G. (1976a) A voltage clamp study of the sodium, calcium and chloride spikes of chick skeletal muscle cells grown in tissue culture. *Dev. Biol. 49*, 412-24.
Fukada, J., Henkart, M.P., Fischbach, G.D. and Smith, T.G. (1976b) Physiological and structural properties of colchicine-treated chick skeletal muscle cells grown in tissue culture. *Dev. Biol. 49*, 395-411.
Fukada, J. and Kameyama, M. (1980) A tissue-culture of nerve cells from adult mammalian ganglia and some electrophysiological pro-

perties of the nerve cells *in vitro*. *Brain Res. 202*, 249-55.
Furmanski, P., Silverman, D.J. and Lubin, M. (1971) Expression of differentiated functions in mouse neuroblastoma mediated by dibutyryl-cyclic adrenosine monophosphate. *Nature 233*, 413-15.
Furshpan, E.J., MacLeish, P.R., O'Lague, P.H. and Potter, D.D. (1976) Chemical transmission between rat sympathetic neurons and cardiac myocytes developing in microcultures: evidence for cholinergic, adrenergic and dual-function neurons. *Proc. Nat. Acad. Sci.* USA *73*, 4225-9.
Gähwiler, B.H. (1975) The effects of GABA, picrotoxin and bicuculline on the spontaneous bioelectric activity of cultured cerebellar Purkinje cells. *Brain Res. 99*, 85-95.
Gähwiler, B.H. (1980) Excitatory action of opioid peptides and opiates on cultured hippocampal pyramidal cells. *Brain Res. 194*, 193-203.
Gähwiler, B.H. (1981) Organotypic monolayer cultures of nervous tissue. *J. Neurosci. Meth. 4*, 329-42.
Gähwiler, B.H. and Dreifuss, J.J. (1979) Hypothalamic neurones in culture. II. A progress report. *J. Physiol. (Paris) 75*, 23-6.
Gähwiler, B.H. and Dreifuss, J.J. (1982) Multiple actions of acetylcholine on hippocampal pyramidal cells in organotypic explant cultures. *Neuroscience 7*, 1243-56.
Gähwiler, B.H., Mamoon, A.M. and Tobias, C.A. (1973) Spontaneous bioelectric activity of cultured cerebellar Purkinje cells during exposure to agents which prevent synaptic transmission. *Brain Res. 53*, 71-9.
Galper, J.B. and Catterall, W.A. (1978) Developmental changes in the sensitivity of embryonic heart cells to tetrodotoxin and D600. *Dev. Biol. 65*, 216-27.
Galper, J.B. and Smith, T.W. (1978) Properties of muscarinic acetylcholine receptors in heart cell cultures. *Proc. Nat. Acad. Sci.* USA *75*, 5831-5.
Galper, J.B. and Smith, T.W. (1980) Agonist and guanine nucleotide modulation of muscarinic cholinergic receptors in cultured heart cells. *J. Biol. Chem. 255*, 9571-9.
Galper, J.B., Dziekan, L.C., O'Hara, D.S. and Smith, T.W. (1982) The biphasic response of muscarinic cholinergic receptors in cultured heart cells to agonists. Effects on receptor number and affinity in intact cells and homogenates. *J. Biol. Chem. 257*, 10344-56.
Galper, J.B., Klein, W. and Catterall, W.A. (1977) Muscarinic acetylcholine receptors in developing chick heart. *J. Biol. Chem. 252*, 8692-9.
Gardner, J.M. and Fambrough, D.M. (1979) Acetylcholine receptor degradation measured by density labelling: effects of cholinergic ligands and evidence against recycling. *Cell 16*, 661-74.
Geller, H.M. (1981) Histamine actions on activity of cultured hypothalamic neurons: evidence for mediation by H_1- and H_2- histamine receptors. *Dev. Brain Res. 1*, 89-101.
Geller, H.M. and Woodward, D.J. (1974) Responses of cultured cerebellar neurons to iontophoretically applied amino acids. *Brain Res.*

74, 67-80.

Gilman, A.G. and Minna, J.D. (1973) Expression of genes for metabolism of cyclic adenosine 3′,5′-monophosphate in somatic cells. I. Responses to catecholamines in parental and hybrid cells. *J. Biol. Chem. 248,* 6610-17.

Gilman, A.G. and Nirenberg, M. (1971) Regulation of adenosine 3′, 5′-cyclic monophosphate metabolism in cultured neuroblastoma cells. *Nature*, *234*, 356-8.

Gimbourne, M.A. and Cotran, R.S. (1975) Human vascular smooth muscle in culture. Growth and ultrastructure. *Lab. Invest. 33*, 16-27.

Godfrey, E.W., Nelson, P.G., Schrier, B.K., Breuer, A.C. and Ransom, B.R. (1975) Neurons from fetal rat brain in a new cell culture system: a multidisciplinary analysis. *Brain Res. 90,* 1-21.

Goshima, K. (1974) Initiation of beating in quiescent myocardial cells by norepinephrine, by contact with beating cells and by electrical stimulation of adjacent FL cells. *Exp. Cell Res. 84*, 223-34.

Goshima, K. (1976) Arrhythmic movements of myocardial cells in culture and their improvement with antiarrhythmic drugs. *J. Mol. Cell. Cardiol. 8*, 217-38.

Goshima, K. (1977) Ouabain-induced arrhythmias of single isolated myocardial cells and cell clusters cultured *in vitro* and their improvement by quinidine. *J. Mol. Cell. Cardiol. 9*, 7-23.

Goshima, K., Masuda, A., Matsui, Y. and Yoshino, S. (1979) Beating of multinucleated giant myocardial cells in culture. *Exp. Cell Res. 120*, 285-93.

Goshima, K. and Wakabayashi, S. (1981a) Inhibitions of ouabain-induced arrhythmias of ouabain-sensitive cells (quail) by contact with ouabain-resistant cells (mouse) and its mechanism. *J. Mol. Cell. Cardiol. 13*, 75-92.

Goshima, K. and Wakabayashi, S. (1981b) Involvement of an Na^+-Ca^{2+} exchange system in genesis of ouabain-induced arrhythmias of cultured myocardial cells. *J. Mol. Cell. Cardiol. 13*, 489-509.

Goshima, K., Wakabayashi, S. and Masuda, A. (1980) Ionic mechanism of morphological changes of cultured myocardial cells on successive incubation in media without and with Ca^{2+}. *J. Mol. Cell. Cardiol. 12*, 1135-57.

Greene, L.A. and Rein, G. (1977a) Release of [^{3}H] norepinephrine from a clonal line of pheochromocytoma cells (PC12) by nicotinic cholinergic stimulation. *Brain Res. 138*, 521-8.

Greene, L.A. and Rein, G. (1977b) Synthesis, storage and release of acetylcholine by a noradrenergic pheochromocytoma cell line. *Nature 268*, 349-51.

Greene, L.A. and Rein, G. (1978) Release of norepinephrine from neurons in dissociated cell cultures of chick sympathetic ganglia via stimulation of nicotinic and muscarinic acetylcholine receptors. *J. Neurochem. 30*, 579-86.

Greene, L.A., Shain, W., Chalazonitis, A., Breakfield, X., Minna, J., Coon, H.G. and Nirenberg, M. (1975) Neuronal properties of hybrid neuroblastoma × sympathetic ganglion cells. *Proc. Nat. Acad. Sci.*

USA *72*, 4923-7.

Greene, L.A., Sytkowski, A.J., Vogel, Z. and Nirenberg, M.W. (1973) α-Bungarotoxin used as a probe for acetylcholine receptors of cultured neurones. *Nature 243*, 163-6.

Greene, L.A. and Tischler, A.S. (1976) Establishment of a noradrenergic clonal line of rat adrenal pheochromocytoma cells which respond to nerve growth factor. *Proc. Nat. Acad. Sci.* USA *73*, 2424-8.

Grinvald, A., Ross, W.N. and Farber, I. (1981) Simultaneous optical measurements of electrical activity from multiple sites on processes of cultured neurons. *Proc. Nat. Acad. Sci.* USA *78*, 3245-9.

Gröschel-Stewart, U., Chamley, J.H., Campbell, G.R. and Burnstock, G. (1975) Changes in myosin distribution in dedifferentiating and redifferentiating smooth muscle cells in tissue culture. *Cell Tiss. Res. 165*, 13-22.

Gruener, R. and Kidokoro, Y. (1982) Acetylcholine sensitivity of innervated and noninnervated *Xenopus* muscle cells in culture. *Dev. Biol. 91*, 86-92.

Gruol, D.L., Siggins, G.R., Padjen, A.L. and Forman, D.S. (1981) Explant cultures of adult amphibian sympathetic ganglia: electrophysiological and pharmacological investigation of neurotransmitter and nucleotide action. *Brain Res. 223*, 81-106.

Guharay, G. and Usherwood, P.N.R. (1981) Characterization of the effect of 5-hydroxytryptamine on N1E-115 neuroblastoma cells. *Br. J. Pharmac. 74*, 294P-5P.

Gullis, R.J. (1977) Statement. *Nature 265*, 764.

Guroff, G., Dickens, G., End, D. and Landos, C. (1981) The action of adenosine analogs on PC12 cells. *J. Neurochem. 37*, 1431-9.

Gutmann, E. (1976) Neurotrophic relations. *Ann. Rev. Physiol. 38*, 177-216.

Gwynn, G.J. and Costa, E. (1982) Opioids regulate cGMP formation in cloned neuroblastoma cells. *Proc. Nat. Acad. Sci.* USA *79*, 690-4.

Hagiwara, S. and Ohmori, H. (1982) Studies of calcium channels in rat clonal pituitary cells with patch electrode voltage clamp. *J. Physiol. 331*, 231-52.

Hagiwara, S. and Ohmori, H. (1983) Studies of single channel currents in rat clonal pituitary cells. *J. Physiol. 336*, 649-61.

Halbert, S.P., Bruderer, R. and Thompson, A. (1973) Growth of dissociated beating human heart cells in tissue culture. *Life Sci. 13*, 969-75.

Hamill, O.P., Marty, A., Neher, E., Sakmann, B. and Sigworth, F.J. (1981) Improved patch-clamp techniques for high-resolution current recording from cells and cell-free membrane patches. *Pflügers Arch. 381*, 85-100.

Hamill, O.P. and Sakmann, B. (1981) Multiple conductance states of single acetylcholine receptor channels in embryonic muscle cells. *Nature 294*, 462-4.

Hammonds, R.G. and Li, C.H. (1981) Human β-endorphin: specific binding in neuroblastoma N18TG2 cells. *Proc. Nat. Acad. Sci.* USA

78, 6764-5.

Hamprecht, B. (1974) Cell cultures as model systems for studying the biochemistry of differentiated functions in nerve cells, in *Biochemistry of Sensory Functions*, L. Jaenicke (ed.), Springer-Verlag, Berlin, pp. 391-421.

Hamprecht, B. (1977a) Structural, electrophysiological, biochemical and pharmacological properties of neuroblastoma-glioma cell hybrids in cell culture. *Int. Rev. Cytol. 49*, 99-170.

Hamprecht, B. (1977b) Statement. *Nature 265*, 764.

Harary, I. and Farley, B. (1963) *In vitro* studies on single beating rat heart cells. I. Growth and organization. *Exp. Cell Res. 29*, 451-65.

Harris, A.J. and Dennis, M.J. (1970) Acetylcholine sensitivity and distribution on mouse neuroblastoma cells. *Science 167*, 1253-5.

Harris, A.J., Heinemann, S., Schubert, D. and Tarikis, H. (1971) Trophic interaction between cloned tissue culture lines of nerve and muscle. *Nature 231*, 296-301.

Harris, J.B., Marshall, M.W. and Wilson, P. (1973) A physiological study of chick myotubes grown in tissue culture. *J. Physiol. 229*, 751-66.

Harrison, R.G. (1907) Observations on the living developing nerve fiber. *Proc. Soc. Exp. Biol. Med. 4*, 140-3.

Harrison, R.G. (1910) The outgrowth of the nerve fiber as a mode of protoplasmic movement. *J. Exp. Zool. 9*, 787-848.

Hartzell, H.C. (1980) Distribution of muscarinic acetylcholine receptors and presynaptic nerve terminals in amphibian heart. *J. Cell Biol. 86*, 6-20.

Hartzell, H.C. and Fambrough, D.M. (1973) Acetylcholine receptor production and incorporation into membranes of developing muscle fibers. *Dev. Biol. 30*, 153-65.

Harvey, A.L. (1980) Actions of drugs on developing skeletal muscle. *Pharmac. Ther. 11*, 1-41.

Harvey, A.L. and Dryden, W.F. (1974a) Depolarization, desensitization and the effects of tubocurarine and neostigmine in cultured skeletal muscle. *Eur. J. Pharmac. 27*, 5-13.

Harvey, A.L. and Dryden, W.F. (1974b) Studies on the pharmacology of skeletal muscle in culture: specificity of receptors. *Eur. J. Pharmac. 28*, 125-30.

Harvey, A.L. and Dryden, W.F. (1974c) The actions of some anticholinesterase drugs on skeletal muscle in culture. *J. Pharm. Pharmac. 26*, 865-70.

Harvey, A.L., Paul, D. and Singh, H. (1975) Actions of chandonium iodide on skeletal muscle in culture. *J. Pharm. Pharmac. 27*, Suppl., 62P.

Harvey, A.L., Robertson, J.G. and Witkowski, J.A. (1979) Maturation of human skeletal muscle fibres in explant tissue culture. *J. Neurol. Sci. 41*, 115-22.

Harvey, A.L., Robertson, J.G. and Witkowski, J.A. (1980) Increased membrane potentials of human skeletal muscle fibres in explant tissue culture. *J. Neurol. Sci. 44*, 273-4.

Harvey, A.L. and Van Helden, D. (1981) Acetylcholine receptors in

singly and multiply innervated skeletal muscle fibres of the chicken during development. *J. Physiol. 317*, 397-411.
Hasin, Y., Shimoni, Y., Stein, O. and Stein, Y. (1980) Effect of cholesterol depletion on the electrical activity of rat heart myocytes in culture. *J. Mol. Cell. Cardiol. 12*, 675-83.
Hauschka, S.D. (1972) Cultivation of muscle tissue, in *Growth, Nutrition and Metabolism of Cells in Culture*, G.H. Rothblat and V.J. Cristofalo (eds.), Vol. 2, Academic Press, New York, pp. 67-130.
Hazum, E., Chang, K.-J. and Cuatrecasas, P. (1980) Cluster formation of opiate (enkephalin) receptors in neuroblastoma cells: differences between agonists and antagonists and possible relationships to biological functions. *Proc. Nat. Acad. Sci.* USA *77*, 3038-41.
Heinemann, S., Bevan, S., Kullberg, R., Lindstrom, J. and Rice, J. (1977) Modulation of acetylcholine receptor by antibody against the receptor. *Proc. Nat. Acad. Sci.* USA *74*, 3090-4.
Helfand, S.L., Smith, G.A. and Wessells, N.K. (1976) Survival and development in culture of dissociated parasympathetic neurons from ciliary ganglia. *Dev. Biol. 50*, 541-7.
Hermsmeyer, K. (1976) Cellular basis for increased sensitivity of vascular smooth muscle in spontaneously hypertensive rats. *Circ. Res. 38*, Suppl. II, 53-7.
Hermsmeyer, K., De Cino, P. and White, R. (1976) Spontaneous contractions of dispersed vascular muscle in cell culture. *In Vitro 12*, 628-34.
Hermsmeyer, K. and Robinson, R.B. (1977) High sensitivity of cultured cardiac muscle cells to autonomic agents. *Am. J. Physiol. 233*, C172-9.
Heyer, E.J. and MacDonald, R.L. (1982a) Calcium- and sodium-dependent action potentials of mouse spinal cord and dorsal root ganglion neurons in cell culture. *J. Neurophysiol. 47*, 641-55.
Heyer, E.J. and MacDonald, R.L. (1982b) Barbiturate reduction of calcium-dependent action potentials: correlation with anaesthetic action. *Brain Res. 236*, 157-71.
Heyer, E.J., MacDonald, R.L., Bergey, G.K. and Nelson, P.G. (1981) Calcium-dependent action potentials in mouse spinal cord neurons in cell culture. *Brain Res. 220*, 408-15.
Higgins, D. and Burton, H. (1982) Electrotonic synapses are formed by fetal rat sympathetic neurons maintained in a chemically-defined culture medium. *Neuroscience 7*, 2241-53.
Hild, W., Chang, J.J. and Tasaki, I. (1958) Electrical responses of astrocytic glia from the mammalian central nervous system cultivated *in vitro*. *Experientia 14*, 220-1.
Hild, W. and Tasaki, I. (1962) Morphological and physiological properties of neurons and glial cells in tissue culture. *J. Neurophysiol. 25*, 277-304.
Hill, C.E., Purves, R.D., Watanabe, H. and Burnstock, G. (1976) Specificity of innervation of iris musculature by sympathetic nerve fibres in tissue culture. *Pflügers Arch. 361*, 127-34.
Hogg, B.M., Goss, C.M. and Cole, K.S. (1934) Potentials in embryo rat

heart muscle cultures. *Proc. Soc. Exp. Biol. Med. 32*, 304-7.

Holz, R.W., Senter, R.A. and Frye, R.A. (1982) Relationship between Ca^{2+} uptake and catecholamine secretion in primary dissociated cultures of adrenal medulla. *J. Neurochem. 39*, 635-46.

Hooisma, J., Slaaf, D.W., Meeter, E. and Stevens, W.F. (1975) The innervation of chick striated muscle fibers by the chick ciliary ganglion in tissue culture. *Brain Res. 85*, 79-85.

Horn, R. and Bordwick, M.S. (1980) Acetylcholine-induced current in perfused rat myoballs. *J. Gen. Physiol. 75*, 297-321.

Horn, R., Bordwick, M.S. and Dickey, W.D. (1980) Asymmetry of the acetylcholine channel revealed by quaternary anesthetics. *Science, 210*, 205-7.

Horn, R. and Patlak, J. (1980) Single channel currents from excised patches of muscle membrane. *Proc. Nat. Acad. Sci.* USA *77*, 6930-4.

Horowitz, J.D., Barry, W.H. and Smith, T.W. (1982) Lack of interaction between digoxin and quinidine in cultured heart cells. *J. Pharmac. Exp. Ther. 220*, 488-93.

Hösli, L., Andrès, P.F. and Hösli, E. (1971) Effects of glycine on spinal neurones grown in tissue culture. *Brain Res. 34*, 399-402.

Hösli, L., Andrès, P.F. and Hösli, E. (1972) Effects of potassium on the membrane potential of spinal neurones in tissue culture. *Pflügers Arch. 333*, 362-5.

Hösli, L., Andrès, P.F. and Hösli, E. (1976) Ionic mechanisms associated with the depolarization by glutamate and aspartate on human and rat spinal neurones in tissue culture. *Pflügers Arch. 363*, 43-8.

Hösli, L. and Hösli, E. (1978) Action and uptake of neurotransmitters in CNS tissue culture. *Rev. Physiol. Biochem. Pharmac. 81*, 135-88.

Hösli, L., Hösli, E. and Andrès, P.F. (1973) Nervous tissue culture – a model to study action and uptake of putative neurotransmitters such as amino acids. *Brain Res. 62,* 597-602.

Huang, L.-Y.M., Catterall, W.A. and Ehrenstein, G. (1978) Selectivity of cations and nonelectrolytes for acetylcholine-activated channels in cultured muscle cells. *J. Gen. Physiol. 71*, 397-410.

Huang, L.-Y.M., Catterall, W.A. and Ehrenstein, G. (1979) Comparison of ionic selectivity of batrachotoxin-activated channels with different tetrodotoxin dissociation constants. *J. Gen. Physiol. 73*, 839-54.

Huang, L.-Y.M., Moran, N. and Ehrenstein, G. (1982) Batrachotoxin modifies the gating kinetics of sodium channels in internally perfused neuroblastoma cells. *Proc. Nat. Acad. Sci.* USA *79*, 2082-5.

Hubbard, J.I., Llinàs, R. and Quastel, D.M. (1969) *Electrophysiological Analysis of Synaptic Transmission.* Edward Arnold, London.

Hugues, M., Romey, G., Duval, D., Vincent, J.P. and Lazdunski, M. (1982) Apamin as a selective blocker of the calcium-dependent potassium channel in neuroblastoma cells: voltage-clamp and biochemical characterization of the toxin receptor. *Proc. Nat. Acad. Sci.* USA *79*, 1308-12.

Iacovitti, L., Joh, T.H., Park, D.H. and Bunge, R.P. (1981) Dual expression of neurotransmitter synthesis in cultured autonomic neurons.

J. Neurosci. 1, 685-90.

Iacovitti, L., Johnson, M.I., Joh, T.H. and Bunge, R.P. (1982) Biochemical and morphological characterization of sympathetic neurons grown in a chemically-defined medium, *Neuroscience 7*, 2225-39.

Iijima, T. and Pappano, A.J. (1979) Ontogenetic increase of the maximal rate of rise of the chick embryonic heart action potential: relationship to voltage, time and tetrodotoxin. *Circulation Res. 44*, 358-67.

Ingebrigtsen, R. (1913) Studies of the degeneration and regeneration of axis cylinders *in vitro*. *J. Exp. Med. 17*, 182-91.

Ives, H.E., Schultz, G.S., Galardy, R.E. and Jamieson, J.D. (1978) Preparation of functional smooth muscle cells from the rabbit aorta. *J. Exp. Med. 148*, 1400-13.

Jackson, M.B. and Lecar, H. (1979) Single postsynaptic channel currents in tissue cultured muscle. *Nature 282*, 863-4.

Jackson, M.B., Lecar, H., Askanas, V. and Engel, W.K. (1982a) Single cholinergic receptor channel currents in cultured human muscle. *J. Neurosci. 2*, 1465-73.

Jackson, M.B., Lecar, H., Mathers, D.A. and Barker, J.L. (1982b) Single channel currents activated by GABA, muscimol and (–) pentobarbital in cultured mouse spinal neurons. *J. Neurosci. 2*, 889-94.

Jessell, T.M., Siegel, R.E. and Fischbach, G.D. (1979) Induction of acetylcholine receptors on cultured skeletal muscle by a factor extracted from brain and spinal cord. *Proc. Nat. Acad. Sci.* USA *76*, 5397-401.

Jessen, K.R., McConnell, J.D., Purves, R.D., Burnstock, G. and Chamley-Campbell, J. (1978) Tissue culture of mammalian enteric neurons. *Brain Res. 152*, 573-9.

Jessen, K.R., Saffrey, M.J. and Burnstock, G. (1983) The enteric nervous system in tissue culture. I. Cell types and their interactions in explants of the myenteric and submucous plexuses from guinea pig, rabbit and rat. *Brain Res. 262*, 17-35.

Johnson, M., Ross, D., Meyers, M., Rees, R., Bunge, R., Wakshull, E. and Burton, H. (1976) Synaptic vesicle cytochemistry changes when cultured sympathetic neurones develop cholinergic interactions. *Nature 262*, 308-10.

Jones, J.L., Lepeschkin, E., Jones, R.E. and Rush, S. (1978) Response of cultured myocardial cells to countershock-type electric field stimulation. *Am. J. Physiol. 235*, H214-22.

Josephson, I. and Sperelakis, N. (1977) Ouabain blockade of inward slow current in cardiac muscle. *J. Mol. Cell. Cardiol. 9*, 409-18.

Josephson, I. and Sperelakis, N. (1978) 5′-Guanylimidodiphosphate stimulation of slow Ca^{2+} current in myocardial cells. *J. Mol. Cell. Cardiol. 10*, 1157-66.

Jourdon, P. and Sperelakis, N. (1980) Electrical properties of cultured heart cell reaggregates from newborn rat ventricles: comparison with intact non-cultured ventricles. *J. Mol. Cell. Cardiol. 12*, 1441-58.

Jumblatt, J.E. and Tischler, A.S. (1982) Regulation of muscarinic ligand binding sites by nerve growth factor in PC12 phaeochromocytoma cells. *Nature 297*, 152-4.

Kalcheim, C., Duksin, D. and Vogel, Z. (1982) Aggregation of acetylcholine receptors in nerve-muscle cocultures is decreased by inhibitors of collagen production. *Neurosci. Lett. 31*, 265-70.

Kano, M. (1975) Development of excitability in embryonic chick skeletal muscle cells. *J. Cell. Physiol. 86*, 503-10.

Kano, M. and Shimada, Y. (1971) Innervation and acetylcholine sensitivity of skeletal muscle cells differentiated *in vitro* from chick embryo. *J. Cell. Physiol. 78*, 233-42.

Kano, M. and Shimada, Y. (1973) Tetrodotoxin-resistant electric activity in chick skeletal muscle cells differentiated *in vitro*. *J. Cell. Physiol. 81*, 85-90.

Kano, M., Shimada, Y. and Ishikawa, K. (1971) Acetylcholine sensitivity of skeletal muscle cells differentiated *in vitro* from chick embryo. *Brain Res. 25*, 216-19.

Kano, M., Shimada, Y. and Ishikawa, K. (1972) Electrogenesis of embryonic chick skeletal muscle cells differentiated *in vitro*. *J. Cell. Physiol. 79*, 363-6.

Kano, M. and Yamamoto, M. (1977) Development of spike potentials in skeletal muscle cells differentiated *in vitro* from chick embryo. *J. Cell. Physiol. 90*, 439-41.

Kano, M., Susuki, N. and Ojima, H. (1979) Neurotrophic effect of nerve extract on development of tetrodotoxin-sensitive spike potential in skeletal muscle cells in culture. *J. Cell. Physiol. 99*, 327-32.

Kao, I. and Drachman, D.B. (1977) Myasthenic immunoglobulin accelerates acetylcholine receptor degradation. *Science 196*, 527-9.

Karlsson, E. (1979) Chemistry of protein toxins in snake venoms, in *Snake Venoms*. Vol. 52, *Handbk. Exp. Pharmac.* C.-Y. Lee (ed.), Springer-Verlag, Berlin, pp. 159-212.

Kato, E. and Narahashi, T. (1982a) Low sensitivity of the neuroblastoma cell cholinergic receptors to erabutoxin and α-bungarotoxin. *Brain Res. 245*, 159-62.

Kato, E. and Narahashi, T. (1982b) Characteristics of the electrical response to dopamine in neuroblastoma cells. *J. Physiol. 333*, 213-26.

Katz, B. and Thesleff, S. (1957) A study of the "desensitization" produced by acetylcholine at the motor end-plate. *J. Physiol. 138*, 63-80.

Kaumann, A.J., McInerny, T.K., Gilmour, D.P. and Blinks, J.R. (1980) Comparative assessment of β-adrenoceptor blocking agents as simple competitive antagonists in isolated heart muscle: similarity of inotropic and chronotropic blocking potencies against isoproterenol. *Naunyn-Schmiedebergs Arch. Pharmac. 311*, 219-36.

Kenimer, J.G. and Nirenberg, M. (1981) Desensitization of adenylate cyclase to prostaglandin E_1 or 2-chloroadenosine. *Mol. Pharmac. 20*, 585-91.

Kidokoro, Y. (1973) Development of action potentials in a clonal rat

skeletal cell line. *Nature New Biol.* *241*, 158-9.
Kidokoro, Y. (1975a) Sodium and calcium components of the action potential in a developing skeletal muscle cell line. *J. Physiol.* *244*, 145-59.
Kidokoro, Y. (1975b) Developmental changes of membrane electrical properties in a rat skeletal muscle cell line. *J. Physiol.* *244*, 129-43.
Kidokoro, Y., Anderson, M.J. and Gruener, R. (1980) Changes in synaptic potential properties during acetylcholine receptor accumulation and neurospecific interactions in *Xenopus* nerve-muscle cell culture. *Dev. Biol.* *78*, 464-83.
Kidokoro, Y. and Gruener, R. (1982) Distribution and density of α-bungarotoxin binding sites on innervated and noninnervated *Xenopus* muscle cells in culture. *Dev. Biol.* *91*, 78-85.
Kidokoro, Y. and Heinemann, S. (1974) Synapse formation between clonal muscle cells and rat spinal cord explants. *Nature* *252*, 593-4.
Kidokoro, Y., Miyazaki, S. and Ozawa, S. (1982) Acetylcholine-induced membrane depolarization and potential fluctuations in the rat adrenal chromaffin cell. *J. Physiol.* *324*, 203-20.
Kidokoro, Y. and Patrick, J. (1978) Correlation between miniature endplate potential amplitudes and acetylcholine receptor densities in the neuromuscular contact formed *in vitro*. *Brain Res.* *142*, 368-73.
Kidokoro, Y. and Yeh, E. (1981) Synaptic contacts between embryonic *Xenopus* neurons and myotubes formed from a rat skeletal muscle cell line. *Dev. Biol.* *86*, 12-18.
Kidokoro, Y. and Yeh, E. (1982) Initial synaptic transmission at the growth cone in *Xenopus* nerve-muscle cultures. *Proc. Nat. Acad. Sci.* USA *79*, 6727-31.
Kimes, B.W. and Brandt, B.L. (1976a) Characterization of two putative smooth muscle cell lines from rat thoracic aorta. *Exp. Cell Res.* *98*, 349-66.
Kimes, B.W. and Brandt, B.L. (1976b) Properties of a clonal muscle cell line from rat heart. *Exp. Cell Res.* *98*, 367-81.
Kimhi, Y. (1981) Nerve cells in clonal systems, in *Excitable Cells in Tissue Culture*, P.G. Nelson and M. Lieberman (eds), Plenum Press, New York, pp. 173-245.
Kimhi, Y., Palfrey, C., Spector, I., Barak, Y. and Littauer, U.Z. (1976) Maturation of neuroblastoma cells in the presence of dimethylsulfoxide. *Proc. Nat. Acad. Sci.* USA *73*, 462-6.
King, K.L., Boder, G.B., Williams, D.C. and Harley, R.J. (1978) Chronotropic effect of tyramine on rat heart cells cultured with sympathetic neurons. *Eur. J. Pharmac.* *51*, 331-5.
Kitzes, M.C. and Berns, M.W. (1979) Electrical activity of rat myocardial cells in culture: La^{3+}-induced alterations. *Am. J. Physiol.* *237*, C87-95.
Klee, W.A. and Nirenberg, M. (1974) A neuroblastoma × glioma hybrid cell line with morphine receptors. *Proc. Nat. Acad. Sci.* USA *71*, 3474-7.
Klein, W.L., Nathanson, N. and Nirenberg, M. (1979) Muscarinic acetylcholine receptor regulation by accelerated rate of receptor loss.

Biochem. Biophys. Res. Comm. 90, 506-12.

Kleinfeld, M., Schade, O. and Gruen, F. (1969) Action of diphenylhydantoin on rhythmicity and contactility of cultured embryonic heart cells. *J. Pharmac. Exp. Ther. 179*, 84-90.

Ko, C.-P., Burton, H. and Bunge, R.P. (1976a) Synaptic transmission between rat spinal cord explants and dissociated superior cervical ganglion neurons in tissue culture. *Brain Res. 117*, 437-60.

Ko, C.-P., Burton, H., Johnson, M.I. and Bunge, R.P. (1976b) Synaptic transmission between rat superior cervical ganglion neurons in dissociated cell cultures. *Brain Res. 117*, 461-85.

Koidl, B., Tritthart, H.A. and Erkinger, S. (1980) Cultured embryonic chick heart cells: photometric measurement of the cell pulsation and the effects of calcium ions, electrical stimulation and temperature. *J. Mol. Cell Cardiol. 12*, 165-78.

Koike, T. and Miyake, M. (1977) Effect of concanavalin A on the cholinergic responses of mouse neuroblastoma cells. *Neurosci. Lett. 5*, 209-13.

Kondo, K., Shimizu, T. and Hayashi, O. (1981) Effects of prostaglandin D_2 on membrane potential in neuroblastoma × glioma hybrid cells as determined with a cyanine dye. *Biochem. Biophys. Res. Comm. 98*, 648-55.

Koski, G. and Klee, W.A. (1981) Opiates inhibit adenylate cyclase by stimulating GTP hydrolysis. *Proc. Nat. Acad. Sci.* USA *78*, 4185-9.

Kouvelas, E.D., Dichter, M.A. and Greene, L.A. (1978) Chick sympathetic neurons develop receptors for α-bungarotoxin *in vitro*, but the toxin does not block nicotinic receptors. *Brain Res. 154*, 83-93.

Kullberg, R.W., Brehm, P. and Steinbach, J.H. (1981) Nonjunctional acetylcholine receptor channel open time decreases during development of *Xenopus* muscle. *Nature 289*, 411-13.

Kumakura, K., Guidotti, A., Yang, H.-Y.T., Saiani, L. and Costa, E. (1980) A role for the opiate peptides that presumably co-exist with acetylcholine in splanchnic nerves, in *Neural Peptides and Neuronal Communication*. E. Costa and M. Trabucci (eds.), *Adv. Biochem. Psychopharmac. 22*, Raven Press, New York, pp. 571-80.

Kuramoto, T., Perez-Polo, J.R. and Haber, B. (1977) Membrane properties of a human neuroblastoma. *Neurosci. Lett. 4*, 151-9.

Kuramoto, T., Werrbach-Perez, K., Perez-Polo, J.R. and Haber, B. (1981) Membrane properties of a human neuroblastoma. II. Effects of differentiation. *J. Neurosci. Res. 6*, 441-9.

Kuromi, H., Gonoi, T. and Hasegawa, S. (1981) Neurotrophic substance develops tetrodotoxin-sensitive action potential and increases curare-sensitivity of acetylcholine response in cultured rat myotubes. *Dev. Brain Res. 1*, 369-79.

Lake, N.C. (1915/16) Observations upon the growth of tissues *in vitro* relating to the origin of the heart beat. *J. Physiol. 50*, 364-9.

Land, B.R., Podleski, T.R., Salpeter, E.E. and Salpeter, M.M. (1977) Acetylcholine receptor distribution on myotubes in culture correlated to acetylcholine sensitivity. *J. Physiol. 269*, 155-76.

Land, B.R., Sastre, A. and Podleski, T.R. (1973) Tetrodotoxin-sensitive and -insensitive action potentials in myotubes. *J. Cell. Physiol. 82*, 497-510.

Lane, M.-A., Sastre, A., Law, M. and Salpeter, M.M. (1977) Cholinergic and adrenergic receptors on mouse cardiocytes *in vitro*. *Dev. Biol. 57*, 254-69.

Lantz, L.C., Elsas, L.J. and DeHaan, R.L. (1980) Ouabain-resistant hyperpolarization induced by insulin in aggregates of embryonic heart cells. *Proc. Nat. Acad. Sci.* USA *77*, 3062-6.

Laqueur, E. (1914) Zur Überlebensdauer von Säugetierorganen mit Automatie. *Zentrbl. Physiol. 28*, 728.

Larno, S., Lhoste, F., Auclair, M.-C. and Lechat, P. (1980) Interaction between parathyroid hormone and the β-adrenoceptor system in cultured rat myocardial cells. *J. Mol. Cell. Cardiol. 12*, 955-64.

Lasher, R.S. and Zagon, I.S. (1972) The effect of potassium on neuronal differentiation in cultures of dissociated newborn rat cerebellum. *Brain Res. 41*, 482-8.

Lau, Y.H., Robinson, R.B., Rosen, M.R. and Bilezikian, J.P. (1980) Subclassification of β-adrenergic receptors in cultured rat cardiac myoblasts and fibroblasts. *Circulation Res. 47*, 41-8.

Law, P.Y., Hertz, A. and Loh, H.H. (1979) Demonstration and characterization of a stereospecific opiate receptor in the neuroblastoma N18TG2 cells. *J. Neurochem. 33*, 1177-87.

Lawrence, J.C. and Catterall, W.A. (1981a) Tetrodotoxin-insensitive sodium channels. Ion flux studies of neurotoxin action in a clonal rat muscle line. *J. Biol. Chem. 256*, 6213-22.

Lawrence, J.C. and Catterall, W.A. (1981b) Tetrodotoxin-insensitive sodium channels. Binding of polypeptide neurotoxins in primary cultures of rat muscle cells. *J. Biol. Chem. 256*, 6623-9.

Lawson, S.N. and Biscoe, T.J. (1973) Electrophysiological observations on immature rat nerve cells grown in tissue culture. *J. Cell. Physiol. 82*, 285-98.

Lawson, S.N., Biscoe, T.J. and Headley, P.M. (1976) The effect of electrophoretically applied GABA on cultured dissociated spinal cord and sensory ganglion neurones of the rat. *Brain Res. 117*, 493-7.

Lecar, H. and Sachs, F. (1981) Membrane noise analysis, in *Excitable Cells in Tissue Culture*, P.G. Nelson and M. Lieberman (eds.), Plenum Press, New York, pp. 137-72.

Le Douarin, G., Renaud, J.F., Renaud, D. and Coraboeuf, E. (1974) Influence of insulin on sensitivity to tetrodotoxin of isolated chick embryo heart cells in culture. *J. Mol. Cell. Cardiol. 6*, 523-9.

Lefkowitz, R.J., O'Hara, D. and Warshaw, J.B. (1974) Surface interaction of [^{3}H]norepinephrine with cultured chick embryo myocardial cells. *Biochim. Biophys. Acta 332*, 317-28.

Lehmkuhl, D. and Sperelakis, N. (1963) Transmembrane potential of trypsin-dispersed chick heart cells cultured *in vitro*. *Am. J. Physiol. 205*, 1213-20.

Leiman, A.L. and Seil, F.J. (1973) Spontaneous and evoked bioelectric activity in organised cerebellar tissue cultures. *Exp. Neurol. 40*,

748-58.

Leiman, A.L., Seil, F.J. and Kelly, J.M. (1975) Maturation of electrical activity of cerebral neocortex in tissue culture. *Exp. Neurol. 48*, 275-91.

Lewis, M.R. (1915) Rhythmical contraction of the skeletal muscle tissue observed in tissue culture. *Am. J. Physiol. 38*, 153-61.

Lewis, M.R. (1924) Spontaneous rhythmical contraction of the muscles of the bronchial tubes and air sacs of the chick embryo. *Am. J. Physiol. 68*, 385-8.

Lewis, M.R. and Lewis, W.H. (1917a) The contraction of smooth muscle cells in tissue cultures. *Am. J. Physiol. 44*, 67-74.

Lewis, M.R. and Lewis, W.H. (1917b) Behaviour of cross-striated muscle in tissue cultures. *Am. J. Anat. 22*, 169-194.

Lewis, W.H. and Lewis, M.R. (1912) The cultivation of sympathetic nerves from the intestine of chick embryos in saline solutions. *Anat. Rec. 6*, 7-31.

Libby, P., Bursztajn, S. and Goldberg, A.L. (1980) Degradation of the acetylcholine receptor in cultured muscle cells: selective inhibitors and the fate of undegraded receptors. *Cell 19*, 481-91.

Libby, P. and Goldberg, A.L. (1981) Comparison of the control and pathways for degradation of the acetylcholine receptor and average protein in cultured muscle cells. *J. Cell. Physiol. 107*, 185-94.

Lieberman, M. (1967) Effects of cell density and low K on action potentials of cultured chick heart cells. *Circulation Res. 21*, 879-88.

Lieberman, M., Horres, C.R., Shigeto, N., Ebihara, L., Aiton, J.F. and Johnson, E.A. (1981) Cardiac muscle with controlled geometry. Application to electrophysiological and ion transport studies in *Excitable Cells in Tissue Culture*, P.G. Nelson and M. Lieberman (eds.), Plenum Press, New York, pp. 379-408.

Lipshultz, S., Shanfeld, J. and Chacko, S. (1981) Emergence of β-adrenergic sensitivity in the developing chicken heart. *Proc. Nat. Acad. Sci.* USA *78*, 288-92.

Lombet, A., Frelin, C., Renaud, J.-F. and Lazdunski, M. (1982) Na^+ channels with binding sites of high and low affinity for tetrodoxin in different excitable and non-excitable cells. *Eur. J. Biochem. 124*, 199-203.

Lömo, T. (1976) The role of activity in the control of membrane and contractile properties of skeletal muscle. In *Motor Innervation of Muscle*, S.Thesleff (ed.), Academic Press, London, pp. 289-321.

Lompre, A.M., Poggioli, J. and Vassort, G. (1979) Maintenance of fast Na-channels during primary culture of embryonic chick heart cells. *J. Mol. Cell. Cardiol. 11*, 813-25.

Lumsden, C.E. (1968) Nervous tissue in culture, in *The Structure and Function of Nervous Tissue*, G.H. Bourne (ed.), Academic Press, New York, pp. 67-140.

McCall, D. (1976) Effect of quinidine and temperature on sodium uptake and contraction frequency of cultured rat myocardial cells. *Circulation Res. 39*, 730-5.

McCall, D. (1979) Cation exchange and glycoside binding in cultured rat heart cells. *Am. J. Physiol. 236*, C87-95.

McCarl, R.L., Szuhaj, B.F. and Houlihan, R.T. (1965) Steroid stimulation of beating cultured rat heart cells. *Science 150*, 1611-13.

McCarthy, K.D. and Harden, T.K. (1981) Identification of two benzodiazepine binding sites on cells cultured from rat cerebral cortex. *J. Pharmac. Exp. Ther. 216*, 183-91.

MacDermot, J., Higashida, H., Wilson, S.P., Matsuzawa, H., Minna, J. and Nirenberg, M. (1979) Adenylate cyclase and acetylcholine release regulated by separate serotonin receptors of somatic cell hybrids. *Proc. Nat. Acad. Sci.* USA *76*, 1135-39.

MacDonald, J.F. and Barker, J.L. (1982) Multiple actions of picomolar concentrations of flurazepam on the excitability of cultured mouse spinal neurons. *Brain Res. 246*, 257-64.

MacDonald, J.F., Barker, J.L., Paul, S.M., Marangos, P.J. and Skolnick, P. (1979) Inosine may be an endogenous ligand for benzodiazepine receptors on cultured spinal neurons. *Science 205*, 715-17.

MacDonald, J.F., Porietis, A.V. and Wojtowicz, J.M. (1982) L-Aspartic acid induces a region of negative slope conductance in the current-voltage relationship of cultured spinal cord neurons. *Brain Res. 237*, 248-53.

MacDonald, J.F. and Wojtowicz, J.M. (1980) Two conductance mechanisms activated by applications of L-glutamic, L-aspartic, DL-homocysteic, *N*-methyl-D-aspartic, and DL-kainic acids to cultured mammalian central neurones. *Can. J. Physiol. Pharmac. 58*, 1393-7.

MacDonald, J.F. and Wojtowicz, J.M. (1982) The effects of L-glutamate and its analogues upon the membrane conductance of central murine neurones in culture. *Can. J. Physiol. Pharmac. 60*, 282-96.

MacDonald, R.L. and Barker, J.L. (1977) Pentylenetetrazol and penicillin are selective antagonists of GABA-mediated post-synaptic inhibition in cultured mammalian neurones. *Nature 267*, 720-1.

MacDonald, R.L. and Barker, J.L. (1978) Different actions of anticonvulsant and anesthetic barbiturates revealed by use of cultured mammalian neurons. *Science 200*, 775-7.

MacDonald, R.L. and Barker, J.L. (1979) Enhancement of GABA-mediated postsynaptic inhibition in cultured mammalian spinal cord neurons: a common mode of anticonvulsant action. *Brain Res. 167*, 323-36.

MacDonald, R.L. and Barker, J.L. (1981) Neuropharmacology of spinal cord neurons in primary dissociated cell culture, in *Excitable Cells in Tissue Culture*, P.G. Nelson and M. Lieberman (eds.), Plenum Press, New York, pp. 81-110.

MacDonald, R.L. and Bergey, G.K. (1979) Valproic acid augments GABA-mediated postsynaptic inhibition in cultured mammalian neurons. *Brain Res. 170*, 558-62.

MacDonald, R.L., Moonen, G., Neale, E.A. and Nelson, P.G. (1982) Cerebellar macroneurons in microexplant cell culture. Postsynaptic amino acid pharmacology. *Dev. Brain Res.* 5, 75-88.

MacDonald, R.L. and Nelson, P.G. (1978) Specific-opiate-induced depression of transmitter release from dorsal root ganglion cells in culture. *Science 199*, 1449-51.

MacDonald, R.L. and Young, A.B. (1981) Pharmacology of GABA-mediated inhibition of spinal cord neurons *in vivo* and in primary dissociated cell culture. *Mol. Cell. Biochem. 38*, 147-62.

McDonald, T.F., Sachs, H.G. and DeHaan, R.L. (1972) Development of sensitivity to tetrodotoxin in beating chick embryo hearts, single cells, and aggregates. *Science 176*, 1248-50.

McKay, R.D.G. and Hockfield, S.J. (1982) Monoclonal antibodies distinguish antigenically discrete neuronal types in the vertebrate central nervous system. *Proc. Nat. Acad. Sci.* USA *79*, 6747-51.

McLawhon, R.W., West, R.E., Miller, R.J. and Dawson, G. (1981) Distinct high-affinity binding sites for benzomorphan drugs and enkephalin in a neuroblastoma-brain hybrid cell line. *Proc. Nat. Acad. Sci.* USA *78*, 4309-13.

McLean, M.J., Pelleg, A. and Sperelakis, N. (1979) Electrophysiological recordings from spontaneously contracting reaggregates of cultured smooth muscle cells from guinea pig vas deferens. *J. Cell Biol. 80*, 539-52.

McLean, M.J., Renaud, J.-F., Niu, M.C. and Sperelakis, N. (1977) Membrane differentiation of cardiac myoblasts induced *in vitro* by an RNA-enriched fraction from adult heart. *Exp. Cell Res. 110*, 1-14.

McLean, M.J., Renaud, J.-F., Sperelakis, N. and Niu, M.C. (1976) Messenger RNA induction of fast sodium ion channels in cultured cardiac myoblasts. *Science 191*, 297-9.

McLean, M.J. and Sperelakis, N. (1974) Rapid loss of sensitivity to tetrodotoxin by chick ventricular myocardial cells after separation from the heart. *Exp. Cell Res. 86*, 351-64.

McLean, M.J. and Sperelakis, N. (1976) Retention of fully differentiated electrophysiological properties of chick embryonic heart cells in culture. *Dev. Biol. 50*, 134-41.

McLean, M.J. and Sperelakis, N. (1977) Electrophysiological recordings from spontaneously contracting reaggregates of cultured vascular smooth muscle cells from chick embryos. *Exp. Cell Res. 104*, 309-18.

McManaman, J.L., Blosser, J.C. and Appel, S.H. (1982) Inhibitors of membrane depolarization regulate acetylcholine receptor synthesis by a calcium-dependent, cyclic nucleotide-independent mechanism. *Biochim. Biophys. Acta 720*, 28-35.

Magazanik, L.G. and Vyskočil, F. (1970) Dependence of acetylcholine desensitization on the membrane potential of frog muscle fibre and on the ionic changes in the medium. *J. Physiol. 210*, 507-18.

Magchielse, T. and Meeter, E. (1982) Reduction of polyneuronal innervation of muscle cells in tissue culture after long-term indirect stimulation. *Dev. Brain Res. 3*, 130-3.

Mains, R.E. and Patterson, P.H. (1973) Primary cultures of dissociated sympathetic neurons. I. Establishment of long-term growth in culture and studies of differentiated properties. *J. Cell Biol. 59*, 329-45.

Margiotta, J.F. and Berl, D.K. (1982) Functional synapses are established between ciliary ganglion neurones in dissociated cell culture. *Nature 296*, 152-4.

Mark, G.E., Chamley, J.H. and Burnstock, G. (1973) Interactions between autonomic nerves and smooth and cardiac muscle cells in tissue culture. *Dev. Biol. 32*, 194-200.

Markowitz, C. (1931) Response of explanted embryonic cardiac tissue to epinephrine and acetylcholine. *Am. J. Physiol. 97*, 271-5.

Marshall, K.C., Wojtowicz, J.M. and Hendelman, W.J. (1980) Patterns of functional synaptic connections in organised cultures of cerebellum. *Neuroscience 5*, 1847-57.

Marty, A. (1981) Ca-dependent K channels with large unitary conductance in chromaffin cell membranes. *Nature 291*, 497-500.

Maruyama, Y., Yamashita, E. and Inomata, R. (1980) Effects of intracellular or extracellular application of tetraethylammonium on the action potential in cultured chick embryonic heart muscle cell. *Experientia 36*, 557-8.

Masson-Pévet, M., Jongsma, H.J. and De Bruijne, J. (1976) Collagenase- and trypsin-dissociated heart cells: a comparative ultrastructural study. *J. Mol. Cell. Cardiol. 8*, 747-57.

Masuko, S., Kuromi, H. and Shimada, Y. (1979) Isolation and culture of motoneurons from embryonic chicken spinal cords. *Proc. Nat. Acad. Sci.* USA *76*, 3537-41.

Masurovsky, E.B. and Benitez, H.H. (1967) Apparent innervation of chick cardiac muscle by sympathetic neurons in organised culture. *Anat. Rec. 157*, 285.

Mathers, D.A. and Barker, J.L. (1980) (–) Pentobarbital opens ion channels of long duration in cultured mouse spinal neurons. *Science 209*, 507-9.

Mathers, D.A. and Barker, J.L. (1981) GABA- and glycine-induced Cl^{1-} channels in cultured mouse spinal neurons require the same energy to close. *Brain Res. 224*, 441-5.

Mathers, D.A., Jackson, M.B., Lecar, H. and Barker, J.L. (1981) Single channel currents activated by GABA, muscimol, and (–) pentobarbital in cultured mouse spinal neurons. *Biophys. J. 33*, 14a.

Matsuda, Y., Yoshida, S. and Yonezawa, T. (1976) A Ca-dependent regenerative response in rodent dorsal root ganglion cells cultured *in vitro*. *Brain Res. 115*, 334-8.

Matsuda, Y., Yoshida, S. and Yonezawa, T. (1978) Tetrodotoxin sensitivity and Ca component of action potentials of mouse dorsal root ganglion cells cultured *in vitro*. *Brain Res. 154*, 69-82.

Matsuzawa, H. and Nirenberg, M. (1975) Receptor-mediated shifts in cGMP and cAMP levels in neuroblastoma cells. *Proc. Nat. Acad. Sci.* USA *72*, 3472-6.

Mauger, J.P., Worcel, M., Tassin, J. and Courtois, Y. (1975) Contractility of smooth muscle of rabbit aorta in tissue culture. *Nature 255*, 337-8.

Mercer, E.N. and Dower, G.E. (1966) Normal and arrhythmic beating in isolated cultured heart cells and the effects of digoxin, quinidine and procaine amide. *J. Pharmac. Exp. Ther. 153*, 203-10.

Merickel, M., Gray, R., Chauvin, P. and Appel, S. (1981) Electrophysiology of human muscle in culture. *Exp. Neurol. 72*, 281-93.
Messer, A. (1981) Primary monolayer cultures of the rat corpus striatum: morphology and properties related to acetylcholine and γ-aminobutyrate. *Neuroscience 6*, 2677-87.
Messing, A. (1982) Cholinergic agonist-induced down regulation of neuronal α-bungarotoxin receptors. *Brain Res. 232*, 479-84.
Mettler, F.A., Grundfest, H., Crain, S.M. and Murray, M.R. (1952) Spontaneous electrical activity from tissue cultures. *Trans. Am. Neurol. Assoc. 77*, 52-3.
Minna, J.D. and Gilman, A.G. (1973) Expression of genes for metabolism of cyclic adenosine 3′:5′-monophosphate in somatic cells. II. Effects of prostaglandin E_1 and theophylline on parental and hybrid lines. *J. Biol. Chem. 248*, 6618-25.
Minna, J., Nelson, P., Peacock, J., Glazer, D. and Nirenberg, M. (1971) Genes for neuronal properties expressed in neuroblastoma × L cell hybrids. *Proc. Nat. Acad. Sci.* USA *68*, 234-9.
Miyake, M. (1978) The development of action potential mechanism in a mouse neuronal cell line *in vitro*. *Brain Res. 143*, 349-54.
Miyake, M. and Shibata, S. (1981) A novel mode of neurotoxin action. A polypeptide toxin isolated from *Anemonia sulcata* shifts the voltage dependence on the maximal rate of rise of Na^+ action potentials in a mouse neuronal clone. *Mol. Pharmac. 20*, 453-6.
Mϕlstad, P., Bϕhmer, T. and Hovig, T. (1978) Carnitine-induced uptake of L-carnitine into cells from an established cell line from human heart (CCL27). *Biochim. Biophys. Acta 512,* 557-65.
Moody-Corbett, F., Weldon, P.R. and Cohen, M.W. (1982) Cholinesterase localization at sites of nerve contact on embryonic amphibian muscle cells in culture. *J. Neurocytol. 11*, 381-94.
Moolenaar, W.H. and Spector, I. (1977) Membrane currents examined under voltage clamp in cultured neuroblastoma cells. *Science 196*, 331-3.
Moolenaar, W.H. and Spector, I. (1978) Ionic currents in cultured mouse neuroblastoma cells under voltage-clamp conditions. *J. Physiol. 278,* 265-86.
Moolenaar, W.H. and Spector, I. (1979a) The calcium action potential and a prolonged calcium dependent after-hyperpolarization in mouse neuroblastoma cells. *J. Physiol. 292*, 297-306.
Moolenaar, W.H. and Spector, I. (1979b) The calcium current and the activation of a slow potassium conductance in voltage-clamped mouse neuroblastoma cells. *J. Physiol. 292*, 307-23.
Moonen, G., Neale, E.A., MacDonald, R.L., Gibbs, W. and Nelson, P.G. (1982) Cerebellar macroneurons in microexplant cell culture. Methodology, basic electrophysiology and morphology after horseradish peroxidase injection. *Dev. Brain Res. 5*, 59-73.
Mori, J., Ashida, H., Maru, E. and Tatsuno, J. (1982) Effects of Ca ions on action potentials in immature cultured neurons from chick cerebral cortex. *J. Cell. Physiol. 110*, 241-4.
Moscona, A. (1952) Cell suspensions from organ rudiments of chick

embryos. *Exp. Cell Res. 3*, 535-9.

Moura, A.-M. and Simpkins, H. (1975) Cyclic AMP levels in cultured myocardial cells under the influence of chronotropic and inotropic agents. *J. Mol. Cell. Cardiol. 7*, 71-7.

Mudge, A.W., Leeman, S.E. and Fischbach, G.D. (1979) Enkephalin inhibits release of substance P from sensory neurons in culture and decreases action potential duration. *Proc. Nat. Acad. Sci.* USA *76*, 526-30.

Murray, M.R. (1960) Skeletal muscle in tissue culture, in *Structure and Function of Muscle*, G.H. Bourne (ed.), Academic Press, New York, Vol. 2. pp. 111-136.

Murray, M.R. (1965a) Muscle, in *Cells and Tissues in Culture. Methods, Biology and Physiology*. E.N. Willmer (ed.), Academic Press, London, Vol. 2 pp. 311-72.

Murray, M.R. (1965b) Nervous tissue *in vitro*, in *Cells and Tissue Culture, Methods, Biology and Physiology,* E.N. Willmer (ed.), Academic Press, London, Vol. 2. pp. 373-455.

Murray, M.R. and Kopech, G. (1953) *A Bibliography of the Research in Tissue Culture*, 2 Vols. Academic Press, New York.

Myers, P.R., Blosser, J. and Shain, W. (1978) Neurotransmitter modulation of prostaglandin E_1-stimulated increases in cyclic AMP. II. Characterization of a cultured neuronal cell line treated with dibutyryl cyclic *AMP. Biochem. Pharmac. 27,* 1173-7.

Myers, P.R. and Livengood, D.R. (1975) Dopamine depolarising response in a vertebrate neuronal somatic cell hybrid. *Nature 255*, 235-7.

Myers, P.R., Livengood, D.R. and Shain, W. (1977) Characterization of a depolarizing dopamine response in a vertebrate neuronal somatic cell hybrid. *J. Cell. Physiol. 91*, 103-18.

Nakajima, Y., Kidokoro, Y. and Klier, F.G. (1980) The development of functional neuromuscular junctions *in vitro*: an ultrastructural and physiological study. *Dev. Biol. 77*, 52-72.

Nathan, R.D. (1981) Aggregates of fetal rat heart cells: electrophysiology and tetrodotoxin sensitivity. *J. Mol. Cell. Cardiol. 13*, 241-9.

Nathan, R.D. and DeHaan, R.L. (1978) *In vitro* differentiation of a fast Na^+ conductance in embryonic heart cell aggregates. *Proc. Nat. Acad. Sci.* USA *75*, 2776-80.

Nathan, R.D. and DeHaan, R.L. (1979) Voltage clamp analysis of embryonic heart cell aggregates. *J. Gen. Physiol. 73*, 175-98.

Neher, E. and Marty, A. (1982) Discrete changes of cell membrane capacitance observed under conditions of enhanced secretion in bovine adrenal chromaffin cells. *Proc. Nat. Acad. Sci.* USA *79*, 6712-16.

Neher, E. and Sakmann, B (1976a) Single channel currents recorded from membrane of denervated frog muscle fibres. *Nature 260*, 799-802.

Neher, E. and Sakmann, B. (1976b) Noise analysis of drug-induced voltage clamp currents in denervated frog muscle fibres. *J. Physiol. 258*, 705-29.

Neher, E., Sakmann, B. and Steinbach, J.H. (1978) The extracellular patch clamp. A method for resolving currents through individual open channels in biological membranes. *Pflügers Arch. 375*, 219-28.

Nelson, D.J. and Sachs, F. (1979) Single ionic channels observed in tissue-cultured muscle. *Nature 282*, 861-3.

Nelson, P.G. (1973) Electrophysiological studies of normal and neoplastic cells in tissue culture, in *Tissue Culture of the Nervous System*, G. Sato (ed.), Plenum Press, New York, pp. 135-60.

Nelson, P.G. (1975) Nerve and muscle cells in culture. *Physiol. Rev. 55*, 1-61.

Nelson, P.G. and Lieberman, M. (1981) *Excitable Cells in Tissue Culture*. Plenum Press, New York.

Nelson, P.G., Christian, C.N., Daniels, M.P., Henkart, M., Bullock, P., Mullinax, D. and Nirenberg, M. (1978) Formation of synapses between cells of a neuroblastoma X glioma hybrid clone and mouse myotubes. *Brain Res. 147*, 245-59.

Nelson, P., Christian, C. and Nirenberg, M. (1 976) Synapse formation between clonal neuroblastoma X glioma hybrid cells and striated muscle cells. *Proc. Nat. Acad. Sci.* USA *73*, 123-7.

Nelson, P.G., Neale, E.A. and MacDonald, R.L. (1981) Electrophysiological and structural studies of neurons in dissociated cell cultures of the central nervous system, in *Excitable Cells in Tissue Culture*, P.G. Nelson and M. Lieberman (eds.), Plenum Press, New York, pp. 39-80.

Nelson, P.G. and Peacock, J.H. (1973) Electrical activity in dissociated cell cultures from fetal mouse cerebellum, *Brain Res. 61*, 163-74.

Nelson, P.G., Peacock, J.H. and Amano, T. (1971a) Responses of neuroblastoma cells to iontophoretically applied acetylcholine. *J. Cell. Physiol. 77*, 353-62.

Nelson, P.G., Peacock, J.H., Amano, T. and Minna, J. (1971b) Electrogenesis in mouse neuroblastoma cells *in vitro*. *J. Cell. Physiol. 77*, 337-52.

Nelson, P.G., Ransom, B.R., Henkart, M. and Bullock, P.N. (1977) Mouse spinal cord in cell culture. IV. Modulation of inhibitory synaptic function. *J. Neurophysiol. 40*, 1178-87.

Nelson, P., Ruffner, W. and Nirenberg, M. (1969) Neuronal tumor cells with excitable membranes grown *in vitro*. *Proc. Nat. Acad. Sci.* USA *64*, 1004-10.

Nishi, R. and Berg, D.K. (1977) Dissociated ciliary ganglion neurons *in vitro*: survival and synapse formation. *Proc. Nat. Acad. Sci.* USA *74*, 5171-5.

Nishi, R. and Berg, D.K. (1979) Survival and development of ciliary ganglion neurons alone in cell culture. *Nature 277*, 232-4.

Noble, M.D., Brown, T.H. and Peacock, J.H. (1978) Regulation of acetylcholine receptor levels by a cholinergic agonist in mouse cell cultures. *Proc. Nat. Acad. Sci.* USA *75*, 3488-92.

Norwood, C.R., Castaneda, A.R. and Norwood, W.I. (1980) Heterogeneity of rat cardiac cells of defined origin in single cell culture. *J. Mol. Cell. Cardiol. 12*, 201-10.

Nowak, L.M. and MacDonald, R.L. (1982) Substance P: ionic basis for depolarizing responses of mouse spinal cord neurons in cell culture. *J. Neurosci. 2*, 1119-28.

Nowak, L.M. and MacDonald, R.L. (1983) Muscarine-sensitive voltage-dependent potassium current in cultured murine spinal cord neurons. *Neurosci. Lett. 35*, 85-91.

Nowak, L.M., Young, A.B. and MacDonald, R.L. (1982) GABA and bicuculline actions on mouse spinal cord and cortical neurons in cell culture. *Brain Res. 244*, 155-64.

Nurse, C.A. (1981a) Interactions between dissociated rat sympathetic neurons and skeletal muscle cells developing in cell culture. I. Cholinergic transmission, *Dev. Biol. 88*, 55-70.

Nurse, C.A. (1981b) Interactions between dissociated rat sympathetic neurons and skeletal muscle cells developing in cell culture. II. Synaptic mechanisms, *Dev. Biol. 88*, 71-9.

Nurse, C.A. and O'Lague, P.H. (1975) Formation of cholinergic synapses between dissociated sympathetic neurons and skeletal myotubes of the rat in cell culture. *Proc. Nat. Acad. Sci.* USA *72*, 1955-9.

Obata, K. (1974) Transmitter sensitivities of some nerve and muscle cells in culture. *Brain Res. 73*, 71-88.

Obata, K. (1977) Development of neuromuscular transmission in culture with a variety of neurons and in the presence of cholinergic substances and tetrodotoxin. *Brain Res. 119*, 141-53.

Obata, K. and Oide, M. (1980) Development of acetylcholine sensitivity of embryonic chick atria in culture. *Dev. Neurosci. 3*, 28-38.

Obata, K., Oide, M. and Tanaka, H. (1978) Excitatory and inhibitory actions of GABA and glycine on embryonic chick spinal neurons in culture. *Brain Res. 144*, 179-84.

Oh, T.H. (1975) Neurotrophic effects: characterization of the nerve extract that stimulates muscle development in culture. *Exp. Neurol. 46*, 432-8.

Oh, T.H. (1976) Neurotrophic effects of sciatic nerve extracts on muscle development in culture. *Exp. Neurol. 50*, 376-86.

Ohmori, H., Yoshida, S. and Hagiwara, S. (1981) Single K^+ channel currents of anomalous rectification in cultured rat myotubes. *Proc. Nat. Acad. Sci.* USA *78*, 4960-4.

Okarma, T.B., Tramell, P. and Kalman, S.M. (1972) The surface interaction between digoxin and cultured heart cells. *J. Pharmac. Exp. Ther. 183*, 559-76.

Okun, L.M. (1972) Isolated dorsal root ganglion neurons in culture: cytological maturation and extension of electrically active processes. *J. Neurobiol. 3*, 111-51.

O'Lague, P.H., Furshpan, E.J. and Potter, D.D. (1978a) Studies on rat sympathetic neurons developing in cell culture. II. Synaptic mechanisms. *Dev. Biol. 67*, 404-23.

O'Lague, P.H., Obata, K., Claude, P., Furshpan, E.J. and Potter, D.D. (1974) Evidence for cholinergic synapses between dissociated rat sympathetic neurons in cell culture. *Proc. Nat. Acad. Sci.* USA *71*,

3602-6.
O'Lague, P.H., Potter, D.D. and Furshpan, E.J. (1978b) Studies on rat sympathetic neurons developing in cell culture. I. Growth characteristics and electrophysiological properties. *Dev. Biol. 67*, 384-403.
O'Lague, P.H., Potter, D.D. and Furshpan, E.J. (1978c) Studies on rat sympathetic neurons developing in cell culture. III. Cholinergic transmission. *Dev. Biol. 67*, 424-43.
Olsen, R.W. (1982) Drug interactions at the GABA receptor-ionophore complex. *Ann. Rev. Pharmac. Toxicol. 22*, 245-77.
Olsen, M.I. and Bunge, R.P. (1973) Anatomical observations on the specificity of synapse formation in tissue culture. *Brain Res. 59*, 19-33.
Orida, N. and Poo, M.-M. (1981) Maintenance and dissolution of acetylcholine receptor clusters in the embryonic muscle cell membrane. *Dev. Brain Res. 1*, 293-8.
Pace, C.S., Murphy, M., Conant, S. and Lacy, P.E. (1977) Somatostatin inhibition of glucose-induced electrical activity in cultured rat islet cells. *Am. J. Physiol. 233*, C164-71.
Pado, C.H., Munson, R., Glaser, L. and Gottlieb, D.I. (1980) Evidence for ionic channels in cultured chick embryonic CNS cells. *Brain Res. 185*, 187-91.
Pappano, A.J. and Skrowonek, C.A. (1974) Reactivity of chick embryo heart to cholinergic agonists during ontogenesis: decline in desensitization at the onset of cholinergic transmission. *J. Pharmac. Exp. Ther. 191*, 109-18.
Pappano, A.J. and Sperelakis, N. (1969) Low K^+ conductance and low resting membrane potentials of isolated single cultured heart cells. *Am. J. Physiol. 217*, 1076-82.
Patrick, J., Heinemann, S.F., Lindstrom, J., Schubert, D. and Steinbach, J.H. (1972) Appearance of acetylcholine receptors during differentiation of a myogenic cell line. *Proc. Nat. Acad. Sci.* USA *69*, 2762-6.
Patrick, J., McMillan, J., Wolfson, H. and O'Brien, J.C. (1977) Acetylcholine receptor metabolism in a nonfusing muscle cell line. *J. Biol. Chem. 252*, 2143-53.
Patrick, J. and Stallcup, B. (1977) α-Bungarotoxin binding and cholinergic receptor function on a rat sympathetic nerve line. *J. Biol. Chem. 252*, 8629-33.
Paul, J. (1975) *Cell and Tissue Culture*, Churchill Livingstone, Edinburgh.
Peacock, J.H. (1979) Electrophysiology of dissociated hippocampal cultures from fetal mice. *Brain Res. 169*, 247-60.
Peacock, J., Minna, J., Nelson, P. and Nirenberg, M. (1972) Use of aminopterin in selecting electrically active neuroblastoma cells. *Exp. Cell Res. 73*, 367-77.
Peacock, J.H., McMorris, F.A. and Nelson, P.G. (1973a) Electrical excitability and chemosensitivity of mouse neuroblastoma × mouse or human fibroblast hybrids. *Exp. Cell Res. 79*, 199-212.
Peacock, J.H. and Nelson, P.G. (1973) Chemosensitivity of mouse

neuroblastoma cells *in vitro*. *J. Neurobiol.* *4*, 363-74.
Peacock, J.H., Nelson, P.G. and Goldstone, M.W. (1973b) Electrophysiologic study of cultured neurons dissociated from spinal cords and dorsal root ganglia of fetal mice. *Dev. Biol. 30*, 137-52.
Peacock, J.H., Rush, D.F. and Mathers, L.H. (1979) Morphology of dissociated hippocampal cultures from fetal mice. *Brain Res. 169*, 231-46.
Pearce, B.R. and Dutton, G.R. (1982) L-glutamate increases the spontaneous release of [^{3}H]GABA from cultured cerebellar neurons. *Dev. Brain Res. 3*, 492-6.
Peng, H.B. and Nakajima, Y. (1978) Membrane particle aggregates in innervated and noninnervated cultures of *Xenopus* embryonic muscle cells. *Proc. Nat. Acad. Sci.* USA *75*, 500-4.
Percy, V.A., Lamm, M.C.L. and Taljaard, J.J.F. (1981) Effects of δ-aminolaevulinic acid, porphobilinogen, amino acids and barbiturates on calcium accumulation by cultured neurons. *Biochem. Pharmac. 30*, 665-6.
Podleski, T.R., Axelrod, D., Ravdin, P., Greenberg, I., Johnson, M.M. and Salpeter, M.M. (1978) Nerve extract induces increase and redistribution of acetylcholine receptors on cloned muscle cells. *Proc. Nat. Acad. Sci.* USA *75*, 2035-9.
Polinger, I.S. (1970) Separation of cell types in embryonic heart cell cultures. *Exp. Cell Res. 63*, 78-82.
Polinger, I.S. (1973) Growth and DNA synthesis in embryonic chick heart cells, *in vivo* and *in vitro*. *Exp. Cell Res. 76*, 253-62.
Powell, J.A. and Fambrough, D.M. (1973) Electrical properties of normal and dysgenic mouse skeletal muscle in culture. *J. Cell. Physiol. 82*, 21-38.
Prasad, K.N. (1981) Cell Culture, in *Methods in Neurobiology*, R. Lahue (ed.), Plenum Press, New York, Vol. 1, pp. 245-63.
Prasad, K.N. and Gilmer, K.N. (1974) Demonstration of dopamine-sensitive adenylate cyclase in malignant neuroblastoma cells and change in sensitivity of adenylate cyclase to catecholamines in "differentiated" cells. *Proc. Nat. Acad. Sci.* USA *71*, 2525-9.
Prives, J.M. and Paterson, B.M. (1974) Differentiation of cell membranes in cultures of embryonic chick breast muscle. *Proc. Nat. Acad. Sci.* USA *71*, 3208-11.
Prives, J., Hoffman, L., Tarrab-Hazdai, R., Fuchs, S. and Amsterdam, A. (1979) Ligand induced changes in stability and distribution of acetylcholine receptors on surface membranes of muscle cells. *Life Sci. 24*, 1713-18.
Prives, J., Silman, I. and Amsterdam, A. (1976) Appearance and disappearance of acetylcholine receptor during differentiation of chick skeletal muscle *in vitro*. *Cell 7*, 543-50.
Purdy, J.E., Lieberman, M., Roggeveen, A.E. and Kirk, R.G. (1972) Synthetic strands of cardiac muscle. Formation and ultrastructure. *J. Cell Biol. 55*, 563-78.
Puro, D.G. and Nirenberg, M. (1976) On the specificity of synapse formation. *Proc. Nat. Acad. Sci.* USA *73*, 3544-8.

Purves, R.D. (1974) Muscarinic excitation: a microelectrophoretic study on cultured smooth muscle cells. *Br. J. Pharmac. 52*, 77-86.
Purves, R.D. and Vrbová, G. (1974) Some characterstics of myotubes cultured from slow and fast chick muscles. *J. Cell. Physiol. 84*, 97-100.
Purves, R.D., Mark, G.E. and Burnstock, G. (1973) The electrical activity of single isolated smooth muscle cells. *Pflügers Arch. 341*, 325-30.
Purves, R.D., Hill, C.E., Chamley, J.J., Mark, G.E., Fry, D.M. and Burnstock, G. (1974) Functional autonomic neuromuscular junctions in tissue culture. *Pflügers Arch. 350*, 1-7.
Quandt, F.N. and Narahashi, T. (1982) Modification of single Na^+ channels by batrachotoxin. *Proc. Nat. Acad. Sci.* USA *79*, 6732-6.
Rang, H.P. (1981) The characteristics of synaptic currents and responses to acetylcholine of rat submandibular ganglion cells. *J. Physiol. 311*, 23-55.
Ransom, B.R. and Barker, J.L. (1975) Pentobarbital modulates transmitter effects on mouse spinal neurones grown in tissue culture. *Nature 254*, 703-5.
Ransom, B.R. and Barker, J.L. (1981) Physiology and pharmacology of mammalian central neurons in cell culture. *Adv. Cell. Neurobiol. 2*, 84-114.
Ransom, B.R., Barker, J.L. and Nelson, P.G. (1975) Two mechanisms for poststimulus hyperpolarisations in cultured mammalian neurones. *Nature 256*, 424-5.
Ransom, B.R., Bullock, P.N. and Nelson, P.G. (1977a) Mouse spinal cord in cell culture. III. Neuronal chemosensitivity and its relationship to synaptic activity. *J. Neurophysiol. 40*, 1163-77.
Ransom, B.R., Christian, C.N., Bullock, P.N. and Nelson, P.G. (1977b) Mouse spinal cord in cell culture. II. Synaptic activity and circuit behaviour. *J. Neurophysiol. 40*, 1151-62.
Ransom, B.R. and Holz, R.W. (1977) Ionic determinants of excitability in cultured mouse dorsal root ganglion and spinal cord cells. *Brain Res. 136*, 445-53.
Ransom, B.R., Neale, E., Henkart, M., Bullock, P.N. and Nelson, P.G. (1977c) Mouse spinal cord in cell culture. I. Morphology and intrinsic neuronal electrophysiologic properties. *J. Neurophysiol. 40*, 1132-50.
Ravdin, P.M. and Berg, D.K. (1979) Inhibition of neuronal acetylcholine sensitivity by α-toxins from *Bungarus multicinctus* venom. *Proc. Nat. Acad. Sci.* USA *76*, 2072-6.
Reiser, G., Heumann, R., Kemper, W., Lautenschlager, E. and Hamprecht, B. (1977) Influence of cations on the electrical activity of neuroblastoma × glioma hybrid cells. *Brain Res. 130*, 495-504.
Reiser, G., Scholz, F. and Hamprecht, B. (1982) Pharmacological and electrophysiological characterization of lithium ion flux through the action potential sodium channel in neuroblastoma × glioma hybrid cells. *J. Neurochem. 39*, 228-34.
Renaud, J.F., Scanu, A.M., Kazazoglou, T., Lombet, A., Romey, G.

and Lazdunski, M. (1982) Normal serum and lipoprotein-deficient serum give different expressions of excitability, corresponding to different stages of differentiation, in chicken cardiac cells in culture. *Proc. Nat. Acad. Sci.* USA *79*, 7768-72.

Renaud, J.-F., Sperelakis, N. and Le Douarin, G. (1978) Increase of cyclic AMP levels induced by isoproterenol in cultured and non-cultured chick embryonic hearts. *J. Mol. Cell Cardiol. 10*, 281-6.

Repke, H. and Maderspach, K. (1982) Muscarinic acetylcholine receptors on cultured glial cells. *Brain Res. 232*, 206-11.

Reuter, H. (1983) Calcium channel modulation by neurotransmitters, enzymes and drugs. *Nature 301*, 569-74.

Reuter, H., Stevens, C.F., Tsien, R.W. and Yellen, G. (1982) Properties of single calcium channels in cardiac cell culture. *Nature 297*, 501-4.

Richelson, E. (1977a) Lithium entry through the sodium channel of mouse neuroblastoma cells: a biochemical study. *Science 196*, 1001-2.

Richelson, E. (1977b) Antipsychotics block muscarinic acetylcholine receptor-mediated cyclic GMP formation in cultured mouse neuroblastoma cells. *Nature 266*, 371-3.

Richelson, E. (1978a) Desensitization of muscarinic receptor-mediated cyclic GMP formation by cultured nerve cells. *Nature 272*, 366-8.

Richelson, E. (1978b) Tricyclic antidepressants block histamine H_1 receptors of mouse neuroblastoma cells. *Nature 274*, 176-7.

Richelson, E. (1978c) Histamine H_1 receptor-mediated guanosine $3',5'$-monophosphate formation by cultured mouse neuroblastoma cells. *Science*, *201*, 69-71.

Richelson, E., Prendergast, F.G. and Divinetz-Romero, S. (1978) Muscarinic receptor-mediated cyclic GMP formation by cultured nerve cells - ionic dependence and effect of local anesthetics. *Biochem. Pharmac. 27*, 2039-48.

Rifas, L., Fant, J., Makman, M.M. and Seifter, S. (1979) The characterization of human uterine smooth muscle cells in culture. *Cell Tiss. Res. 196*, 385-95.

Rinaldini, L.M. (1959) An improved method for the isolation and quantitative cultivation of embryonic cells. *Exp. Cell Res. 16*, 477-505.

Ritchie, A.K. (1979) Cathcholamine secretion in a rat pheochromocytoma cell line: two pathways for calcium entry. *J. Physiol. 286*, 541-61.

Ritchie, A.K. and Fambrough, D.M. (1975a) Ionic properties of the acetylcholine receptor in cultured rat myotubes. *J. Gen. Physiol. 65*, 751-67.

Ritchie, A.K. and Fambrough, D.M. (1975b) Electrophysiological properties of the membrane and acetylcholine receptor in developing rat and chick myotubes. *J. Gen. Physiol. 66*, 327-55.

Robbins, R.J., Sutton, R.E. and Reichlin, S. (1982) Effects of neurotransmitters and cyclic AMP on somatostatin release from cultured cerebral cortical cells. *Brain Res. 234*, 377-86.

Robbins, N. and Yonezawa, T. (1971) Developing neuromuscular junctions: first signs of chemical transmission during formation in tissue culture. *Science 172,* 395-8.

Robinson, R.B. and Legato, M.J. (1980) Maintained differentiation in rat cardiac monolayer cultures: tetrodotoxin sensitivity and ultrastructure. *J. Mol. Cell. Cardiol. 12*, 493-8.

Roeske, W.R. and Wildenthal, K. (1981) Responsiveness to drugs and hormones in the murine model of cardiac ontogenesis. *Pharmac. Ther. 14*, 55-66.

Romey, G. and Lazdunski, M. (1982) Lipid-soluble toxins thought to be specific for Na^+ channels block Ca^{++} channels in neuronal cells. *Nature 297,* 79-80.

Romijn, H.J., Habets, A.M.M.C., Mud, M.T. and Wolters, P.S. (1982) Nerve outgrowth, synaptogenesis and bioelectric activity in fetal rat cerebral cortex tissue cultured in serum-free chemically-defined medium. *Dev. Brain Res. 2,* 583-9.

Ross, R. (1971) The smooth muscle cell. II. Growth of smooth muscle in culture and formation of elastic fibers. *J. Cell Biol. 50*, 172-86.

Rothblat, G.H. and Cristofalo, V.J. (1972) *Growth, Nutrition and Metabolism of Cells in Culture*, 3 Vols, Academic Press, New York.

Rotundo, R.L. and Fambrough, D.M. (1980a) Synthesis, transport and fate of acetylcholinesterase in cultured chick embryo muscle cells. *Cell 22*, 583-94.

Rotundo, R.L. and Fambrough, D.M. (1980b) Secretion of acetylcholinesterase : relation to acetylcholine receptor metabolism. *Cell 22*, 595-602.

Rous, P. and Jones, F.S. (1916) A method for obtaining suspensions of living cells from the fixed tissues, and for the plating out of individual cells. *J. Exp. Med. 23*, 549-55.

Rubin, L.L., Schuetze, S.M. and Fischbach, G.D. (1979) Accumulation of acetylcholinesterase at newly formed nerve-muscle synapses. *Dev. Biol. 69*, 46-58.

Rubin, L.L., Schuetze, S.M., Weill, C.L. and Fischbach, G.D. (1980) Regulation of acetylcholinesterase appearance at neuromuscular junctions *in vitro*. *Nature 283*, 264-7.

Sabol, S.L. and Nirenberg, M. (1979a) Regulation of adenylate cyclase neuroblastoma × glioma hybrid cells by α-adrenergic receptors. I. Inhibition of adenylate cyclase mediated by α-receptors. *J. Biol. Chem. 254*, 1913-20.

Sabol, S.L. and Nirenberg, M. (1979b) Regulation of adenylate cyclase of neuroblastoma × glioma hybrid cells by α-adrenergic receptors. II. Long-lived increase of adenylate cyclase activity mediated by α-receptors. *J. Biol. Chem. 254*, 1921-6.

Sacerdote De Lustig, E. (1942) Estudio del Automatismo Muscular en Cultivos *in vitro* con Eserina, Acetilcolina, Adrenalina y Atropina. *Rev. Soc. Argent. Biol. 18*, 524-31.

Sacerdote De Lustig, E. (1943) Accion de Sustancias Curarizantes sobre Cultivos de Musculo de Embrion de Pollo. *Rev. Soc. Argent. Biol. 19*, 159-69.

Sachs, F. and Lecar, H. (1977) Acetylcholine-induced current fluctuations in tissue-cultured muscle cells under voltage clamp. *Biophys. J. 17*, 129-43.
Sachs, H.G., McDonald, T.F. and DeHaan, R.L. (1973) Tetrodotoxin sensitivity of cultured embryonic heart cells depends on cell interactions. *J. Cell Biol. 56*, 255-8.
Saji, M. and Miura, M. (1982) Transmitter sensitivities among various types of cultured brain stem neurons. *Brain Res. 233*, 83-96.
Sakmann, B. and Brenner, H.R. (1978) Change in synaptic channel gating during neuromuscular development. *Nature 276*, 401-2.
Salpeter, M.M., Spanton, S., Holley, K. and Podleski, T.R. (1 982) Brain extract causes acetylcholine receptor redistribution which mimics some early events at developing neuromuscular junctions. *J. Cell Biol. 93*, 417-25.
Sand, O., Haug, E. and Kautvik, K.M. (1980) Effects of thyroliberin and 4-aminopyridine on action potentials and prolactin release and synthesis in rat pituitary cells in culture. *Acta Physiol. Scand. 198*, 247-52.
Sand, O., Ozawa, S. and Kautvik, K.M. (1981a) Sodium and calcium action potentials in cells derived from a rat medullary thyroid carcinoma. *Acta Physiol. Scand. 112,* 287-91.
Sand, O., Ozawa, S. and Hove, K. (1981b) Electrophysiology of cultured parathyroid cells from the goat. *Acta Physiol. Scand. 113*, 45-50.
Sastre, S. and Podleski, T.R. (1976) Pharmacologic characterization of Na^+ ionophores in L6 myotubes. *Proc. Nat. Acad. Sci.* USA *73*, 1355-9.
Schlapfer, W.T. (1981) Tissue and organ culture, in *Methods in Neurobiology*, R. Lahue (ed.), Plenum Press, New York. Vol. 1., pp. 183-244.
Schlapfer, W.T., Mamoon, A.-M. and Tobias, C.A. (1972) Spontaneous bioelectric activity of neurons in cerebellar cultures: evidence for synaptic interactions. *Brain Res. 45*, 345-63.
Schlesinger, H.R., Frazer, A., Friedman, R., Mendels, J. and Hummeler, K. (1979) Lithium ion uptake associated with the stimulation of action potential ionophores of cultured human neuroblastoma cells. *Life Sci. 25,* 957-68.
Schubert, D., Harris, A.J., Devine, C.E. and Heinemann, S. (1974a) Characterization of a unique muscle cell line. *J. Cell Biol. 61*, 398-413.
Schubert, D., Harris, A.J., Heinemann, S., Kidokoro, Y., Patrick, J. and Steinbach, J.H. (1973) Differentiation and interaction of clonal cell lines of nerve and muscle, in *Tissue Culture of the Nervous System*, G. Sato (ed.), Plenum Press, New York, pp. 55-86.
Schubert, D., Heinemann, S., Carlisle, W., Tarikas, H., Kimes, B., Patrick, J., Steinbach, J.H., Culp, W. and Brandt, B.L. (1974b) Clonal cell lines from the rat central nervous system. *Nature 249*, 224-7.
Schubert, D., Humphreys, S., Baroni, C. and Cohn, M. (1969) *In vitro*

differentiation of a mouse neuroblastoma. *Proc. Nat. Acad. Sci.* USA *64*, 316-23.
Schuetze, S.M. (1980) The acetylcholine channel open time in chick muscle is not decreased following innervation. *J. Physiol. 303*, 111-24.
Schuetze, S.M., Frank, E.F. and Fischbach, G.D. (1978) Channel open time and metabolic stability of synaptic and extrasynaptic acetylcholine receptors on cultured chick myotubes. *Proc. Nat. Acad. Sci.* USA *75*, 520-3.
Schwarzfeld, T.A. and Jacobson, S.L. (1981) Isolation and development in cell culture of myocardial cells of the adult rat. *J. Mol. Cell. Cardiol. 13*, 563-75.
Scott, B.S. (1982) Adult neurons in cell culture: electrophysiological characterization and use in neurobiological research. *Prog. Neurobiol. 19*, 187-211.
Scott, B.S., Engelbert, V.E. and Fisher, K.C. (1969) Morphological and electrophysiological characteristics of dissociated chick embryonic spinal ganglion cells in culture. *Exp. Neurol. 23*, 230-48.
Scott, B.S. and Fisher, K.C. (1970) Potassium concentration and number of neurons in cultures of dissociated ganglia. *Exp. Neurol. 27*, 16-22.
Scott, B.S., Petit, T.L., Becker, L.E. and Edwards, B.A.V. (1979) Electric membrane properties of human DRG neurons in cell culture and the effect of high K medium. *Brain Res. 178*, 529-44.
Shainberg, A. and Brik, H. (1978) The appearance of acetylcholine receptors triggered by fusion of myoblasts *in vitro. FEBS Lett. 88*, 327-31.
Shainberg, A., Cohen, S.A. and Nelson, P.G. (1976) Induction of acetylcholine receptors in muscle cultures. *Pflügers Arch. 361*, 255-61.
Sharma, S.K., Nirenberg, M. and Klee, W.A. (1975) Morphine receptors as regulators of adenylate cyclase activity. *Proc. Nat. Acad. Sci.* USA *72*, 590-4.
Shifrin, G.S. and Klein, W.L. (1980) Regulation of muscarinic acetylcholine receptor concentration in cloned neuroblastoma cells. *J. Neurochem. 34*, 993-9.
Shigenobu, K. and Sperelakis, N. (1972) Calcium current channels induced by catecholamines in chick embryonic hearts whose fast sodium channels are blocked by tetrodotoxin or elevated potassium. *Circulation Res. 31*, 932-52.
Shigenobu, K. and Sperelakis, N. (1974) Failure of development of fast Na^+ channels during organ culture of young embryonic chick hearts. *Dev. Biol. 39*, 326-30.
Shimada, Y. (1968) Supression of myogenesis by heterotypic and heterospecific cells in monolayer culture. *Exp. Cell Res. 51*, 564-78.
Shimada, Y. and Fischman, D.A. (1973) Morphological and physiological evidence for the development of functional neuromuscular junctions *in vitro*. *Dev. Biol. 31*, 200-25.
Shimada, Y., Fischman, D.A. and Moscona, A.A. (1969) Formation of

neuromuscular junctions in embryonic cell cultures. *Proc. Nat. Acad. Sci.* USA *62*, 715-21.

Shimizu, T., Mizuno, N., Amano, T. and Hayaishi, O. (1979) Prostaglandin D_2, a neuromodulator. *Proc. Nat. Acad. Sci.* USA *76*, 6231-4.

Shrier, A. and Clay, J.R. (1980) Pacemaker currents in chick embryonic heart cells change with development. *Nature 283*, 670-1.

Shrier, A. and Clay, J.R. (1982) A comparison of the pacemaker properties of chick embryonic atrial and ventricular heart cells. *J. Memb. Biol. 69*, 49-56.

Sigworth, F.J. and Neher, E. (1980) Single Na^+ channel currents observed in cultured rat muscle cells. *Nature 287*, 447-9.

Simantov, R. and Sachs, L. (1973) Regulation of acetylcholine receptors in relation to acetylcholinesterase in neuroblastoma cells. *Proc. Nat. Acad. Sci.* USA *70*, 2902-5.

Sinback, C.N. and Coon, H.G. (1982) Electrophysiological and pharmacological properties of cultured rat thyroid cells. *J. Cell. Physiol. 112*, 391-402.

Sinback, C.N. and Shain, W. (1979) Electrophysiological properties of human oviduct smooth cells in dissociated cell culture. *J. Cell. Physiol. 98*, 377-94.

Sinback, C.N. and Shain, W. (1980) Chemosensitivity of single smooth muscle cells to acetylcholine, noradrenaline, and histamine *in vitro*. *J. Cell. Physiol. 102*, 99-112.

Sine, S. and Taylor, P. (1979) Functional consequences of agonist-mediated state transitions in the cholinergic receptor. Studies in cultured muscle cells. *J. Biol. Chem. 254*, 3315-25.

Singer, J.J. and Walsh, J.V. (1980a) Passive properties of the membrane of single freshly isolated smooth muscle cells. *Am. J. Physiol. 239*, C153-61.

Singer, J.J. and Walsh, J.V. (1980b) Rectifying properties of the membrane of single freshly isolated smooth muscle cells. *Am. J. Physiol. 239*, C175-81.

Skaper, S.D., Adler, R. and Varon, S. (1979) A procedure for purifying neuron-like cells in cultures from central nervous tissue with a defined medium. *Dev. Neurosci. 2*, 233-7.

Smilowitz, H. and Fischbach, G.D. (1978) Acetylcholine receptors on chick mononucleated muscle precursor cells. *Dev. Biol. 66*, 539-49.

Smith, T.G., Barker, J.L., Smith, B.M. and Colburn, T.R. (1981) Voltage clamp techniques applied to cultured skeletal muscle and spinal neurons, in *Excitable Cells in Tissue Culture*, P.G. Nelson and M. Lieberman (eds.), Plenum Press, New York, pp. 111-36.

Spector, I. (1981) Electrophysiology of clonal nerve cell lines, in *Excitable Cells in Tissue Culture*, P.G. Nelson and M. Lieberman (eds.), Plenum Press, New York, pp. 247-77.

Spector, I., Kimhi, Y. and Nelson, P.G. (1973) Tetrodotoxin and cobalt blockade of neuroblastoma action potentials. *Nature New Biol. 246*, 124-6.

Spector, I., Palfrey, C. and Littauer, U.Z. (1975) Enhancement of the electrical excitability of neuroblastoma cells by valinomycin. *Nature* *254*, 121-4.

Spector, I. and Prives, J.M. (1977) Development of electrophysiological and biochemical membrane properties during differentiation of embryonic skeletal muscle in culture. *Proc. Nat. Acad. Sci.* USA *74*, 5166-70.

Sperelakis, N. and Pappano, A.J. (1969) Increase in P_{Na} and P_K of cultured heart cells produced by veratridine. *J. Gen. Physiol.* *53*, 97-114.

Sperelakis, N. and Shigenobu, K. (1972) Changes in membrane properties of chick embryonic hearts during development. *J. Gen. Physiol.* *60*, 430-53.

Sperelakis, N. and Shigenobu, K. (1974) Organ-cultured chick embryonic hearts of various ages. I. Electrophysiology. *J. Mol. Cell. Cardiol.* *6*, 449-71.

Spitzer, N.C. (1979) Ion channels in development. *Ann. Rev. Neurosci.* *2*, 363-97.

Spitzer, N.C. and Lamborghini, J.E. (1976) The development of the action potential mechanism of amphibian neurons isolated in culture. *Proc. Nat. Acad. Sci.* USA *73*, 1641-5.

Stallcup, W.B. (1977) Comparative pharmacology of voltage-dependent sodium channels. *Brain Res.* *135*, 37-53.

Stallcup, W.B. (1979) Sodium and calcium fluxes in a clonal nerve cell line. *J. Physiol.* *286*, 525-40.

Stallcup, W.B. and Patrick, J. (1980) Substance P enhances cholinergic receptor desensitization in a clonal nerve cell line. *Proc. Nat. Acad. Sci.* USA *77*, 634-8.

Stammati, A.P., Silano, V. and Zucco, F. (1981) Toxicology investigations with cell culture systems. *Toxicology* *20*, 91-153.

Steinbach, J.H. (1974) Role of muscle activity in nerve-muscle interactions *in vitro*. *Nature* *248*, 70-1.

Steinbach, J.H. (1975) Acetylcholine responses on clonal myogenic cells *in vitro*. *J. Physiol.* *247*, 393-405.

Steinbach, J.H., Merlie, J., Heinemann, S. and Bloch, R. (1979) Degradation of junctional and extrajunctional acetylcholine receptors by developing rat skeletal muscle. *Proc Nat. Acad. Sci.* USA *76*, 3547-51.

Strange, P.G. (1978) Effect of a phosphodiesterase inhibitor on cyclic GMP changes induced by muscarinic agonists in mouse neuroblastoma cells. *Br. J. Pharmac.* *64*, 450P-1P.

Study, R.E. (1980) Phenytoin inhibition of cyclic guanosine 3′:5′-monophosphate (cGMP) accumulation in neuroblastoma cells by calcium channel blockade. *J. Pharmac. Exp. Ther.* *215*, 575-81.

Study, R.E. and Barker, J.L. (1981) Diazepam and (−)-pentobarbital: fluctuation analysis reveals different mechanisms for potentiation of γ-aminobutyric acid responses in cultured central neurons. *Proc. Nat. Acad. Sci.* USA *78*, 7180-4.

Study, R.E., Breakefield, X.O., Bartfai, T. and Greengard, P. (1978) Voltage-sensitive calcium channels regulate guanosine 3′,5′-cyclic monophosphate levels in neuroblastoma cells. *Proc. Nat. Acad. Sci.* USA *75*, 6295-9.

Stya, N. and Axelrod, D. (1983) Diffusely distributed acetylcholine receptors can participate in cluster formation on cultured rat myotubes. *Proc. Nat. Acad. Sci.* USA *80*, 449-53.

Sugiyama, H., Yamashita, Y. and Murakami, F. (1982) Multiple molecular forms of acetylcholine receptors in cultured skeletal muscle cells: subcellular localization and characterization. *J. Neurochem. 39*, 1038-46.

Swaiman, K.F., Neale, E.A., Fitzgerald, S.C. and Nelson, P.G. (1982) A method for large-scale production of mouse brain cortical cultures, *Dev. Brain Res. 3*, 361-9.

Syapin, P.J., Salvaterra, P.M. and Engelhardt, J.K. (1982) Neuronal-like features of TE671 cells: presence of a functional nicotinic cholinergic receptor. *Brain Res. 231,* 365-77.

Sytkowski, A.J., Vogel, Z. and Nirenberg, M. (1973) Development of acetylcholine receptor clusters on cultured muscle cells. *Proc. Nat. Acad. Sci.* USA *70*, 270-4.

Szepsenwol, J. (1946) Comparison of growth, differentiation, activity and action currents of heart and skeletal muscle in tissue culture. *Anat. Rec. 95*, 125-46.

Szepsenwol, J. (1947) Electrical excitability and spontaneous activity in explants of skeletal and heart muscle of chick embryos. *Anat. Rec. 98*, 67-85.

Taylor, J.E., El-Fakahany, E. and Richelson, E. (1979) Long-term regulation of muscarinic acetylcholine receptors on cultured nerve cells. *Life Sci. 25*, 2181-7.

Taylor, J.E. and Richelson, E. (1979) Desensitization of histamine H_1 receptor-mediated cyclic GMP formation in mouse neuroblastoma cells. *Mol. Pharmac., 15*, 462-71.

Teerapong, P., Marshall, I.G., Harvey, A.L., Singh, H., Paul, D., Bhardwaj, T.R. and Ahuja, N.K. (1979) The effects of dihydrochandonium and other chandonium analogues on neuromuscular and autonomic transmission. *J. Pharm. Pharmac. 31,* 521-8.

Teng, N.N.H. and Fiszman, M.Y. (1976) Appearance of acetylcholine receptors in cultured myoblasts prior to fusion. *J. Supramol. Structure 4*, 381-7.

Thoenen, H. and Barde, Y.-A. (1980) Physiology of nerve growth factor. *Physiol. Rev. 60*, 1284-335.

Thomson, C.M. and Dryden, W.F. (1980) Development of membrane conductance of chick skeletal muscle in culture. *Can. J. Physiol. Pharmac. 58*, 600-5.

Ticku, M.K., Huang, A. and Barker, J.L. (1980a) Characterization of γ-aminobutyric acid receptor binding in cultured brain cells. *Mol. Pharmac. 17*, 285-9.

Ticku, M.K., Huang, A. and Barker, J.L. (1980b) GABA receptor binding in cultured mammalian spinal cord neurons. *Brain Res. 182*, 201-6.

Tourneur, Y., Romey, G. and Lazdunski, M. (1982) Phencyclidine blockade of sodium and potassium channels in neuroblastoma cells. *Brain Res. 245*, 154-8.

Traber, J., Fischer, K., Latzin, S. and Hamprecht, B. (1974) Morphine antagonizes the action of prostaglandin in neuroblastoma cells but not of prostaglandin and noradrenaline in glioma and glioma × fibroblast hybrid cells. *FEBS Lett. 49,* 260-3.

Traber, J., Fischer, K., Buchen, C. and Hamprecht, B. (1975a) Muscarinic response to acetylcholine in neuroblastoma × glioma hybrid cells. *Nature 255*, 558-60.

Traber, J., Reiser, G., Fischer, K. and Hamprecht, B. (1975b) Measurements of adenosine 3′:5′-cyclic monophosphate and membrane potential in neuroblastoma × glioma hybrid cells: opiates and adrenergic agonists cause effects opposite to those of prostaglandin E_1. *FEBS Lett. 52,* 327-32.

Trautmann, A. (1982) Curare can open and block ionic channels associated with cholinergic receptors. *Nature 298,* 272-5.

Tsai, J.S. and Chen, A. (1978) Effect of L-triiodothyronine on $(-)^3$H-dihydroalprenolol binding and cyclic AMP response to (−)adrenaline in cultured heart cells. *Nature 275,* 138-40.

Tuttle, J.B. and Richelson, E. (1979) Phenytoin action on the excitable membrane of mouse neuroblastoma. *J. Pharmac. Exp. Ther. 211*, 632-7.

Tuttle, J.B., Suszkiw, J.B. and Ard, M. (1980) Long-term survival and development of dissociated parasympathetic neurons in culture. *Brain Res. 183*, 161-80.

Undrovinas, A.I., Yushmanova, A.V., Hering, S. and Rosenshtraukh, L.V. (1980) Voltage clamp method on single cardiac cells from adult rat heart. *Experientia 36*, 572-4.

Varon, S. and Raiborn, C.W. (1969) Dissociation, fractionation and culture of embryonic brain cells. *Brain Res. 12*, 180-99.

Varon, S. and Raiborn, C. (1971) Excitability and conduction in neurons of dissociated ganglionic cell cultures. *Brain Res. 30*, 83-98.

Varon, S. and Raiborn, C. (1972) Dissociation, fractionation and culture of chick embryo sympathetic ganglionic cells. *J. Neurocytol. 1*, 211-21.

Vincent, J.-D. and Barker, J.L. (1979) Substance P: evidence for diverse roles in neuronal function from cultured mouse spinal neurons. *Science 205*, 1409-12.

Vogel, Z., Daniels, M.P. and Nirenberg, M. (1976) Synapse and acetylcholine receptor synthesis by neurons dissociated from retina. *Proc. Nat. Acad. Sci.* USA *73*, 2370-4.

Vogel, Z., Sytkowski, A.J. and Nirenberg, M.W. (1972) Acetylcholine receptors of muscle grown *in vitro*. *Proc. Nat. Acad. Sci.* USA *69*, 3180-4.

Wahlström, A., Brandt, M., Moroder, L., Wünsch, E., Lindeberg, G., Ragnarsson, U., Terenius, L. and Hamprecht, B. (1977) Peptides related to β-lipotropin with opioid activity. Effects on levels of adenosine 3′:5′-cyclic monophosphate in neuroblastoma × glioma hybrid cells. *FEBS Lett. 77,* 28-32.

Wakade, A.R. and Wakade, T.D. (1982) Relationship between membrane depolarization, calcium influx and norepinephrine release in sympathetic neurons maintained in culture. *J. Pharmac. Exp. Ther. 223*, 125-9.

Wakshull, E., Johnson, M.I. and Burton, H. (1979a) Postnatal rat sympathetic neurons in culture. I. A comparison with embryonic neurons. *J. Neurophysiol. 42*, 1410-25.

Wakshull, E., Johnson, M.I. and Burton, H. (1979b) Postnatal rat sympathetic neurons in culture. II. Synaptic transmission by postnatal neurons. *J. Neurophysiol. 42*, 1426-36.

Walsh, J.V. and Singer, J.J. (1980a) Calcium action potentials in single freshly isolated smooth muscle cells. *Am. J. Physiol. 239*, C162-74.

Walsh, J.V. and Singer, J.J. (1980b) Penetration-induced hyperpolarization as evidence for Ca^{2+}-activation of K^{+} conductance in isolated smooth muscle cells. *Am. J. Physiol. 239*, C182-9.

Walsh, J.V. and Singer, J.J. (1981) Voltage clamp of single freshly dissociated smooth muscle cells: current-voltage relationships for three currents. *Pflügers Arch. 390*, 207-10.

Wastek, G.J., Lopez, J.R. and Richelson, E. (1981) Demonstration of a muscarinic receptor-mediated cyclic GMP-dependent hyperpolarization of the membrane potential of mouse neuroblastoma cells using [^{3}H] tetraphenylphosphonium. *Mol. Pharmac. 19*, 15-20.

Wehner, J.M., Feinman, R.D. and Sheppard, J.R. (1982) β-Adrenergic response in mouse CNS reaggregate cultures. *Dev. Brain Res. 3*, 207-17.

Wekerle, H., Paterson, B., Ketelson, U.-P. and Feldman, M. (1975) Striated muscle fibres differentiate in monolayer cultures of adult thymus reticulum. *Nature 256*, 493-4.

Werz, M.A. and MacDonald, R.L. (1982a) Opioid peptides decrease calcium-dependent action potential duration of mouse dorsal root ganglion neurons in cell culture. *Brain Res. 239*, 315-21.

Werz, M.A. and MacDonald, R.L. (1982b) Opiate alkaloids antagonize postsynaptic glycine and GABA responses: correlation with convulsant action. *Brain Res. 236*, 107-19.

Werz, M.A. and MacDonald, R.L. (1982c) Heterogeneous sensitivity of cultured dorsal root ganglion neurones to opioid peptides selective for μ- and δ-opiate receptors. *Nature 299*, 730-3.

Whitsett, J.A., Noguchi, A., Neely, J.E., Johnson, C.L. and Moore, J.J. (1981) β_1-Adrenergic receptors in human neuroblastoma. *Brain Res. 216*, 73-87.

Wildenthal, K. (1970) Factors promoting the survival and beating of intact foetal mouse hearts in organ culture. *J. Mol. Cell. Cardiol. 1*, 101-4.

Wildenthal, K. (1974) Studies of fetal mouse hearts in organ culture: influence of prolonged exposure to triiodothyronine on cardiac responsiveness to isoproterenol, glucagon, theophylline, acetylcholine and dibutytyl cyclic 3′,5′-adenosine monophosphate. *J. Pharmac. Exp. Ther. 190*, 272-9.

Wilkinson, M., Gibson, C.J., Bressler, B.H. and Inman, D.R. (1974)

Hypothalamic neurons in dissociated cell culture. *Brain Res. 82* 129-38.

Witkowski, J.A. (1977) Diseased muscle cells in culture. *Biol. Rev. 52*, 431-76.

Witkowski, J.A. (1979) Alexis Carrel and the mysticism of tissue culture. *Med. Hist. 23*, 279-96.

Wojtowicz, J.M., Gysen, M. and MacDonald, J.F. (1981) Multiple reversal potentials for responses to L-glutamic acid. *Brain Res. 213*, 195-200.

Wojtowicz, J.M., Marshall, K.C. and Hendelman, W.J. (1978) Electrophysiological and pharmacological studies of the inhibitory projection from the cerebellar cortex to the deep cerebellar nuclei in tissue culture. *Neuroscience 3*, 607-18.

Wollenberger, A. (1964) Rhythmic and arrhythmic contactile activity of single myocardial cells cultured *in vitro*. *Circulation Res. 15*, Suppl. II, 184-201.

Wollenberger, A. and Irmler, R. (1978) Effects of adrenaline and methylisobutylxanthine on adenosine 3′,5′-monophosphate levels in cultures of beating heart cells of the newborn rat, in *Cardiac Adaptation*, Rec. Adv. Studies Cardiac Structure Metab., Vol. 12, T. Kobayashi,, Y. Ito and G. Rona (eds.), University Park Press, Baltimore, pp. 689-95.

Yaffe, D. (1968) Retention of differentiation potentialities during prolonged cultivation of myogenic cells. *Proc. Nat. Acad. Sci.* USA *61*, 477-83.

Yasin, R., Van Beers, G., Nurse, K.C.E., Al-Ani, S., Landon, D.N. and Thompson, E.J. (1977) A quantitative technique for growing adult human skeletal muscle in culture starting from mononucleated cells. *J. Neurol. Sci. 32*, 347-60.

Yavin, E. and Yavin, Z. (1974) Attachment and culture of dissociated cells from rat embryo cerebral hemispheres on polylysine-coated surface. *J. Cell Biol. 62*, 540-6.

Yee, A.G., Fischbach, G.D. and Karnovsky, M.J. (1978) Clusters of intramembranous particles on cultured myotubes at sites that are highly sensitive to acetylcholine. *Proc. Nat. Acad. Sci.* USA *75*, 3004-8.

Yellen, G. (1982) Single Ca^{2+}-activated nonselective cation channels in neuroblastoma. *Nature 296*, 357-9.

Zelcer, E. and Sperelakis, N. (1981) Angiotensin induction of active responses in cultured reaggregates of rat aortic smooth muscle cells. *Blood Vessels 18*, 263-79.

Zipser, B., Crain, S.M. and Bornstein, M.B. (1973) Directly evoked "paroxysmal" depolarizations of mouse hippocampal neurons in synaptically organised explants in long-term culture. *Brain Res. 60*, 489-95.

Ziskind, L. and Dennis, M.J. (1978) Depolarising effect of curare on embryonic rat muscles. *Nature 276*, 622-3.

INDEX